# PREFACE

*Principles of Anatomy and Physiology in the Laboratory* has been prepared to accompany the authors' *Principles of Anatomy and Physiology* textbook, which was designed for use in introductory courses. However, we have provided a comprehensive and flexible sequence of modules, and this Manual may readily be used with other anatomy and physiology textbooks as well.

The Manual contains numerous modules for the introductory course. Many of the modules can be completed in a single laboratory period. In some instances, several modules can be completed in a single period. In other cases, a double laboratory period may be devoted to one module. We feel that such a design provides the instructor with considerable flexibility in selecting appropriate modules, varied sequences, and specific quantities of work that best suit his needs and are within the limits of practical considerations. Moreover, such a design affords utilization of the Manual in either one-semester or one-year courses in anatomy and physiology.

Each module is a self-contained unit consisting of three principal divisions. The *OBJECTIVES* provide both the student and instructor with the desired outcomes of the module. The *CON-CLUSIONS* are a series of questions designed to review and apply the principles and concepts studied in the module. The *STUDENT ACTIVITY* is designed to help the student reinforce one or more of the concepts under investigation.

Throughout the Manual, space is provided for students to answer questions. In addition, there is space for students to make drawings and label diagrams. There are also many Tables for recording data.

We would like to emphasize that the Manual is both a laboratory workbook and a study guide. The laboratory modules require students to make microscopic examinations and evaluations of cells and tissues, to observe and interpret chemical reactions, to record data, to make gross examinations of organs and systems, and to conduct physiological experiments and interpret and apply the results of the experiments. The study modules have been carefully developed to reinforce lecture and laboratory work, and to introduce clinical applications.

This Manual is distinctive in its use of clinical applications, its comprehensiveness, its flexibility, and the quality of its illustrations. The Manual contains 44 separate modules that cover virtually every aspect of the human body that might be treated in an introductory course. In this regard, we feel that the instructor is less burdened with preparation since all he has to do is select the appropriate modules from the Manual without having to supplement his syllabus from several sources. This

Manual also provides the instructor with a great deal of flexibility regarding assignments. Many of the modules have been specifically designed so that the instructor, at his discretion, may assign a module to be completed out of class to supplement lecture or laboratory work. These assignments require the student to research specific areas of study. Or, the instructor may wish to assign some modules in class for evaluation. The authors' textbook is clinically oriented and, accordingly, the Manual is also clinically oriented. We feel that most laboratory work in anatomy and physiology courses neglects the clinical aspects. We have, therefore, incorporated many modules that deal only with clinical applications. Here again, the instructor may assign these modules as out-of-class projects or use them as a means of evaluation. We are especially pleased with the outstanding quality of the illustrations in the modules. They have been reproduced from the authors' textbook. We hope that the use of these illustrations will greatly assist the student to cross-refer to the illustrations in the textbook should the instructor decide to use the authors' textbook.

Several appendixes have been included. These are Appendix A, Preparations of Specimens for Experimentation, Appendix B, Use of Physiologic Equipment, and Appendix C, Metric Units of Length and Some English Equivalents.

A complimentary *Teachers' Guide* is offered for instructors who use *Principles of Anatomy and Physiology in the Laboratory*. It is designed to assist in the selection of appropriate audio-visual materials and other materials required to complete each module, to present instructional concepts, and to provide students with correct responses to questions asked in the modules.

Gerard J. Tortora
Nicholas P. Anagnostakos

# CONTENTS

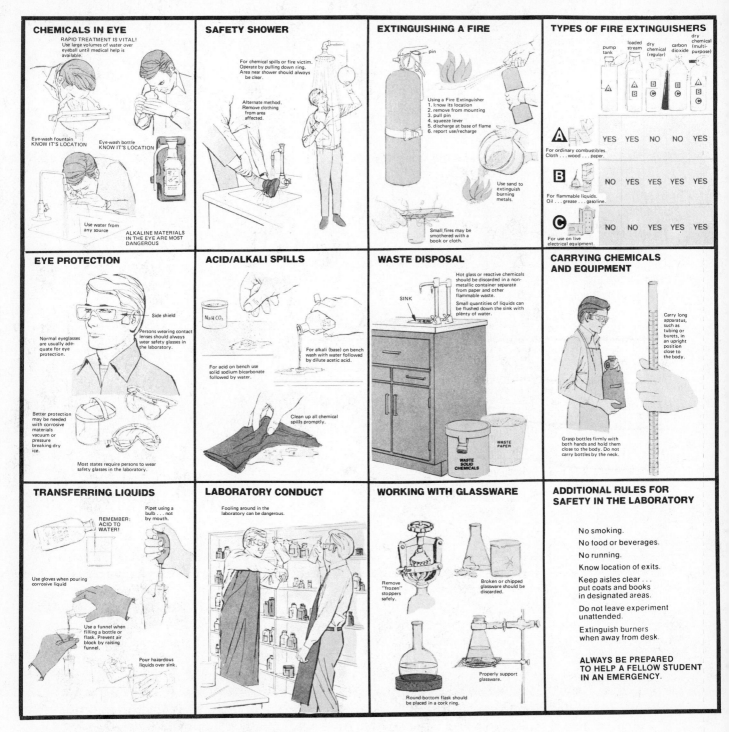

## CHEMICALS IN EYE

RAPID TREATMENT IS VITAL! Use large volumes of water over eyeball until medical help is available.

Eye-wash fountain KNOW IT'S LOCATION

Eye-wash bottle KNOW IT'S LOCATION

Use water from any source

ALKALINE MATERIALS IN THE EYE ARE MOST DANGEROUS

## SAFETY SHOWER

For chemical spills or fire victim. Operate by pulling down ring. Area near shower should always be clear.

Alternate method. Remove clothing from area affected.

## EXTINGUISHING A FIRE

pin

Using a Fire Extinguisher
1. know its location
2. remove from mounting
3. pull pin
4. squeeze lever
5. discharge at base of flame
6. report use/recharge

Use sand to extinguish burning metals.

Small fires may be smothered with a book or cloth.

## TYPES OF FIRE EXTINGUISHERS

| | pump tank | loaded stream | dry chemical (regular) | carbon dioxide | dry chemical (multi-purpose) |
|---|---|---|---|---|---|
| **A** For ordinary combustibles. Cloth . . . wood . . . paper. | YES | YES | NO | NO | YES |
| **B** For flammable liquids. Oil . . . grease . . . gasoline. | NO | YES | YES | YES | YES |
| **C** For use on live electrical equipment. | NO | NO | YES | YES | YES |

## EYE PROTECTION

Side shield

Normal eyeglasses are usually adequate for eye protection.

Persons wearing contact lenses should always wear safety glasses in the laboratory.

Better protection may be needed with corrosive materials vacuum or pressure breaking dry ice.

Most states require persons to wear safety glasses in the laboratory.

## ACID/ALKALI SPILLS

$Na_2CO_3$

For alkali (base) on bench wash with water followed by dilute acetic acid.

For acid on bench use solid sodium bicarbonate followed by water.

Clean up all chemical spills promptly.

## WASTE DISPOSAL

SINK

Hot glass or reactive chemicals should be discarded in a non-metallic container separate from paper and other flammable waste.

Small quantities of liquids can be flushed down the sink with plenty of water.

WASTE SOLID CHEMICALS

WASTE PAPER

## CARRYING CHEMICALS AND EQUIPMENT

Carry long apparatus, such as tubing or burets, in an upright position close to the body.

Grasp bottles firmly with both hands and hold them close to the body. Do not carry bottles by the neck.

## TRANSFERRING LIQUIDS

REMEMBER: ACID TO WATER!

Pipet using a bulb . . . not by mouth.

Use gloves when pouring corrosive liquid

Use a funnel when filling a bottle or flask. Prevent air block by raising funnel.

Pour hazardous liquids over sink.

## LABORATORY CONDUCT

Fooling around in the laboratory can be dangerous.

## WORKING WITH GLASSWARE

Remove "frozen" stoppers safely.

Broken or chipped glassware should be discarded.

Properly support glassware.

Round-bottom flask should be placed in a cork ring.

## ADDITIONAL RULES FOR SAFETY IN THE LABORATORY

No smoking.

No food or beverages.

No running.

Know location of exits.

Keep aisles clear . . . put coats and books in designated areas.

Do not leave experiment unattended.

Extinguish burners when away from desk.

**ALWAYS BE PREPARED TO HELP A FELLOW STUDENT IN AN EMERGENCY.**

(Courtesy Science Related Materials, Janesville, Wis. 53545.)

# MODULE 1
# MICROSCOPY

**OBJECTIVES**

1. To identify the different parts of the compound microscope

2. To learn the proper use and care of the compound microscope

3. To calculate the approximate sizes of materials under investigation

4. To compare the resolution of a photomicrograph and an electron micrograph

5. To contrast the principles employed in light microscopy and electron microscopy

One of the most important instruments that you will use in your anatomy and physiology course is the compound microscope. With this instrument, objects too small to be seen clearly with the naked eye can become highly magnified and their minute details will be revealed. This delicate and expensive instrument must be cared for properly, if you are to be completely successful in your understanding of it, and if you wish to make accurate observations of the various specimens you examine. In this module you will also be introduced to some of the principles employed in electron microscopy.

There is little doubt that the single most important instrument that led to the systematic study of cells has been the **compound microscope.** This instrument, initially developed by the Dutch lens maker Zacharias Janssen at the close of the sixteenth century, has undergone considerable modification over the years. A photograph of a modern compound microscope is shown in Figure 1-1a. Any optical instrument is limited in the structural details it can make visible by its resolution. The *resolution* of a microscope is the ability of a microscope to distinguish the smallest distance between 2 points on an object. Resolution is determined, among other factors, by the wavelength of radiation used to illuminate the object. The smaller the wavelength, the smaller the limit of resolution and the more structural detail is visible. Since the source of radiation for the compound microscope is light and since light has a relatively long wavelength, the smallest objects that can be seen in detail are about 0.3 micrometers ($\mu$m). Parts smaller than 0.3 $\mu$m cannot be resolved with the light microscope. One *micrometer,* symbolized $\mu m$, is a microscopic unit of measurement equal to 1/1,000 millimeter (mm.) or about 1/25,000 inch. A photomicrograph of ciliated epithelium is shown in Figure 1-2a. A *photomicrograph* is a photograph of an object taken under a light microscope.

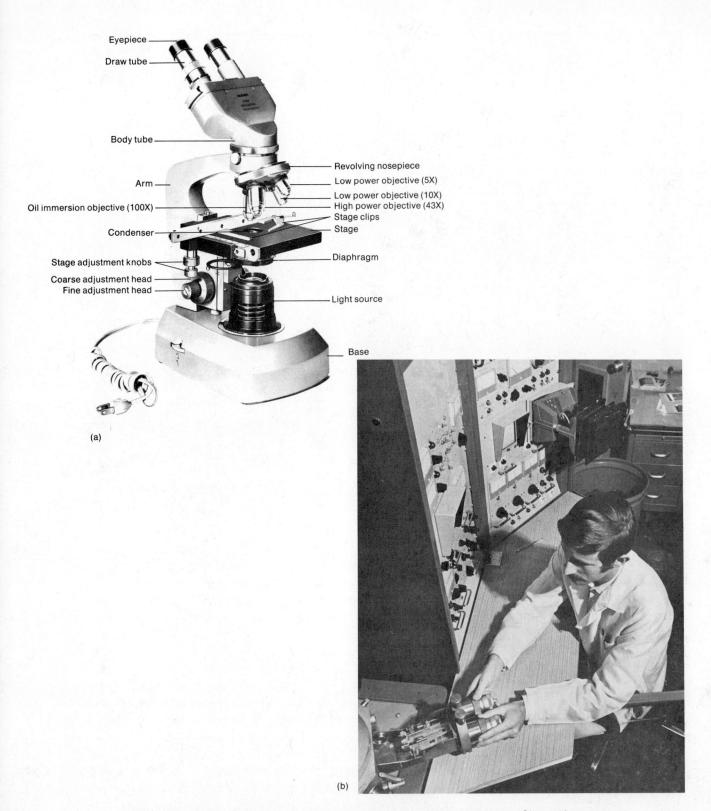

Eyepiece
Draw tube
Body tube
Arm
Oil immersion objective (100X)
Condenser
Stage adjustment knobs
Coarse adjustment head
Fine adjustment head

Revolving nosepiece
Low power objective (5X)
Low power objective (10X)
High power objective (43X)
Stage clips
Stage
Diaphragm
Light source

Base

(a)

(b)

**Figure 1–1.** Modern microscopes. (a) Modern compound light microscope. Since microscopes of this kind use light, they cannot resolve structures smaller than 0.3 micrometer. This limitation is the result of the relatively long length of visible light waves. (b) Modern electron microscope. Instead of light, this instrument uses a beam of electrons. Since the wavelengths of electrons are about 1/100,000 that of the normal wavelength of white light, an electron microscope can magnify objects up to 200,000 times (a: Courtesy of Tasco Sales, Inc., Miami, Fla., b: Courtesy of RCA Laboratories, Princeton, N.J.).

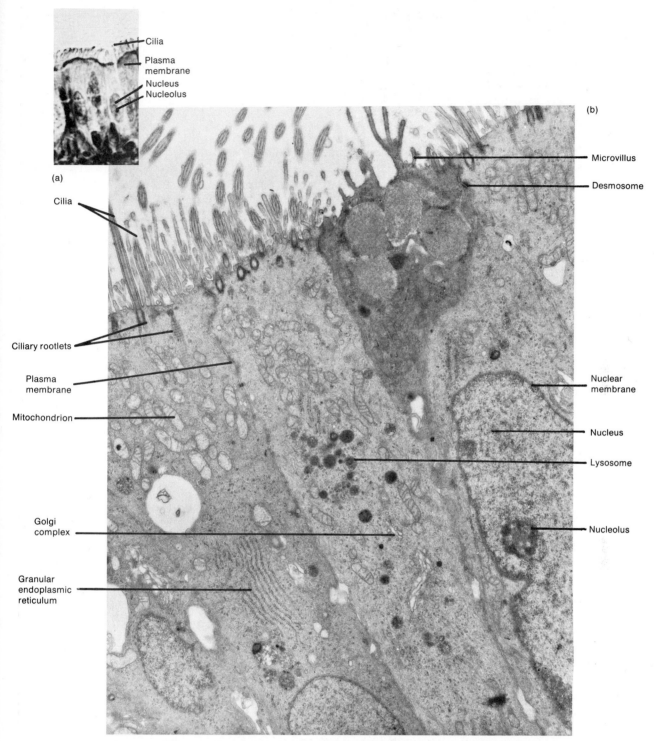

Cilia

Plasma
membrane

Nucleus
Nucleolus

(a)

(b)

Cilia

Ciliary rootlets

Plasma
membrane

Mitochondrion

Golgi
complex

Granular
endoplasmic
reticulum

Microvillus

Desmosome

Nuclear
membrane

Nucleus

Lysosome

Nucleolus

**Figure 1–2.** Comparison between the resolution of a light microscope and an electron microscope. (a) Photomicrograph of ciliated epithelial cells, 450X. Note the cilia at the surfaces of the cells, the plasma membrane between cells, the nucleus, and the nucleolus inside the nucleus. (b) Electron micrograph of ciliated epithelial cells, 54,000X. The resolution of the electron microscope becomes readily apparent. Note the cilia and their rootlets, the plasma membrane, the nucleus, and the nucleolus. In addition, observe the mitochondrion, granular endoplasmic reticulum, Golgi complex, lysosome, microvillus, and desmosome. These structures are rendered visible by the resolution of the electron microscope (a: Courtesy of Carolina Biological Supply Company, Burlington, N.C. b: Courtesy of Dr. John A. Terzakis, Lenox Hill Hospital, New York, N.Y.).

Many parts of cells are much smaller than 0.3 $\mu$m, and if they are to be seen in any detail another kind of microscope must be used. This instrument, which came into use in the 1940s, is called an **electron microscope** (Figure 1-1b). Instead of light, an electron microscope employs high electrical voltages to drive a beam of electrons. As a result, an electron microscope increases resolution because the wavelength of electrons is much smaller than that of visible light (about 1/100,000 of visible light). The present limit of resolution is approximately 3Å. One *angstrom*, symbolized Å, is equal to 1/10,000 of a micrometer, or 1/250,000,000 of an inch. Whereas maximum magnifications with the light microscope are 1,000-2,000X, an electron microscope can magnify an object up to 200,000X.

A listing of commonly used metric units of length is presented in Appendix C.

When using a light microscope, the observer looks directly at the object. In the electron microscope, however, the observer either views a fluorescent screen which lights up as the beam of electrons hits it, or the image is directed onto a photographic emulsion (film or glass plate) which is then developed, enlarged, and examined. The resulting photograph is called an *electron micrograph* (Figure 1-2b). If you compare the photomicrograph of the ciliated epithelium in Figure 1-2a with the electron micrograph of the ciliated epithelium in Figure 1-2b, you will see considerable differences in resolution.

## PART I  THE COMPOUND MICROSCOPE

### IDENTIFICATION OF THE PARTS OF THE MICROSCOPE

The microscope should be carefully carried from the cabinet to your desk by placing one hand around the arm and the other hand firmly under the base. Gently place it on your desk, directly in front of you, with the arm facing you. Locate the following parts on the microscope, and as each is discussed, label Figure 1-3. Although different microscopes vary somewhat in design, they consist of the same basic parts.

1. *Base*—the heavy U-shaped structure upon which the microscope rests.
2. *Body tube*—the main cylindrical part.
3. *Arm*—angular or curved part of the frame.
4. *Inclination joint*—some microscopes have a movable hinge allowing the instrument to be tilted to a comfortable viewing position.
5. *Stage*—a platform upon which slides or other material to be studied is placed. The opening in the center allows light to pass from below through the material being examined.
6. *Stage (spring) clips*—two clips mounted on the stage which hold slides to be examined securely in place.
7. *Mirror*—found in some microscopes below the stage, functioning to direct light from its source through the stage opening and through the lenses. If the light source is built in, then no mirror is found.
8. *Diaphragm*—located beneath the stage functioning to regulate light intensity passing through the lenses to the observer's eyes. One of 2 types of diaphragms are usually used. An *iris diaphragm,* as found in cameras, is a series of sliding leaves which vary the size of the opening. Another type, a *disc diaphragm,* consists of a plate with a graded series of holes any of which may be rotated into position.
9. *Condenser*—a lens that controls the light-beam size that is located beneath the stage opening.
10. *Condenser adjustment knob*—a knob that functions to raise and lower the condenser for optimum viewing. When it is in its highest position it allows full illumination.
11. *Coarse adjustment knob*—when it is turned, it functions to raise and lower the body tube for focusing the microscope.
12. *Fine adjustment knob*—usually found below the coarse adjustment and used for fine or final focusing. Some microscopes have both coarse and fine adjustments combined into one.
13. *Nosepiece*—a plate at the bottom of the body tube which is usually circular.

4

**Figure 1–3.** Parts of a compound light microscope (Courtesy of American Optical Corporation, Buffalo, N.Y.).

14. *Revolving nosepiece*—the lower, movable part of the nosepiece that contains the various objective lenses.
15. *Scanning objective*—a lens, marked 5X on most instruments.
16. *Low power objective*—a lens, marked 10X on most instruments.
17. *High power objective*—a lens, marked 43X or 45X on most instruments.
18. *Oil-immersion objective*—a lens marked 100X on most instruments, and distinguished by an etched red circle (special instructions for this objective are discussed later).
19. *Ocular (eyepiece)*—a lens that is removable, at the top of the body tube, marked 10X on most microscopes.

## RULES OF MICROSCOPY

There are certain basic rules of microscopy that must be observed at all times, in order to obtain maximum efficiency and provide proper care for your microscope.

1. All parts of the microscope must be kept clean, especially the lenses of the ocular, objectives, condenser, and mirror. Only the special lens paper that is provided should be used, and never paper towels or cloths since these tend to scratch the delicate glass surfaces. When using lens paper, do not use the same area on the paper for cleaning the lens. As you go across the lens, change the position of the paper each time.
2. Do not permit the objectives to get wet, especially when observing a wet mount (described later). These preparations must always be examined by using a cover slip or the image becomes distorted.

3. Your instructor should be consulted if any mechanical or optical difficulties arise. *Do not try to solve these problems yourself.*
4. It is important that you keep *both* eyes open at all times while observing materials through the microscope. At the start this will be difficult but with practice it becomes a natural procedure. This technique will help you to draw and observe microscopic specimens without moving your head. Only your eyes will move.
5. The low power objective should always be used first to locate an object; then, if necessary, switch to a higher power.
6. If you are using the high power objectives, never focus with the coarse adjustment. The distance between these objectives and the slide is very small and you may break the cover glass and the slide and scratch the lens system.
7. When looking through the microscope, never focus downward. By observing from one side you can see that the objectives do not make contact with the cover slip and slide.
8. Make sure that you raise the body tube before placing a slide on the stage or before removing a slide.
9. At the end of a laboratory session when the microscope is to be returned to the cabinet, leave it with the scanning or low power objective aligned, the diaphragm open, and the condenser raised to its highest fixed position.

## SETTING UP THE MICROSCOPE

1. The microscope is placed on the table with the arm toward you, and with the back of the base at least one inch from the edge of the table.
2. Position yourself and the microscope so that you can look into the ocular comfortably.
3. Wipe the objectives, the top lens of the eyepiece, the condenser, and the mirror with lens paper. This sequence should be the most delicate and the least dirty lens is cleaned first. Xylol or ethanol may be applied to the lens paper in order to remove grease and oil from the lenses and microscope slides.
4. Position the low power objective in line with the body tube. When it is in its proper place, it will click. Lower the body tube with the coarse adjustment until the bottom of the lens is approximately ¼ inch from the stage.
5. The minimum amount of light is then admitted by opening the diaphragm if it is an iris, or turning the disc to its largest opening.
6. Place your eye to the eyepiece (ocular), and adjust the concave surface of the mirror to the light source.
7. The mirror is then adjusted until a uniform circle of light appears without any shadows. This is called the *microscopic field* and the microscope is ready for use.

## PREPARING MATERIALS FOR STUDY

The materials for microscopic examination can be prepared in one of 2 basic ways: (1) the temporary (wet) mount or (2) the permanent mount. Whichever method is used, the material to be studied must be thin enough for light to pass through it.

1. *Temporary (wet) mount.* In this method, the preparation will be studied for only a short period of time. This mount is prepared as follows (Figure 1–4):
   a. A small drop of water is placed near the center of a clean glass microscope slide.
   b. Then a *very small* amount of the material to be examined is placed in this water, making sure that the liquid covers the material.
   c. A glass or plastic cover slip is then held at a 45° angle, contacting the drop of water, and gradually allowed to fall into place over the material. When the cover slip is dropped from too flat an angle, many air bubbles are produced, interfering with accurate observations.

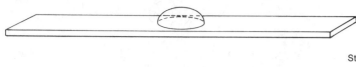

Step 1.  A drop of water is placed on a clean microscope slide.

Step 2.  The object is added to the drop of water.

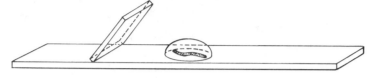

Step 3.  Cover slip is held at a 45° angle against slide.

Step 4.  Cover slip is moved against drop of water at same angle.

Step 5.  Cover slip is dropped slowly over water.

**Figure 1–4.** Preparing a wet mount (Modified from Lawrence S. Dillon and William A. Cooper, *A Laboratory Survey of Biology,* 2nd ed., 1969, The Macmillan Company, New York, N.Y.).

2. *Permanent mount* (optional). This method is a tedious, time-consuming process that is used for slides that will be examined on many different occasions, weeks or months apart.
   a. The material is killed by placing it in an FAA solution (10 ml. of 40% formaldehyde, 50 ml. of 95% ethanol, 2 ml. of glacial acetic acid, and 40 ml. of water).
   b. The killed material is then dehydrated by passing it through a number of grades of alcohol. Gradations of 50, 70, 85, 95, and 100 percent are usually used.
   c. The dehydrated material is then imbedded in a block of paraffin wax.
   d. A microtome is then used to cut very thin sections of the material which are then mounted on a microscope slide with an adhesive.
   e. The paraffin is then dissolved with an organic solvent (toluene or xylene), and the slide is passed back down the graded alcohols to 50 percent.
   f. The slide is then passed through certain specific dye solutions; the material is dehydrated once again.
   g. The cover slip is added and sealed to the slide with resin.

## USING THE MICROSCOPE

1. Raise the body tube to its highest fixed position, using the coarse adjustment.
2. Make a temporary mount using a single letter of newsprint, or use a slide that has been specially prepared with a letter, usually the letter "e." If you prepare such a slide, cut a single letter—"a," "b," or "e"—from the smallest print available and place this letter in the correct position to be read with the naked eye.
3. Place the slide on the stage, making sure that the letter is centered over the opening in the stage, directly over the condenser. Secure the slide in place with the stage clips.

4. The low power objective should be in line with the body tube and the objective should be about ¼ inch from the cover slip.

5. The body tube should then be lowered as far as it will go, while watching it from the side, taking care not to touch the slide. The tube will reach an automatic stop which prevents the low power objective from hitting the slide.

6. While looking through the eyepiece, the coarse adjustment knob is turned counterclockwise, raising the body tube. Watch for the material to suddenly appear in the microscopic field. If it is in proper focus, the low power objective is about ½ inch above the slide. When focusing, always *raise* the body tube.

7. The fine adjustment is used to complete the focusing, using a counterclockwise motion once again.

8. Compare the position of the letter as originally seen with its appearance under the microscope. How has the position of the letter been changed?

9. While looking at the slide through the ocular, change the position of the slide by either moving it with your thumbs, or using the mechanical stage knobs, if the microscope is equipped with them. This exercise is to teach you to quickly and efficiently move your material in various directions. In which direction does the letter move when you move the slide to the left?
   This is called "scanning" a slide and will be useful in the examination of living material or examining slides where the object to be observed is not immediately centered under the tube.
   Make a drawing of the letter as it appears under low power in the space provided on page 9.

10. Change your magnification from low to high power by carrying out the following steps:
    a. Place the letter in the center of the field under low power. This is important because you are now focusing on a smaller area of the microscopic field.
    b. Increase your illumination.
    c. The letter should be in focus, and if the microscope is *parfocal*, ("parfocal" means that when clear focus has been attained using any objective at random, revolving the nosepiece results in a change in magnification but the specimen is still in focus), the high power objective can be switched into line with the body tube without changing focus. If it is not completely in focus after switching the lens, a slight turn of the fine adjustment knob will complete it.
    d. If your microscope is not parfocal, observe the stage from one side and carefully switch the high power objective in line with the body tube.
    e. While still observing from the side and using the coarse adjustment, *carefully* lower the objective until it almost touches the slide.
    f. Look through the ocular and focus up slowly. Finish focusing by turning the fine adjustment toward you.
    g. If your microscope has an oil-immersion objective, special procedures must be followed. A drop of special *immersion oil* is placed on the microscope slide, and this objective is lowered until it just contacts the oil. If you have a parfocal microscope, you do not have to raise or lower the objectives. For example, if you are using the high power objective and the specimen is in focus, just switch the high power objective out of line with the body tube. Then add the oil, switch the oil immersion objective into position, and the specimen should be in focus. The same holds true when you switch from low power to high power. The special light-transmitting properties of the oil are such that light is concentrated at one tiny spot, permitting the use of extra powerful objectives in a relatively narrow field of vision. This objective is extremely close to the slide being examined, so extra precautions must be taken by *never focusing downward* while you are looking through the eyepiece. Whenever you finish working with immersion oil be sure to saturate a piece of lens paper with xylol or alcohol and clean the oil immersion objective and the slide if it is to be used again.

Is as much of the letter visible under high power as under low power?

Explain

Make a drawing of the letter as it appears under high power in the space provided.

LOW POWER DRAWING        HIGH POWER DRAWING
OF LETTER             OF LETTER

Now select a prepared slide containing 3 different colored threads. This will show you that a specimen mounted on a slide has depth as well as length and width. At lower magnification the amount of depth of the specimen that is clearly in focus, the depth of field, is greater than that at higher magnification. You must focus at different depths in order to determine the position (depth) of each thread.

After you make your observation, complete the following with regard to the location of the different threads:

What color is at the bottom, closest to the slide?

On top, closest to the cover glass?

In the middle?

## MAGNIFICATION AND MEASUREMENT OF A MICROSCOPE FIELD

The magnification of your microscope is calculated by multiplying the magnification of the ocular by the magnification of the objectives used. Example: An ocular of 10X used with an objective of 5X would give a total magnification of 50X. Calculate the total magnification of each of the objectives on your microscope:

(1) Ocular _____ X Objective _____ = _____

(2) Ocular _____ X Objective _____ = _____

(3) Ocular _____ X Objective _____ = _____

(4) Ocular _____ X Objective _____ = _____

(5) Why is your microscope called a compound microscope?_____

You can estimate the size of microscopic material by comparing it with the known diameter of the low power field, place a clear plastic millimeter ruler across the stage, with the inner edge of a vertical line just visible at the left edge of the field and the horizontal line across the diameter of the field. The distance between the first and second vertical lines is 1 mm., but it will be necessary to calculate the distance between the second line and right edge of the field.

By using the equation one mm. = 1000 $\mu$m (micrometers), calculate the diameter of your low power field in micrometers.

Calculate the diameter, in micrometers, of some materials like cotton fibers, wool strands, and silk strands under low and high power. Record your observations in the space provided below.

SPECIMEN                                    DIAMETER IN MICROMETERS

_____          _____

_____          _____

_____          _____

_____          _____

_____          _____

**CONCLUSIONS**

1. List some ways to regulate the amount of light that enters a microscope.

2. Which objective gives you your highest total magnification?

   Explain

3. What is the function of each of the following?

   objective

   condenser

   diaphragm

   eyepiece

4. What is the difference between a photomicrograph and an electron micrograph?

5. What is a micrometer?

   An angstrom?

   How many microns are there in one centimeter?

6. Should you focus with one objective, then rotate the nosepiece of your microscope to another objective and find that your material is still in focus, your microscope is said to be

7. Why is oil used with the oil-immersion objective?

8. What is the name of the smaller of the 2 knobs responsible for moving the body tube of a microscope up and down?

9. Which objective needs the most light: scanning, low power or high power?

   Why?

**10**

10. Should you move the body tube of a microscope *upward* or *downward* before changing from a lower to a higher magnification?

Why?

## PART II    THE ELECTRON MICROSCOPE

Compare the electron micrograph with the photomicrograph in Figure 1-2.

1. What differences do you observe?

## CONCLUSIONS

1. What is resolution?

2. What determines resolution?

3. Why is resolution greater with an electron microscope as compared to a light microscope?

4. How do you think electron microscopy has helped scientists to understand the structure of cells?

## STUDENT ACTIVITY

Although compound microscopes vary considerably with regard to design, they operate on the same basic principle. In order to familiarize you with variations in design, we have provided a photograph of another kind of microscope in Figure 1-5 which we would like you to label. Next to the labeled part, please indicate the function of the part.

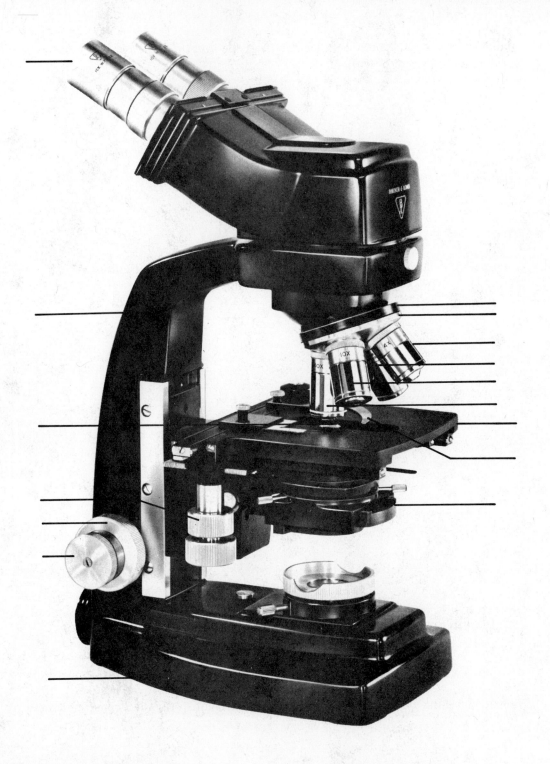

**Figure 1–5.** Compound light microscope (Courtesy of Bausch & Lomb, Rochester, N.Y.).

# MODULE 2
# CELL STRUCTURE

**OBJECTIVES**

1. To identify the basic parts of a cell and describe the function of each part

2. To compare various cells of the body with regard to structure and function

A **cell** is the basic living structural and functional unit of the body. *Cytology* is the specialized branch of science that is concerned with the study of cells. Cells of the body vary considerably with regard to shape, size, and structure. In an attempt to study as many of these parts as possible, we have designed a *generalized cell*, which represents a collection of many different cells combined into one (see Figure 2-1).

## PART I   IDENTIFICATION OF CELL PARTS

Refer to Figure 2-1, a generalized cell based upon recent electron micrograph studies. With the aid of your textbook, and any other materials made available by your instructor, label the parts of the cell. In the space provided below, describe the function of the part indicated.

1. plasma membrane

2. cytoplasm

3. nucleus

4. endoplasmic reticulum

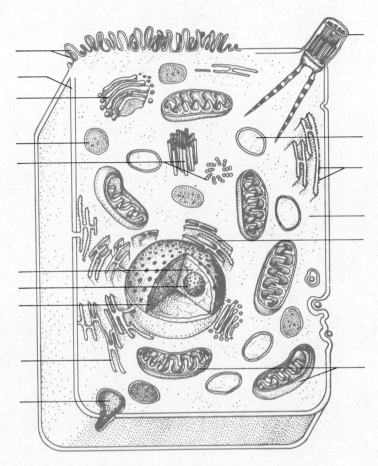

**Figure 2–1.** This generalized cell, based on electron microscope studies, is a composite cell combining parts from many different cells of the body.

5. ribosome

6. Golgi complex

7. mitochondria

8. microvilli

9. vacuole

10. pinocytic vesicle

11. lysosome

12. centriole


13. cilia and flagella


We have provided you with 6 electron micrographs in Figure 2-2. Under each, write the cell parts that you can identify.

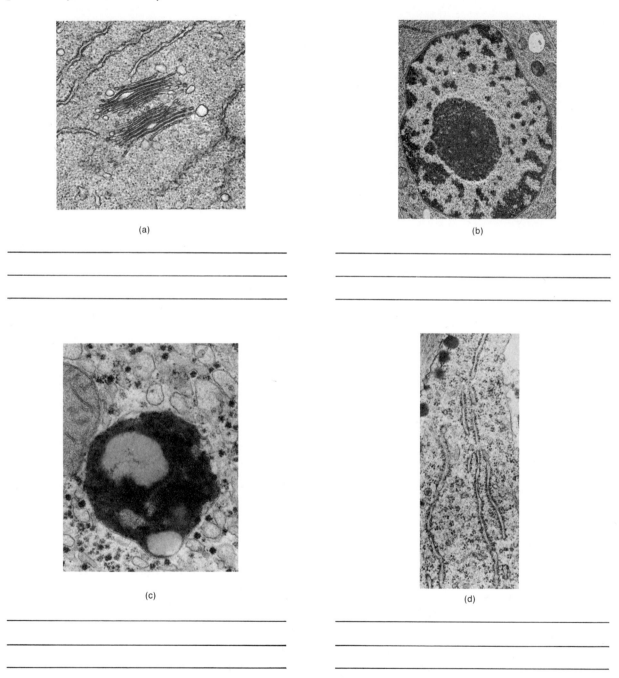

(a)

(b)

(c)

(d)

**Figure 2–2.** Representative electron micrographs. (a), (b), (d) Courtesy of Dr. Myron C. Ledbetter, Brookhaven National Laboratory, Long Island, N.Y. (c) Courtesy of Dr. F. Van Hoof, Université Catholique de Lourain, Montreal, Canada.

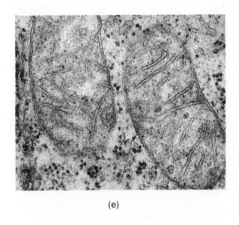

(e)

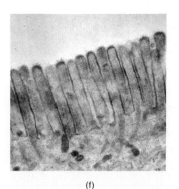

(f)

_____     _____
_____     _____
_____     _____

**Figure 2–2 continued.** (e) Courtesy of Dr. Myron C. Ledbetter, Brookhaven National Laboratory, Long Island, N.Y. (f) Courtesy of Dr. E. B. Sandborn, Université de Montréal, Montreal, Canada.

## PART II   DIVERSITY OF CELLS

Now obtain prepared slides of the following types of cells and examine under the magnifications suggested:

1. ciliated columnar (high power)
2. sperm cell (oil immersion)
3. nerve cell (high power)
4. smooth muscle cells (high power)

After you have made your examination, draw each of the following in the spaces provided and list the parts of the cell you observed and their functions.

CILIATED COLUMNAR                    SPERM CELL

_____     _____
_____     _____
_____     _____
_____     _____
_____     _____

NERVE CELL                                    MUSCLE CELL

_____              _____
_____              _____
_____              _____
_____              _____
_____              _____

**CONCLUSIONS**

1. Define a cell.

2. How are cells related to tissues?

3. What part of the cell is associated with the following activities?

   a. digestion

   b. cell division

   c. protein synthesis

   d. locomotion

   e. ATP synthesis

   f. control center

   g. carbohydrate synthesis

   h. transportation and storage

   i. entrance and exit of materials

4. Based upon your observation and study of cell types, describe the diversity of cells.

**STUDENT ACTIVITY**

Consult your anatomy and physiology textbook and look up 15 different kinds of cells found in the body. (Do not use the same cells that you examined under the microscope.) In the space provided below, list the names of the cells and the specialized function of each.

1.

2.

3.

4.

5.

6.

7.

8.

9.

10.

11.

12.

13.

14.

15.

# MODULE 3
# MOVEMENT OF MATERIALS INTO AND OUT OF CELLS

**OBJECTIVES**

1. To compare the passive processes by which substances move across plasma membranes

2. To contrast the active processes by which materials move across plasma membranes

The plasma membrane not only encloses individual cells but also separates cells from each other and from the external environment. Any materials that enter must pass through the plasma membrane.

In general, substances move across plasma membranes by 2 principal processes—passive and active. In *passive* or *physical processes*, materials move because of differences in concentration (or pressure) from regions of higher concentration (or pressure) to regions of lower concentration (or pressure). The movement continues until an equilibrium or even distribution of substances is accomplished. Passive processes are the result of the kinetic energy (energy of motion) of the substances themselves and do not involve an expenditure of energy by the cell. Examples of passive processes are diffusion, osmosis, filtration, and dialysis.

In *active* or *physiological processes*, substances may move from areas of lower to higher concentration. Moreover, cells must expend energy in order to carry on active processes. Examples of active processes are active transport, phagocytosis, and pinocytosis.

## PART I    PASSIVE PROCESSES

### DIFFUSION

*Diffusion* is the net (greater) movement of molecules or ions from a region of higher concentration to a region of lesser concentration until they are evenly distributed.

In order to demonstrate diffusion, carefully place a large crystal of potassium permanganate into a test tube filled with water. Place the tube in a rack against a white background where it will not be disturbed. Note the diffusion of the crystal through the water at 15-minute intervals. Record the diffusion of the crystal in millimeters per minute at 15-minute intervals. Simply measure the distance of diffusion with a millimeter ruler.

15 minutes

30 minutes

45 minutes

60 minutes

75 minutes

90 minutes

105 minutes

120 minutes

**CONCLUSIONS**

1. Define diffusion.


2. Why do substances diffuse?

3. Cite an example for the following kinds of diffusion:

   a. gas in a gas

   b. liquid in a liquid

**OSMOSIS**

*Osmosis* is the net movement of water through a semipermeable membrane from a region of higher water concentration to a region of lower water concentration.

Refer to the osmosis apparatus in Figure 3-1. Tie a knot very tightly at one end of a 4-inch piece of cellophane dialysis tubing and fill the tubing with a 10 percent sucrose solution that has been colored with congo red (red food coloring may also be used). Close the open end of the tubing with a one-hole cork stopper into which a glass tube has been inserted. Secure and suspend the glass tube and tubing by means of a clamp attached to a ring stand. Insert the cellophane bag into a beaker or flask of water until the water comes up to the bottom of the rubber stopper. As soon as the sugar solution becomes visible in the glass tubing, mark the tube with a wax pencil and note the time. Mark the height of liquid in the tube after 10, 20, and 30-minute intervals by using your millimeter ruler.

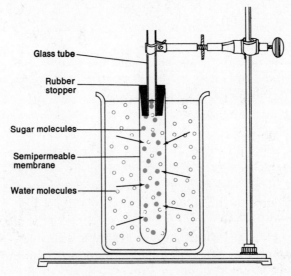

Glass tube

Rubber stopper

Sugar molecules

Semipermeable membrane

Water molecules

**Figure 3-1.** Osmosis apparatus.

**20**

10 minutes

20 minutes

30 minutes

## CONCLUSIONS

1. Compare diffusion and osmosis.



2. Why do water molecules undergo osmosis?


3. Define a semipermeable membrane.


4. Define equilibrium.


5. Define each of the following and relate to red blood cells:
   a. hypotonic solution


   b. hypertonic solution


   c. isotonic solution



6. What is a physiological saline solution?




## FILTRATION

*Filtration* is the movement of solvents and dissolved substances across a semipermeable membrane from regions of higher pressure to lower pressure, or under the influence of gravity. The pressure is called hydrostatic pressure and is the result of the solvent, usually water. In general, any material having a molecular weight that is less than 100 will be filtered. This is because of the size of the pores in the filter paper.

Refer to Figure 3-2, the filtration apparatus. In this experiment, it is the gravity that pulls the particles through the pores of the membrane (filter). Fold a piece of filter paper by folding it in half and then in half again. Open it up into a cone, place in a funnel, and place the funnel over the beaker. Shake, and pour a mixture of a few particles of powdered wood charcoal (black), 1% copper sulfate (blue), boiled starch (white), and water into the funnel until it almost reaches the top of the filter paper.

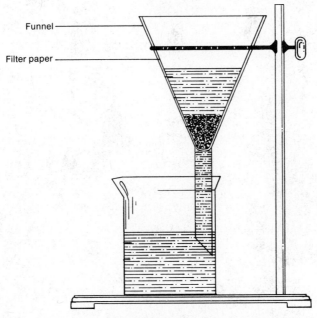

**Figure 3-2.** Filtration apparatus.

Count the number of drops passing through the funnel for the following time intervals:

10 seconds

30 seconds

60 seconds

90 seconds

120 seconds

Observe which substances passed through the filter paper by noting their color. Examine the filter paper to determine whether any colored particles were not filtered. In order to determine if any starch is in the filtrate (liquid in the beaker), add several drops of 0.01M IKI solution. A blue-black color reaction indicates the presence of starch.

## CONCLUSIONS

1. How does filtration differ from osmosis?

2. What types of substances pass through the filter paper?

3. What determines the rate of filtration?

## DIALYSIS

*Dialysis* is the separation of smaller molecules from larger ones by the diffusion of the smaller molecules through a semipermeable membrane.

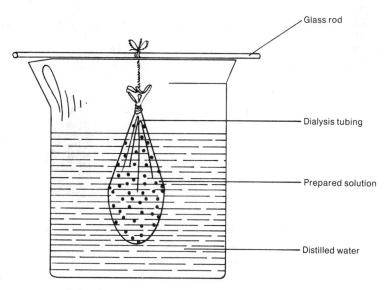

Figure 3-3. Dialysis apparatus.

Refer to the dialysis apparatus in Figure 3-3. Place a prepared solution containing starch, sodium chloride, 5% glucose, and albumin into a piece of dialysis tubing. Tie off the tubing and immerse into a beaker of distilled water.

After one hour, test the water for the presence of each of the substances in the tubing, as follows:

Albumin: Carefully add several drops of concentrated nitric acid to a test tube containing 2 ml. of water from the beaker. Be careful using nitric acid. It can cause severe damage to your skin. Positive reaction = white coagulate

Sugar: Test 5 ml. of the beaker water in a test tube with 5 ml. of Benedict's solution. Heat the solution. Positive reaction = green, yellow, orange, or red precipitate

Starch: Add several drops of IKI solution to 2 ml. of the beaker water in a test tube. Positive reaction = blue-black color

Sodium chloride: Place 2 ml. of beaker water in a test tube and add several drops of 1% silver nitrate. Positive reaction = white precipitate

## CONCLUSIONS

1. Distinguish between filtration and dialysis.

2. What determines whether a substance undergoes dialysis?

## PART II   ACTIVE PROCESSES

### ACTIVE TRANSPORT

*Active transport* is a process by which substances, usually ions, are transported across a plasma membrane from an area of lower to higher concentration.

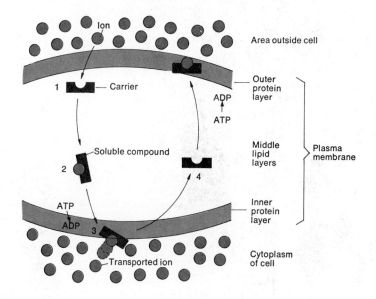

**Figure 3–4.** Proposed mechanism for active transport.

The sequence of events believed to be involved in active transport is shown in Figure 3-4. Describe what is happening in each of the numbered steps.

Step 1.


Step 2.


Step 3.


Step 4.



**CONCLUSIONS**

1. How does active transport differ from passive processes in terms of direction of movement?


2. What is the source of energy for active transport?

**PHAGOCYTOSIS**

*Phagocytosis*, or cell "eating," is the engulfment of solid particles by pseudopodia of the cell. Once the particle is surrounded by the membrane, the membrane folds inward and forms a vacuole around the particle. Enzymes are secreted into the vacuole and the particle is digested. Label the phagocytosis sequence shown in Figure 3-5.

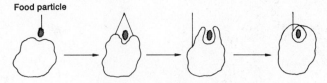

**Figure 3–5.** Phagocytosis.

## CONCLUSIONS

1. What is the source of energy for phagocytosis?

2. Why is the process important to cells?

## PINOCYTOSIS

*Pinocytosis*, or cell "drinking," is the engulfment of a liquid. The liquid is attracted to the surface of the membrane, the membrane folds inward, surrounds the liquid, and detaches from the rest of the intact membrane. Label the pinocytosis sequence shown in Figure 3–6.

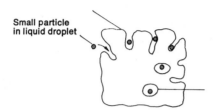

Small particle in liquid droplet

**Figure 3–6.** Pinocytosis.

## CONCLUSIONS

1. What is the source of energy for pinocytosis?

2. How do phagocytosis and pinocytosis differ?

## STUDENT ACTIVITY

For each process listed, indicate where it occurs in the body or why it is important to the body:

1. diffusion

2. osmosis

3. filtration

4. dialysis

5. active transport

6. phagocytosis


7. pinocytosis

# MODULE 4
# CELL DIVISION:
# MITOSIS AND CYTOKINESIS

**OBJECTIVES**

1. To describe and identify the stages of cell division

2. To explain the importance of cell division

**Cell division** is the basic mechanism by which cells reproduce themselves. It provides the body with a means of growth and replacement of diseased or damaged cells. In the overall process, a single parent cell divides and produces 2 daughter cells, each exactly like the starting parent cell.

Cell division is commonly viewed as consisting of 2 principal activities-*mitosis*, or division of the nucleus, and *cytokinesis*, or division of the cytoplasm. Normally, once mitosis is initiated, cytokinesis follows and both continue without interruption until 2 identical daughter cells are formed. Another kind of cell division called meiosis will be studied in module 43. In this process, ova and sperm cells are formed in the ovaries and testes, respectively.

Even though cell division is a continuous process, it may be divided into phases or stages for purposes of study.

Obtain a prepared slide of a whitefish blastula that contains all of the stages of cell division and examine it under high power.

When a cell is carrying on every life process except division, it is said to be in *interphase*. During this time, the cell is between divisions. One of the most important activities of interphase is the replication of DNA so that the 2 daughter cells that eventually form will each have the same kind and amount of DNA. Scan your slide and find a cell in interphase. Such a parent cell is characterized by a clearly defined nuclear membrane. Within the nucleus look for the nucleolus, karyolymph, and the DNA in the form of a granular substance called *chromatin*. Make a labeled diagram of an interphase cell in the space provided in Figure 4-1.

Once a cell completes its interphase activities, mitosis begins. **Mitosis,** or nuclear division, is the distribution of chromosomes into 2 separate and equal nuclei. The 4 stages of mitosis are prophase, metaphase, anaphase, and telophase.

A. *Prophase*. This is the initial stage of mitosis and requires about one half the total time involved in cell division. It is characterized by a series of rather conspicuous events. The paired *centrioles*

**27**

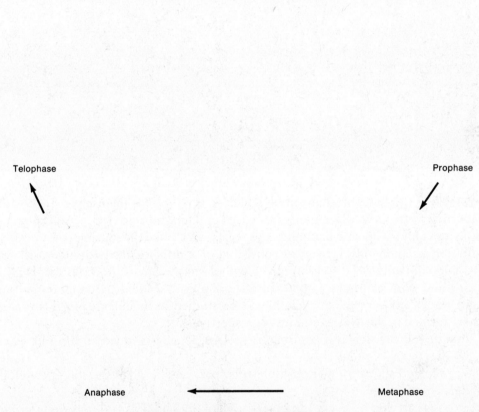

**Figure 4-1.** Mitosis and cytokinesis.

separate and move to opposite poles (ends) of the cell. Once in position they project a series of radiating fibers, the *asters*. The centrioles also form another series of fibers, the *spindle fibers*, which connect the centrioles. At the same time, the chromatin shortens and thickens into distinct rod-shaped bodies, the *chromosomes;* nucleoli become less distinct; and the nuclear membrane disappears. Careful examination of a prophase chromosome reveals that each consists of 2 separate strands of DNA called *chromatids,* joined at a point called a *centromere.* At the end of prophase, the chromatid pairs move toward the equatorial plane (center) of the cell. Isolate a cell in prophase, study it carefully, and draw a labeled diagram in the space provided in Figure 4-1.

**28**

B. *Metaphase*. During the second stage of mitosis, the centromeres line up along the equatorial plane and each chromatid pair becomes attached by its centromere to a spindle fiber. The lengthwise separation of the chromatids takes place, each centromere divides, and independent chromosomes are produced. Find a cell in metaphase and draw a labeled diagram in the space provided in Figure 4-1.

C. *Anaphase*. The third stage of mitosis is characterized by the movement of a complete set of chromosomes to opposite poles of the cell. During this movement, the centromeres attached to spindle fibers lead, while the chromosomes trail like streamers. Draw and label a cell in anaphase in the space provided in Figure 4-1.

D. *Telophase*. The final stage of mitosis consists of a series of events approximately opposite those of prophase. When the chromosome sets reach their respective poles, new nuclear membranes enclose them, chromosomes assume their chromatin form, nucleoli reappear, and the spindle fibers and asters disappear. The formation of 2 nuclei identical to those in cells of interphase terminates telophase and completes a mitotic cycle. Draw and label a telophase cell in the space provided in Figure 4-1.

    **Cytokinesis**, or division of the cytoplasm, begins in late anaphase and terminates in telophase. You can recognize cytokinesis in animal cells by the formation of a *cleavage furrow* which runs around the cell at the equator. The furrow spreads inward as a constricting ring and cuts completely through the cell, forming 2 portions of cytoplasm. At this point, 2 identical daughter cells are formed.

    Following cytokinesis, each daughter cell returns to interphase. Eventually, each cell will grow and undergo mitosis and cytokinesis, and a new divisional cycle begins. Examine your telophase cell again and be sure that it contains a cleavage furrow.

## CONCLUSIONS

1. Define cell division.

2. Why is the process important?

3. Distinguish between mitosis and cytokinesis.

4. Why, under normal circumstances, are daughter cells identical to the parent cells?

5. List the stage of mitosis associated with each of the following:

    a. cytokinesis occurs

    b. nuclear membrane disappears

    c. spindle fibers form

    d. spindle fibers attach to centromeres

    e. centrioles separate

f. chromosomes move to opposite poles

g. nucleoli reappear

## STUDENT ACTIVITY

Below is a series of cells in various stages of division. Under each write the name of the appropriate stage of division.

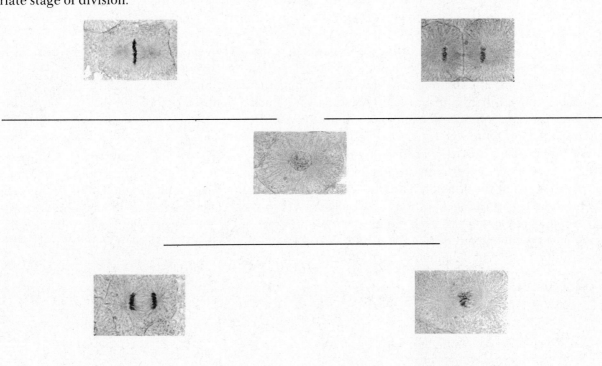

**Figure 4–2.** Photomicrographs courtesy of Carolina Biological Supply Company, Burlington, N.C.

# MODULE 5
# TISSUES

**OBJECTIVES**

1. To define a tissue as a group of specialized cells that perform a specific function

2. To compare the structure, location, and functions of epithelium and selected connective tissue

3. To relate tissue structure to function by observing histological details

4. To define an inflammation and its importance in the repair of damaged tissues

5. To describe the factors that affect tissue repair

A **tissue** is a group of cells operating together to perform a specific function. The cells comprising a tissue are generally derived from the same embryological precursor and are usually found in close physical proximity to one another. The branch of anatomy that deals with the study of tissues is called *histology*. The various body tissues may be categorized into 4 principal kinds: (1) epithelial, (2) connective, (3) muscular, and (4) nervous. In this module you will examine the structure of epithelium and many connective tissues. Other tissues will be studied as parts of the systems to which they belong.

## PART I   EPITHELIUM

*Epithelial tissue* covers body surfaces and lines nearly all cavities of the body.

For each of the types of epithelium listed, obtain a prepared slide and examine it microscopically. Unless otherwise specified by your instructor, examine each under high power. In conjunction with your examination, consult a textbook of anatomy and physiology.

1. *Simple squamous.* After you make your microscopic observation, draw several cells in Table 5-1. In your diagram label the cell membrane, cytoplasm, and nucleus. Also indicate in the Table the location and function of the tissue.

   a. Why is the tissue classified as simple?

b. What does this tell you about the function of the tissue?

c. What is meant by the term squamous?

d. Define endothelium and mesothelium.

2. *Simple cuboidal.* Upon completion of your microscopic examination, fill in the required information in Table 5-1. Once again, label the cell membrane, cytoplasm, and nucleus.
   a. How does the shape of these cells differ from squamous cells?

3. *Simple columnar.* First obtain a slide of simple nonciliated columnar. After examination, complete Table 5-1. In your diagram be sure to label cell membrane, cytoplasm, nucleus, and goblet cell.

   a. What is a goblet cell?

   b. Why is it so called?

   c. What relationship exists between the structure and function of a goblet cell?

   d. What are microvilli?

   e. How is the structure of microvilli related to their function?

   f. Why are goblet cells and microvilli important in the small intestine?

   Now obtain a slide of simple ciliated columnar. Complete Table 5-1 after you make your observation. In your diagram, label cell membrane, cytoplasm, nucleus, goblet cells, and cilia.

   g. What is the function of cilia?

   h. Why are goblet cells and cilia important in the upper respiratory tract?

i. How and where do cilia function in the reproductive system?

4. *Stratified squamous.* Observe your slide and record your observations in Table 5-1. Be sure to indicate the shapes of the cells in the respective layers.

   a. What does the term stratified mean?

   b. What does this tell you about the function of the tissue?

   c. What is meant by keratinization?

   d. Where would you expect to find a keratinized tissue?

5. *Transitional.* Obtain a slide and record your observations in Table 5-1.
   How is the cellular arrangement of this tissue adapted to its function?

6. *Pseudostratified.* After examining your slide, record your observations in Table 5-1.
   Why is this tissue so named?

## CONCLUSIONS

1. Define a tissue.

2. Indicate how epithelium relates to each of the following:

   a. predominance of cells

   b. predominance of intercellular material

   c. nerve supply

   d. blood supply

   e. cellular arrangement

   f. basement membrane

   g. capacity for division

3. For each tissue you examined, relate the structure of the tissue to its function.

4. The secreting portion of glands consists of epithelium specialized for secretion. Distinguish between an exocrine gland and an endocrine gland, and give an example of each.

   a. exocrine

   b. endocrine

5. How is epithelium related to membranes?

6. Distinguish 4 kinds of membranes and indicate where each is found.

   a.

   b.

   c.

   d.

7. Summarize the general functions of epithelium.

TABLE 5–1. EPITHELIUM.

| TISSUE | DIAGRAM | LOCATION | FUNCTION |
|---|---|---|---|
| Simple squamous | | | |
| Simple cuboidal | | | |
| Simple columnar nonciliated | | | |
| ciliated | | | |
| Stratified squamous | | | |
| Transitional | | | |
| Pseudostratified | | | |

## PART II  CONNECTIVE TISSUE

For each of the types of *connective tissue* listed, obtain a prepared slide and examine microscopically. Unless otherwise specified by your instructor, examine each under high power. In conjunction with your examination, consult a textbook of anatomy and physiology.

**35**

1. *Loose (areolar)*. After your microscopic examination, complete Table 5-2. In your diagram, label collagenous and elastic fibers, fibroblasts, macrophages, plasma cells, and mast cells.

   a. What is hyaluronic acid?

   b. What is the clinical significance of injecting hyaluronidase into loose connective tissue?

   c. Compare the following with regard to structure, arrangement, and function.

      1. collagenous fibers

      2. elastic fibers

      3. reticular fibers

   d. Describe the function of each of the following cells in loose connective tissue.

      1. fibroblast

      2. macrophage

      3. plasma cell

      4. mast cell

   e. How is loose connective tissue related to the subcutaneous layer under the skin?

1. *Adipose*. Record your observations in Table 5-2 after examining the slide. Label the plasma membrane, cytoplasm, nucleus, and fat.

   a. Why are these cells called "signet ring" cells?

   b. How are adipose cells adapted to their function?

3. *Collagenous*. This tissue is also called dense fibrous connective tissue. Observe the tissue and complete the appropriate portion of Table 5-2. Label collagenous fibers and fibroblasts.

   a. Why is this tissue the principal component of tendons, ligaments, aponeuroses, and fascia?

4. *Elastic*. Observe the tissue and record the data in Table 5-2. Label elastic fibers and fibroblasts.

   a. How is the arrangement of fibroblasts in this tissue different from that in collagenous connective tissue?

   b. How is the structure of elastic tissue adapted to its function?

   c. Why is the tissue important in arteries?

5. *Reticular.* Observe the tissue and record your observations in Table 5-2. Label the reticular fibers and fibroblasts.

  a. What is the arrangement of fibers?

  b. How is this tissue adapted to its function?

6. *Hyaline cartilage.* Examine the tissue and record your observations in Table 5-2. Label the chondrocytes, lacunae, and matrix.

  a. Are all chondrocytes in lacunae? Explain.

  b. What is another name for hyaline cartilage?

7. *Fibrocartilage.* After examining the tissue, complete the appropriate space in Table 5-2. Label collagenous fibers, chondrocytes, and lacunae.

  a. How are the collagenous fibers arranged?

8. *Elastic cartilage.* Examine the tissue and record your observations in Table 5-2. Label elastic fibers, chondrocytes, and lacunae.

  a. How are the elastic fibers arranged?

  b. How is this tissue adapted to its function?

## CONCLUSIONS

1. Indicate how connective tissue relates to each of the following:

  a. predominance of cells

  b. predominance of matrix

  c. nerve supply

  d. blood supply

  e. cellular arrangement

f. basement membrane

g. capacity for division

2. For each tissue you examined, relate the structure of the tissue to its function.

3. How is connective tissue related to membranes?

4. Distinguish between embryonal and adult connective tissue.
   a. embryonal

   b. adult

5. Summarize the general functions of connective tissue.

TABLE 5–2. CONNECTIVE TISSUES

| TISSUE | DIAGRAM | LOCATION | FUNCTION |
|---|---|---|---|
| Loose | | | |
| Adipose | | | |
| Collagenous | | | |
| Elastic | | | |
| Reticular | | | |
| Hyaline cartilage | | | |
| Fibrocartilage | | | |
| Elastic cartilage | | | |

## PART III   TISSUE DAMAGE AND REPAIR

Consult a textbook of anatomy and physiology and refer to the sections dealing with inflammation. Based upon your reading, answer the following questions:

1.  Define an inflammation.


2.  In what respect is inflammation considered as a homeostatic mechanism?


3.  Describe the responses of the body to tissue damage.


4.  What causes scarring when tissues are damaged?


5.  Why is granulation tissue important in the repair process?


6.  What conditions affect tissue repair?


**STUDENT ACTIVITY**

Compare the regenerative capacities of epithelium, connective tissue, muscle tissue, and nervous tissue.

# MODULE 6
# THE INTEGUMENTARY SYSTEM: STRUCTURE, PHYSIOLOGY, AND DISORDERS

**OBJECTIVES**

1. To identify the structural features of the epidermis and dermis
2. To compare the structure and functions of hair, sebaceous glands, and sweat glands
3. To define selected disorders related to the skin
4. To define medical terminology associated with the integumentary system

The skin, together with all organs derived from it (hair, nails, and glands), and sensory receptors constitute the **integumentary system.** Whereas an *organ* consists of an aggregation of tissues which perform a definite function, a *system* is a group of organs that operate together to perform specialized functions.

## PART I  THE SKIN

The **skin** is the largest organ of the body, occupying a surface area of about 3000 square inches. Among the functions performed by the skin are protection against bacterial invasion, prevention of excessive water loss, protection against harmful light rays, control of body temperature, storage of chemicals, excretion of water and salts, synthesis of several compounds, and the reception of stimuli for touch, pressure, pain, and temperature.

The skin consists of an outer, thinner *epidermis* and an inner, thicker *dermis.* Below the dermis is the *subcutaneous layer* (superficial fascia) that attaches the skin to underlying tissues and organs.

Obtain a prepared slide of human skin and carefully examine the epidermis. Identify the following layers from the outside inward:

*stratum corneum*—25 to 30 rows of flat, dead cells that contain keratin
*stratum lucidum*—3 to 4 rows of clear, flat cells that contain eleidin
*stratum granulosum*—2 to 3 rows of flat cells that contain keratohyaline
*stratum spinosum*—8 to 10 rows of polygonal cells
*stratum basale (germinativum)*—Single layer of columnar cells that constantly undergo division.

**41**

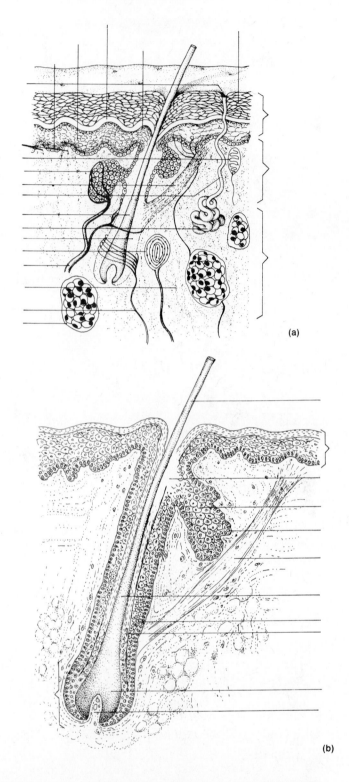

(a)

(b)

**Figure 6–1.** The skin. (a) Structure of the skin. (b) Details of a hair.

Label the 5 epidermal layers you just identified in Figure 6-1a.
Now carefully examine the dermis. Identify the following structures in the dermis:

*collagenous* and *elastic fibers*
*blood vessels*

**42**

*nerves*
*sweat glands*
*sebaceous glands*
*papillary region*—upper 1/5 of dermis
*Meissner's corpuscles*
*Pacinian corpuscles*
*reticular region*—bottom 4/5 of dermis

Label these structures in Figure 6-1a. Study the structures of the skin very carefully on the skin charts and models.

**CONCLUSIONS**

1. Why is the skin an organ?

2. What tissues comprise the subcutaneous layer?

3. What is a subcutaneous injection?

4. What type of epithelium makes up the epidermis?

5. Define each of the following:

   a. keratin

   b. eleidin

   c. keratohyaline

   d. melanin

6. What are papillae?

7. What stimuli are picked up by:

   a. Meissner's corpuscles

   b. Pacinian corpuscles

## PART II   HAIR AND GLANDS

*Hair* is a growth of the epidermis variously distributed over the body. Each hair consists of a shaft, the visible portion above the skin surface, and a root, the portion that penetrates deep into the dermis.

Refer to Figure 6-1b and label the following parts of a hair.

1. *shaft*
2. *root*
3. *follicle*
4. *bulb*
5. *dermal papilla*

6. *matrix*
7. *arrector pili*
8. *sebaceous glands*
9. also label the *sweat glands* in Figure 6-1a.

**CONCLUSIONS**
1. What is the function of hair?


2. List several places where hair is not found.




3. Define the following:
   a. follicle

   b. bulb

   c. dermal papilla

   d. matrix

4. How are lost hairs replaced?


5. How is "gooseflesh" produced?



6. What is a sebaceous gland?


7. Define sebum.




8. Describe the distribution of sebaceous glands.




9. Where are sweat glands located?

10. What is the composition of perspiration?


11. Briefly define the following disorders of the integumentary system:
    a. acne


    b. burn


12. How are burns classified? Describe each.


13. Explain the "rule of nines" for estimating the extent of a burn.


14. Define each of the following terms:
    a. albinism
    b. anhidrosis
    c. callus
    d. carbuncle
    e. comedo
    f. cyst
    g. decubitus ulcer
    h. epidermophytosis
    i. furuncle
    j. hypodermic
    k. intradermal
    l. melanoma
    m. nevi
    n. nodule
    o. papule
    p. polyp
    q. pustule
    r. wart
    s. tinea

**STUDENT ACTIVITY**

Describe, by means of a labeled diagram, how the skin helps to maintain a constant body temperature.

# MODULE 7
# STRUCTURAL PLAN
# OF THE BODY

**OBJECTIVES**

1. To identify by name and location the principal body cavities

2. To identify the major organs located within the principal body cavities

3. To define the membranes of the body

4. To examine the body in the anatomical position

5. To associate common and anatomical terms for various regions of the body

6. To define selected directional terms

7. To describe the common anatomical planes of the body

In this module you will be introduced to the organization of the human body. This will be accomplished through a study of its cavities, membranes, the anatomical position, regional names, directional terms, and planes (or sections).

## PART I   BODY CAVITIES

Examine a torso and charts of the human body and using your textbook as a reference, label the *body cavities* shown in Figure 7-1.

**CONCLUSIONS**

1. Why is man classified as a vertebrate?

2. What is bilateral symmetry?

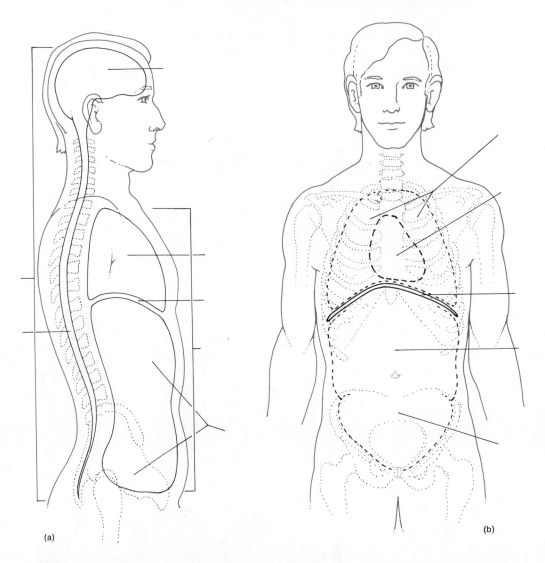

**Figure 7–1.** Body cavities. (a) Median section. (b) Subdivisions of the ventral body cavity.

3. What is a body cavity?

4. What is the makeup of the walls of the dorsal body cavity?

5. What are the 2 subdivisions of the dorsal cavity and what organs are found in each?
   a.
   b.

6. What is the makeup of the walls of the ventral body cavity?

7. What organs are found in each of the following subdivisions of the ventral body cavity?

a. thoracic


b. pleural

c. pericardial

d. mediastinum

e. abdominal



f. pelvic



8. Define viscera.


9. What separates the thoracic from the abdominal cavity?


10. What separates the abdominal from the pelvic cavity?


11. Define retroperitoneal and give an example of retroperitoneal organs.

## PART II   MEMBRANES

The combination of an epithelium and an underlying connective tissue is called a **membrane**. The principal membranes of the body are:

1. *Cutaneous*—skin (Module 6).
2. *Synovial*—line cavities that do not open to exterior and secrete synovia to reduce friction; associated with joints.
3. *Mucous*—line cavities that open to exterior and secrete mucus; associated with the mouth and digestive tract, nose and respiratory passages, reproductive system, and urinary system.
4. *Serous*—line cavities that do not open to the exterior and also cover organs in the cavities; they secrete serous fluid to prevent friction; the portion that lines the cavity is called the *parietal* portion, that which covers the organ in the cavity is called the *visceral* portion.

Consult your textbook and find an example of each of the membranes.

## CONCLUSIONS

1. Define a membrane.


2. Why is the skin a membrane?

3. Name the membrane associated with the following:

    a. covering of heart

    b. lining the heart cavity

    c. covering of lungs

    d. lining the lung cavity

    e. thoracic wall

    f. abdominal wall

    g. covering the small intestine

    h. pelvic wall

    i. lining the knee joint

    j. lining the throat

    h. covering the uterus

## PART III  THE ANATOMICAL POSITION AND REGIONAL NAMES

Figure 7-2 contains a subject in the *anatomical position* in anterior and posterior views. Also shown are selected common names for various regions of the body.

### CONCLUSIONS

1. List the characteristics of a subject in the anatomical position

2. Next to each common name labeled in Figure 7-2, list the anatomical term (examples: skull= cranium, armpit=axilla, head=caput).

## PART IV  DIRECTIONAL TERMS

In order to explain exactly where a structure of the body is located, it is a standard procedure to use *directional terms*. For example, if you want to point out the position of the breastline to someone who knows where the collarbone is, you could say that "the breastbone is inferior and medial to the collarbone." In other words, you have just said that "the breastbone is farther away from the head (inferior) and toward the midline of the body (medial) relative to the collarbone."

Commonly used directional terms are as follows:

1. *Superior (cephalic)*    Toward the head; toward the upper part of a structure.
2. *Inferior (caudal)*    Away from the head; toward the lower part of a structure.
3. *Anterior (ventral)*    Nearer to or at the front of the body.
4. *Posterior (dorsal)*    Nearer to or at the back of the body.
5. *Medial*    Nearer to the midline of the body.
6. *Lateral*    Farther from the midline of the body.
7. *Proximal*    Nearer the attachment of a limb to the trunk.
8. *Distal*    Away from the attachment of a limb to the trunk.
9. *External (superficial)*    Toward the surface of the body.
10. *Internal (deep)*    Away from the surface of the body.

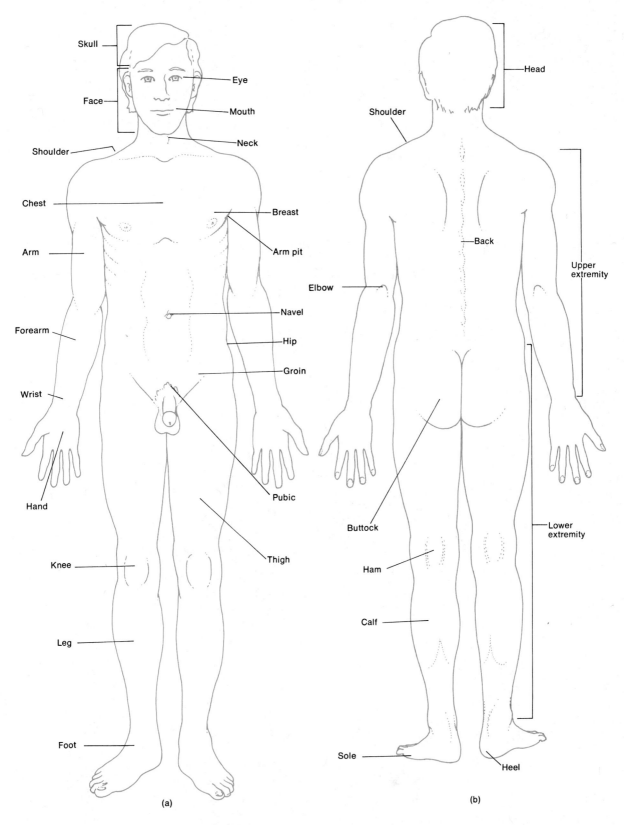

**Figure 7–2.** The anatomical position and regional names. (a) Anterior view. (b) Posterior view.

We would now like you to refer to Figure 7-3, which is a labeled diagram of selected parts of the body. Using only parts labeled, describe the location of the following:

a. The ulna is on the _____ side of the forearm.

b. The humerus is _____ to the radius.

c. The heart is _____ to the liver.

d. The urinary bladder is _____ to the rectum.

e. The lungs are _____ to the heart.

f. The phalanges are _____ to the wrist.

g. The muscles of the arm are _____ to the skin of the arm.

h. The wrist is _____ to the elbow.

i. The kidneys are _____ to the backbone.

j. The muscles of the abdominal wall are _____ to the viscera in the abdominal cavity.

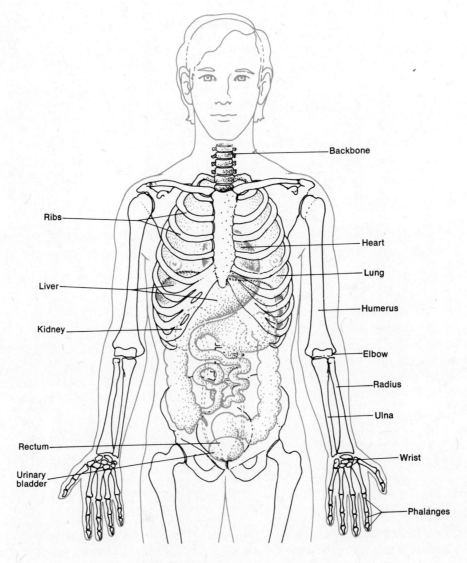

**Figure 7-3.** Directional terms.

**CONCLUSIONS**

1. Why are directional terms important?

2. Write 5 complete sentences using 2 or more different directional terms in each sentence. (Example: The heart is *posterior* to the sternum and *superior* to the liver.)

   a.

   b.

   c.

   d.

   e.

## PART V    PLANES OF THE BODY

The structural plan of the body may also be analyzed with respect to *planes* (or sections) passing through it. Such planes are frequently used to show the relationship to each other of several structures in a region.

Commonly used planes are as follows:

1. *Midsagittal*—runs through the midline of the body vertical to the ground and divides the body into equal right and left sides.
2. *Sagittal*—runs vertical to the ground and divides the body into unequal left and right portions.
3. *Frontal (coronal)*—runs vertical to the ground and divides the body into anterior and posterior portions.
4. *Horizontal (transverse)*—runs parallel to the ground and divides the body into superior and inferior portions.

   Refer to Figure 7–4 and label the planes shown.

**STUDENT ACTIVITIES**

1. Figure 7–5 is a diagram of the 9 regions of the abdomen. Label each region. Next to the name of the region write the name of the entire organ or the major portion of the organ found in the region.
2. Have your partner assume the anatomical position. Point to a region on your partner's body, giving your partner the common name for the region. Have your partner tell you the anatomical name for the region. Reverse roles.

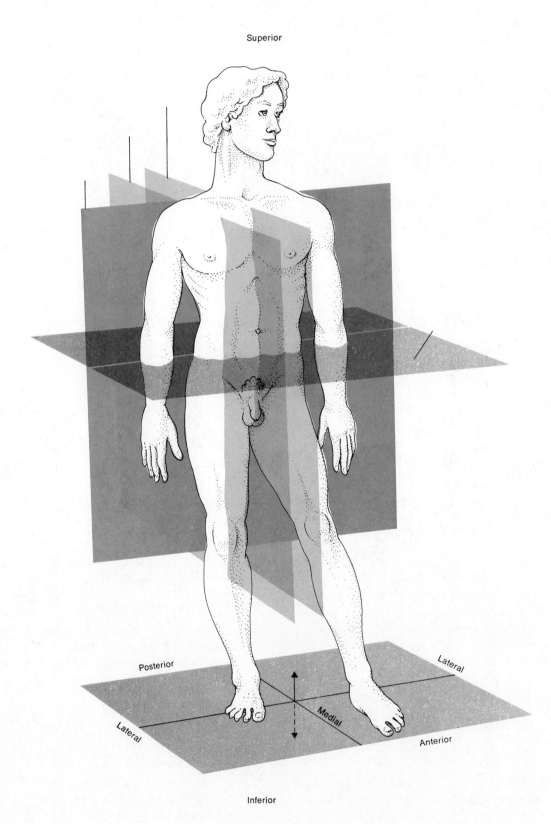

**Figure 7–4.** Planes of the body.

3. Select any 2 parts of the body and have your partner supply the appropriate directional term to describe the relationship of the parts to each other.
4. Using an orange and a knife, make a midsagittal, sagittal, frontal, and transverse plane through the orange.

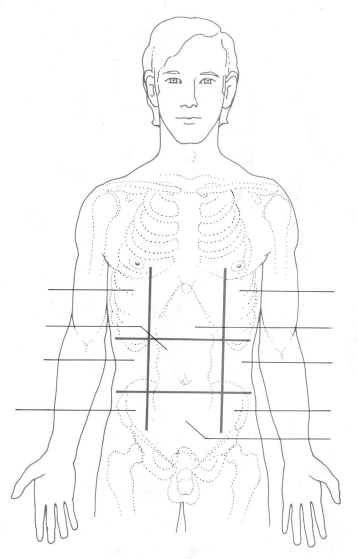

**Figure 7–5.** Abdominal regions.

# MODULE 8
# GROSS STRUCTURE AND HISTOLOGY OF BONES

**OBJECTIVES**

1. To compare the external and internal gross structure of a long bone

2. To examine the histological features of compact bone

Structurally, the **skeletal system** consists of 2 types of connective tissue, cartilage and bone. Since we have already discussed the microscopic structure of cartilage in module 5, we will study only the histology of *bone (osseous) tissue* at this point. We will first, however, examine the gross structure of a typical bone.

## PART I   GROSS STRUCTURE OF A LONG BONE

Examine a fresh long bone and locate the following structures:

1. *periosteum*—a dense white fibrous membrane covering the surface of the bone, except for the areas covered by cartilage.
2. *articular cartilage*—thin layer of hyaline cartilage covering the ends of the bone where joints are formed.
3. *epiphysis*—end or extremity of a bone.
4. *diaphysis*—elongated shaft of a bone.

Now examine a long bone that has been sectioned longitudinally and locate the following structures:

1. *medullary cavity*—cavity within the diaphysis that contains yellow marrow.
2. *endosteum*—a membrane that lines the medullary cavity.
3. *compact (dense) bone*—bone that forms the bulk of the diaphysis and covers the epiphysis.
4. *spongy (cancellous) bone*—bone that forms a small portion of the diaphysis lining the medullary cavity and the bulk of the epiphysis; spongy bone of the epiphysis contains red marrow.

Label the parts of a long bone indicated in Figure 8–1.

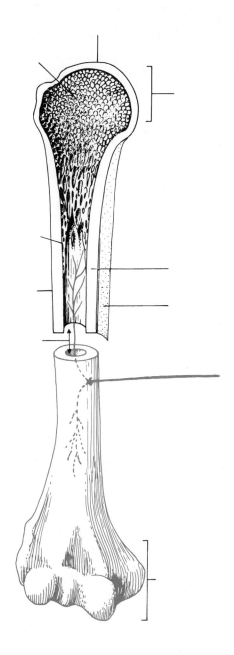

**Figure 8–1.** Parts of a long bone.

## CONCLUSIONS

1. What is the composition of the periosteum?

2. What is the function of the periosteum?

3. Why does cartilage cover the epiphysis of a long bone?

4. What is the function of yellow and red marrow?

   a. yellow

   b. red

5. What is the visible difference between compact and spongy bone?

6. Distinguish between the diaphysis and the epiphysis of a long bone.

## PART II   HISTOLOGY OF COMPACT BONE

Bone tissue, like other connective tissues, contains a large amount of intercellular substance which surrounds widely separated cells. The hard intercellular substance contains calcium salts and collagenous fibers. The cells are referred to as *osteocytes*.

Obtain a prepared slide of compact bone in which several Haversian systems are shown in cross section. Observe under high power. Look for the following structures:

1. *Haversian canal*—circular canal in the center of a Haversian system that runs longitudinally through the bone; the canal contains blood vessels, lymphatics, and nerves.
2. *lamellae*—concentric layers of calcified intercellular substance.
3. *lacunae*—spaces or cavities between lamellae that contain bone cells.
4. *canaliculi*—minute canals that radiate out in all directions from the lacunae and interconnect with each other.
5. *osteocyte*—a bone cell located within a lacuna.
6. *Haversian system*—a Haversian canal plus its surrounding lamellae, lacunae, canaliculi, and osteocytes; the structural unit of compact bone.

Now obtain a prepared slide of a longitudinal section of compact bone and examine under high power. Locate the following:

1. *Volkmann's canals*—canals that extend inward from the periosteum and contain blood vessels, lymphatics and nerves; they extend into the Haversian canals and medullary cavity.
2. *endosteum*
3. *medullary cavity*
4. *lamellae*
5. *lacunae*
6. *canaliculi*
7. *osteocytes*

Label the parts indicated in the microscopic view of bone in Figure 8-2.

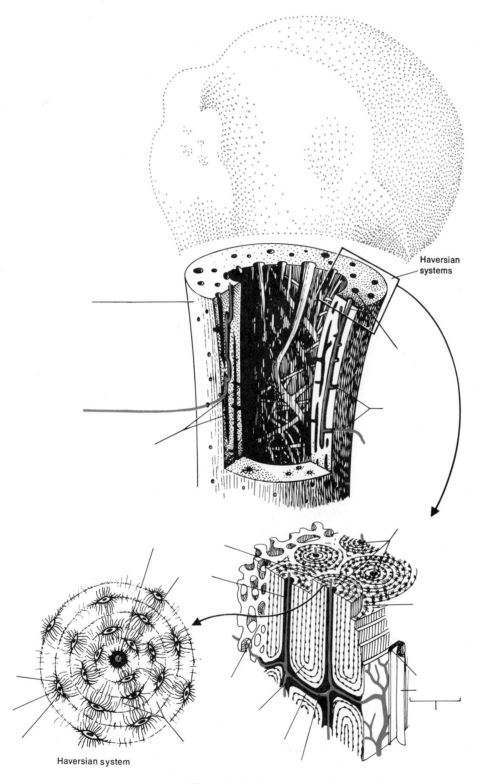

Haversian
systems

Haversian system

**Figure 8–2.** Microscopic structure of bone.

**CONCLUSIONS**

1. What is an osteocyte?

2. Define a Haversian system.

3. Why is bone classified as a connective tissue?

4. Trace the route taken by blood through a bone. Start with the periosteum and end with the medullary cavity.

**STUDENT ACTIVITY**

Support the statement that "bone is a living tissue."

# MODULE 9
# PHYSIOLOGY AND DISORDERS OF BONES

**OBJECTIVES**

1. To define intramembranous (dermal) and endochondral (intracartilaginous) ossification

2. To identify the bones formed by intramembranous and endochondral ossification

3. To explain the sequence of events involved in endochondral ossification

4. To describe selected disorders of bones

5. To define medical terms associated with the skeletal system

In this module you will study some of the functional aspects of bones and some of the clinical conditions associated with disorders of bones.

## PART I   BONE FORMATION

The process by which bone forms is called **ossification**. The process begins at about the sixth week of embryonic life and continues to adulthood. The "skeleton" of an embryo is composed of fibrous membranes and cartilage shaped like bones. The formation of bone directly on or within fibrous membranes is called *intramembranous ossification*. The formation of bone in the cartilage is referred to as *endochondral ossification*. The only basic difference between the two is the method by which the bone forms; the result is the same—bone formation.

Intramembranous ossification is the simpler type of bone formation. Bones formed by this process include the flat bones of the roof of the skull, the mandible, and probably part of the collar bones. Color these bones blue and label them in Figure 9–1.

In endochondral ossification, cartilage is replaced by bone. Most bones of the body form by this process. Color the bones red (Figure 9–1) that form by endochondral ossification. Also label the bones.

Using your textbook as a reference, briefly describe the sequence of events involved in endochondral ossification in Figure 9–2.

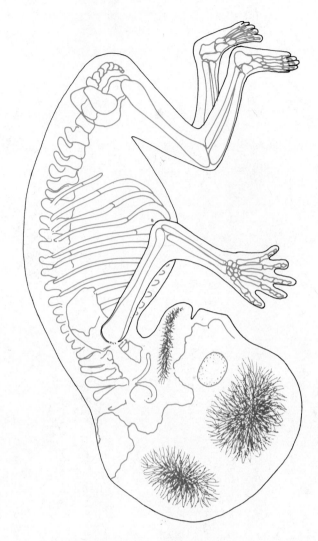

**Figure 9–1.** Ten-week human embryo.

**CONCLUSIONS**

1. Define ossification.


2. Compare intramembranous and endochondral ossification.

    a. intramembranous


    b. endochondral


3. Why is the epiphyseal plate important?


4. What is a primary ossification center?

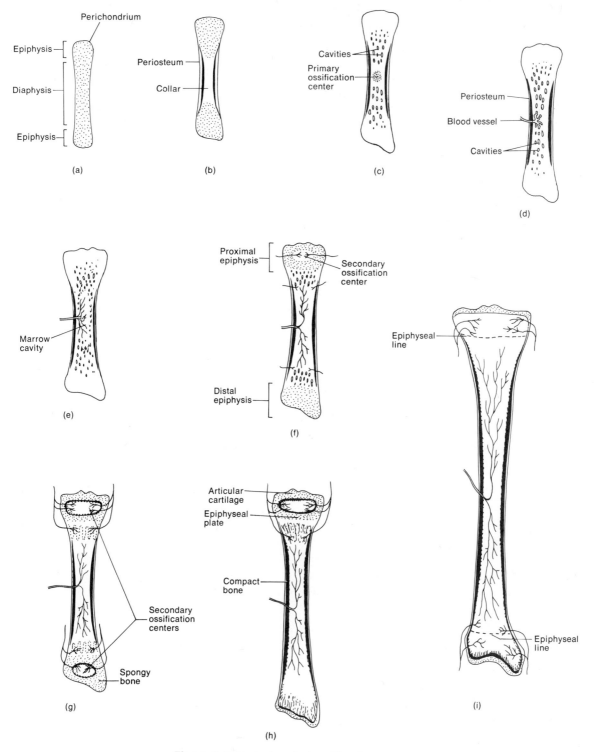

**Figure 9–2.** Endochondral ossification of the tibia.

5. Where is the primary ossification center located?

6. Where are the secondary ossification centers located?

7. Describe briefly how a bone grows in length and diameter.

8. How are bones remodeled?

9. Explain the factors necessary for bone growth.

## PART II   BONE DISORDERS

Consult your textbook and briefly define the following disorders. Be sure to indicate the cause, symptoms, and treatment wherever possible.

1. rickets

2. osteomalacia

3. osteoporosis

4. Paget's disease

5. osteomyelitis

**CONCLUSIONS**

1. How is rickets prevented?

2. How is demineralization related to osteomalacia?

3. Why does osteoporosis primarily affect middle-aged and elderly people?

4. Why is Paget's disease an excellent example of a homeostatic imbalance?

5. Distinguish among the following types of tumors:
   a. osteogenic sarcoma
   b. myeloma
   c. chondrosarcoma

6. Define each of the following medical terms:
   a. achondroplasia
   b. Brodie's abscess
   c. craniotomy
   d. necrosis
   e. metastasis
   f. malignancy
   g. benign tumor
   h. osteitis
   i. osteoarthritis

j. osteoblastoma

k. osteochondroma

l. Pott's disease

**STUDENT ACTIVITY**

If available examine x-rays of the disorders discussed in this module, as well as any others that your instructor may provide.

# MODULE 10
# CLASSIFICATION
# AND STRUCTURAL
# FEATURES OF BONES

**OBJECTIVES**

1. To identify the 4 principal kinds of bones and to note the relationship between structure and function
2. To distinguish the various markings on the surfaces of bones and to relate their structure to function
3. To contrast the principal subdivisions of the axial and appendicular skeletons

In this module you will learn the names and locations of the bones of the body. In addition, you will also identify the principal markings on bones.

## PART I   TYPES OF BONES

The 206 named bones of the body may be classified into 4 principal types:

1. *long*—have greater length than width and consist of a diaphysis and 2 epiphyses; contain more compact than spongy bone. Example: humerus
2. *short*—somewhat cube-shaped and differences in length and width are not important; contain more spongy than compact bone. Example: wrist bones
3. *flat*—generally thin and flat and are composed of 2 more or less parallel plates of compact bone enclosing a layer of spongy bone. Example: sternum
4. *irregular*—very complex shapes; cannot be grouped into any of the 3 categories just described. Example: vertebrae

Other types of bones that are recognized include:

1. *wormian (sutural)*—small clusters of bones between certain cranial bones; variable in number.
2. *sesamoid*—small bones found in tendons; variable in number; the only constant sesamoid bones are the paired knee caps; "sesamoid" means resembling a grain of sesame.

**67**

Examine the disarticulated skeleton, Beauchene skull, and articulated skeleton and find several examples of long, short, flat, and irregular bones. List the bones below.

1. long

2. short

3. flat

4. irregular

Examine the articulated skeleton again and identify the wormian and sesamoid bones.

**CONCLUSIONS**

1. What is the function of:

   a. long bones

   b. short bones

   c. flat bones

   d. irregular bones

2. Between which skull bones did you find the wormian bones?

3. Why are wormian and sesamoid bones (except for the patellae) not classified with the 206 named bones?

## PART II   SURFACE MARKINGS

The surfaces of bones contain various *markings* which serve specific functions. For example, such markings form joints; attach muscles, tendons, and ligaments; and serve as passageways for blood vessels and nerves.

Before you actually observe these markings, consult your textbook and define the following markings. As you look up the definition, be sure to find an example of each. After you have done this, examine the bones made available and list the example you have found. One sample is given.

**68**

| MARKING | DEFINITION | EXAMPLE |
|---|---|---|
| A. Depressions and openings<br>  1. Fissure | Narrow, cleftlike opening between adjacent bones or parts of bones through which blood vessels and nerves pass. | Superior orbital fissure of the sphenoid bone |
|   2. Foramen | | |
|   3. Meatus | | |
|   4. Paranasal sinus | | |
|   5. Groove | | |
|   6. Fossa | | |
| B. Processes that form joints<br>  1. Condyle | | |
|   2. Head | | |
|   3. Facet | | |
| C. Processes that do not form joints<br>  1. Tubercle | | |
|   2. Tuberosity | | |
|   3. Trochanter | | |
|   4. Crest | | |
|   5. Line | | |
|   6. Spine | | |
|   7. Epicondyle | | |

**CONCLUSIONS**

1. Define a surface marking.

2. Select any 10 surface markings just studied and give the specific function of each.

   a.

   b.

   c.

   d.

   e.

   f.

   g.

   h.

   i.

   j.

## PART III   DIVISIONS OF THE SKELETON

The bones of the adult skeleton are grouped into 2 divisions: axial and appendicular. The **axial skeleton** consists of the bones that lie around the axis of the body. The axis is a straight line that runs along the center of gravity of the body, through the head, and down to the space between the feet. The **appendicular skeleton** consists of the bones of the appendages (upper and lower) and the girdles (shoulder and pelvic).

Using your textbook and the articulated skeletons as references, complete the following organizational chart of the skeleton and be sure to include the number of bones in each category. A sample is provided.

| AXIAL SKELETON | NUMBER OF BONES |
|---|---|
| A.  Skull | |
| 1.  cranium | 8 |
| 2.  face | 14 |
| B.  Hyoid | |
| C.  Ear bones | |
| D.  Vertebral column | |
| E.  Thorax | |
| 1. | |
| 2. | |

| APPENDICULAR SKELETON | NUMBER OF BONES |
|---|---|
| A. Shoulder girdles | |
|    1. | |
|    2. | |
| B. Upper extremities | |
|    1. | |
|    2. | |
|    3. | |
|    4. | |
|    5. | |
|    6. | |
| C. Pelvic girdles | |
|    1. | |
| D. Lower extremities | |
|    1. | |
|    2. | |
|    3. | |
|    4. | |
|    5. | |
|    6. | |
|    7. | |

## CONCLUSIONS

1. Distinguish between the axial and appendicular divisions of the skeleton.


2. How many bones comprise each?

   a. axial

   b. appendicular

3. Define the following terms:

   a. cranium

   b. face

   c. hyoid

   d. vertebral column

e. thorax

f. shoulder girdle

g. upper extremity

h. pelvic girdle

i. lower extremity

## STUDENT ACTIVITY

Select a bone from a disarticulated skeleton and hand it to your partner. Ask your partner to do the following:

1. Classify it by type (long, short, flat, irregular)
2. Identify any prominent markings
3. Determine its place in an articulated skeleton (thorax, vertebral column, girdle, extremity, etc.)

Do this with several bones and then reverse roles with your partner.

# MODULE 11
# BONES OF THE BODY

## OBJECTIVES

1. To identify the bones of the skull

2. To locate the principal sutures and fontanels

3. To identify the sinuses of the skull

4. To compare the curves and vertebrae of the spinal column

5. To identify the bones of the thorax

6. To identify the bones of the shoulder girdle

7. To contrast the bones of the upper extremity

8. To identify the pelvic girdle

9. To compare the structural differences between the male and female pelvis

10. To contrast the bones of the lower extremity

11. To identify the arches of the foot

In this module you will study the names and locations of the bones of the body and their markings by examining various regions of the skeleton.

## PART I   THE BONES OF THE ADULT SKULL

The **skull** is composed of 2 sets of bones—*cranial* and *facial*. The 8 cranial bones are: frontal (1), parietal (2), temporal (2), occipital (1), sphenoid (1), and ethmoid (1). The 14 facial bones are: nasal (2), maxillae (2), zygomatic (2), mandible (1), lacrimal (2), palatine (2), inferior conchae (2), and vomer (1).

Figure 11-1 contains 3 views of the skull—anterior, lateral, and medial. In each view, color the cranial bones red and the facial bones blue.

Identify each bone on the Beauchene skull and the articulated skull.

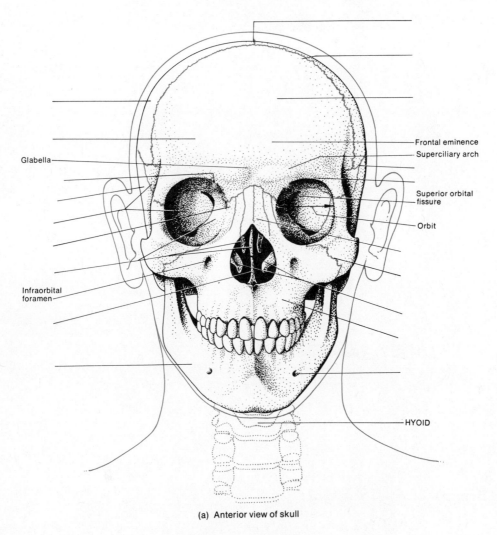

(a) Anterior view of skull

**Figure 11–1.** The skull.

## CONCLUSIONS

1. What is the orbit?

2. What bones form the orbit?

3. How are the skull bones held together?

4. After defining the following, locate each on the Beauchene skull and articulated skull and label those shown in Figure 11–1.

   a. squama

   b. supraorbital margin

   c. supraorbital foramen

   d. zygomatic process

   e. zygomatic arch

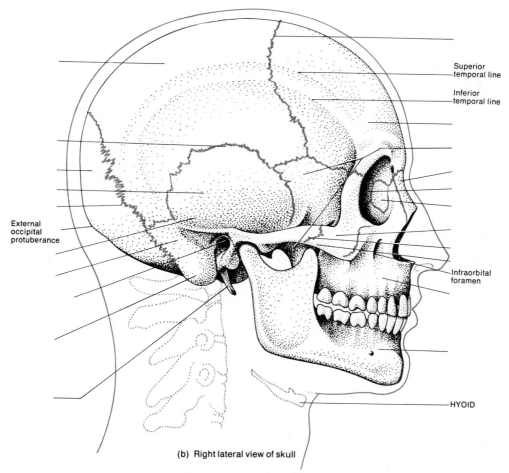

Superior
temporal line

Inferior
temporal line

External
occipital
protuberance

Infraorbital
foramen

HYOID

(b) Right lateral view of skull

**Figure 11–1 continued.**

f.  mandibular fossa

g.  mastoid process

h.  external auditory meatus

i.  styloid process

j.  foramen magnum

k.  occipital condyles

l.  sella turcica

m. greater wings

n.  laminae

o.  perpendicular plate

p.  cribriform plate

q.  crista galli

r.  alveolar processes

s.  temporal process

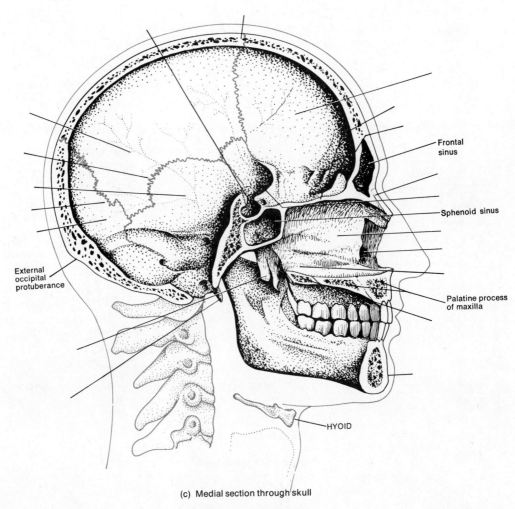

Frontal sinus

Sphenoid sinus

Palatine process of maxilla

External occipital protuberance

HYOID

(c) Medial section through skull

**Figure 11–1 continued.**

    t.  ramus

    u.  condylar process

    v.  mental foramen

    w.  horizontal plates

5. What bones form the nasal septum?


6. Define mastoiditis


7. Why is the sphenoid bone called the "keystone" of the cranial floor?


8. What bones form the hard palate?

9. What is cleft palate?

10. What bones form the nasal cavity?

## PART II SUTURES AND FONTANELS

A **suture** is an immovable joint between skull bones. The 4 prominent sutures are coronal, sagittal, lambdoidal, and squamosal.

Refer to Figure 11-1 and label the 4 sutures in the anterior and lateral views. Locate these sutures on the articulated skeleton.

At birth, membrane-filled spaces called **fontanels** are found between skull bones. The principal fontanels are anterior (frontal), posterior (occipital), anterolateral (sphenoidal), and posterolateral (mastoid).

Refer to Figure 11-2 and label the fontanels, sutures, and the bones. Now locate the fontanels on the skull of a newborn infant.

### CONCLUSIONS

1. Why are sutures immovable?

2. What bones are united at the following sutures?

   a. coronal

   b. sagittal

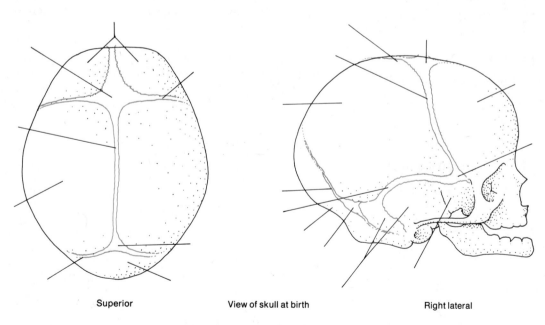

Superior        View of skull at birth        Right lateral

**Figure 11-2.** Fontanels.

c. lambdoidal

d. squamosal

3. What are the functions of fontanels?

4. Describe the location and age of closure of the following fontanels:

a. anterior

b. posterior

c. anterolateral

d. posterolateral

5. How are microcephalus and hydrocephalus related to the anterior fontanel?

## PART III   SINUSES OF THE SKULL

A **sinus** or **paranasal sinus** is a cavity in a bone located near the nasal cavity. Sinuses are found in the frontal, sphenoid, ethmoid, and maxillae.

Label the sinuses shown in Figure 11-3. See if you can locate the sinuses on the Beauchene skull or other demonstration models that may be available.

### CONCLUSIONS

1. How are paranasal sinuses related to the nasal cavity?

2. What is the function of a sinus?

3. Define sinusitis.

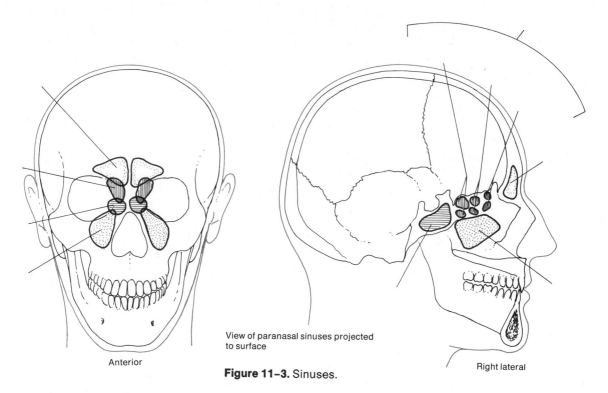

View of paranasal sinuses projected to surface

Anterior

Right lateral

**Figure 11-3.** Sinuses.

## PART IV VERTEBRAL COLUMN

The **vertebral column**, together with the sternum and ribs, constitutes the trunk. The *vertebrae* of the adult column are distributed as follows: 7 cervical, 12 thoracic, 5 lumbar, 5 sacral (fused into one bone, the sacrum), and 3, 4, or 5 coccygeal (fused into one bone, the coccyx). Locate each of these regions on the articulated skeleton. Label the same regions in Figure 11-4, the anterior view.

Examine the vertebral column on the articulated skeleton and identify the cervical, thoracic, lumbar, and sacral (sacrococcygeal) curves. Label the curves in Figure 11-4, the right lateral view.

### CONCLUSIONS

1. What is an intervertebral disc?


2. Define a "slipped disc."


3. What is the purpose of curves in the column?



4. Define a curvature.


5. Define the following curvatures and their possible causes:
   a. scoliosis

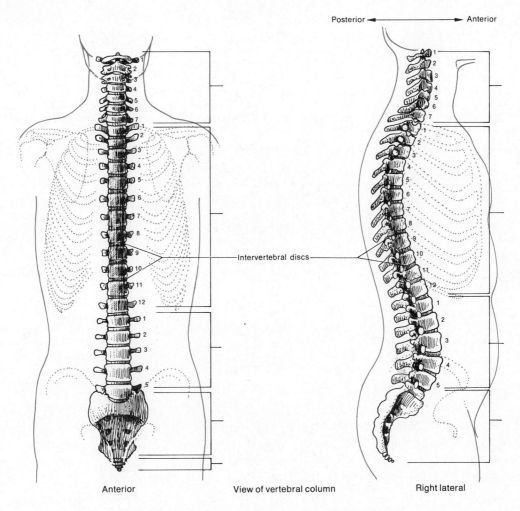

Posterior ◄————————► Anterior

Intervertebral discs

Anterior | View of vertebral column | Right lateral

**Figure 11–4.** The vertebral column.

b. kyphosis

c. lordosis

## PART V   VERTEBRAE

A typical **vertebra** consists of the following portions:

1. *body*—thick, disc-shaped anterior portion.
2. *vertebral (neural) arch*—posterior extension from the body that surrounds the spinal cord and consists of the following parts:
    a. *pedicles*—two short, thick processes that project posteriorly; each has a superior and inferior notch and when successive vertebrae are fitted together the opposing notches form an *intervertebral foramen* through which spinal nerves and blood vessels pass.
    b. *laminae*—form the posterior wall of the vertebral arch.
    c. *vertebral foramen*—opening through which the spinal cord passes; when all the vertebrae are fitted together, the foramina form a canal, the *vertebral canal*.

**80**

3. *processes*—seven processes arise from the vertebral arch:
   a. two *transverse processes*—lateral extensions where the laminae and pedicles join.
   b. one *spinous process*—posterior projection of the lamina.
   c. two *superior articular processes*—articulate with the vertebra above.
   d. two *inferior articular processes*—articulate with the vertebra below.

   Obtain a disarticulated thoracic vertebra and locate each part just described. Now label the vertebra in Figure 11–5a. You should be able to distinguish the general parts on all the types of vertebrae where they are found.

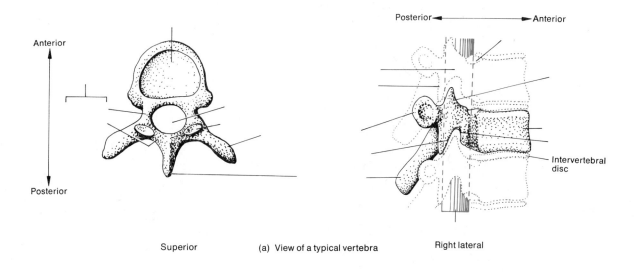

Anterior

Posterior

Superior

(a) View of a typical vertebra

Posterior ◄──────► Anterior

Intervertebral disc

Right lateral

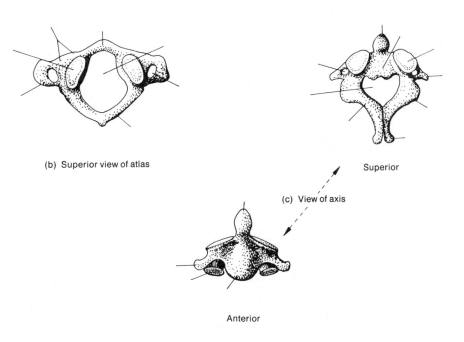

(b) Superior view of atlas

(c) View of axis

Superior

Anterior

**Figure 11-5.** Vertebrae.

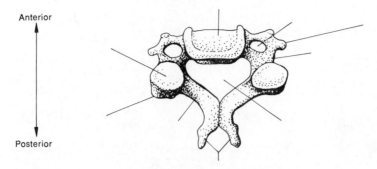

(d) Superior view of a cervical vertebra

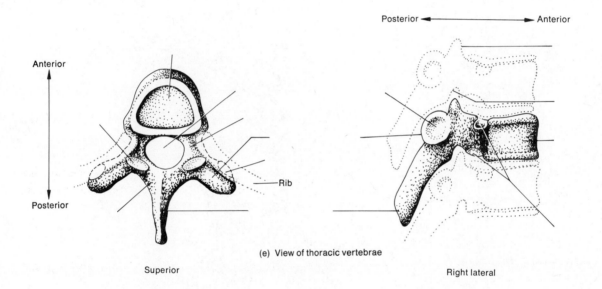

(e) View of thoracic vertebrae

Rib

Superior

Right lateral

Posterior ← → Anterior

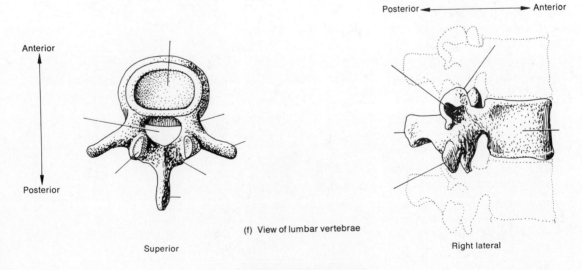

(f) View of lumbar vertebrae

Superior

Right lateral

Posterior ← → Anterior

**Figure 11-5 continued.**

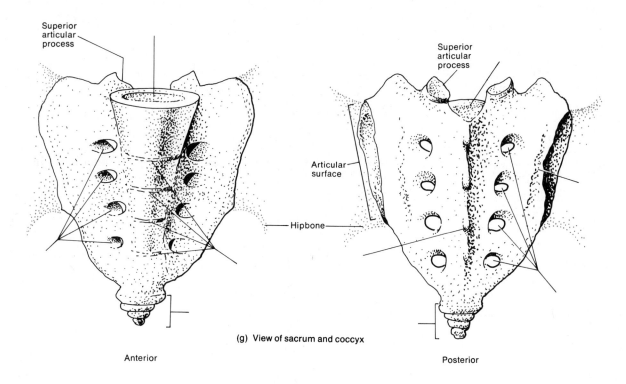

Superior articular process

Superior articular process

Articular surface

Hipbone

(g) View of sacrum and coccyx

Anterior

Posterior

**Figure 11-5 continued.**

Although vertebrae have the same basic design, vertebrae of a given region have special distinguishing features. Obtain the vertebrae listed below and identify the following distinguishing features. After you examine each vertebra, label the appropriate diagrams in Figure 11-5.

1. *cervical vertebrae*
   a. *atlas*—first cervical vertebra (Figure 11-5b).
      *transverse foramen*—opening in transverse process through which an artery, vein, and a spinal nerve pass.
      *anterior arch*—anterior wall of vertebral foramen.
      *posterior arch*—posterior wall of vertebral foramen.
      *lateral mass*—side wall of vertebral foramen.
      Label the other indicated parts.
   b. *axis*—second cervical vertebra (Figure 11-5c).
      *dens*—superior projection of body which articulates with atlas.
      Label the other indicated parts.
   c. *cervical 3-6* (Figure 11-5d).
      *bifid spinous process*—cleft in spinous processes of cervical vertebrae 2-6.
      Label the other indicated parts.
   d. *vertebra prominens*—seventh cervical vertebra; contains a nonbifid spinous process.
2. *thoracic vertebrae* (Figure 11-5e).
   a. *facet*—for articulation with the tubercle of a rib.
   b. *demifacet*—for articulation with the head of a rib.
   c. *spinous processes*—long, pointed, downward-projecting.
      Label the other indicated parts.
3. *lumbar vertebrae* (Figure 11-5f).
   a. *spinous processes*—broad, blunt.
   b. *superior articular processes*—directed medially, not superiorly.

    c. *inferior articular processes*—directed laterally, not inferiorly.
       Label the other indicated parts.
 4. **sacrum** (Figure 11–5g).
    a. *transverse lines*—points where bodies of vertebrae are joined.
    b. *anterior sacral foramina*—4 pairs of foramina that communicate with posterior sacral foramina; passages for blood vessels and nerves.
    c. *median sacral crest*—spinous processes of fused vertebrae.
    d. *lateral sacral crest*—transverse processes of fused vertebrae.
    e. *posterior sacral foramina*—4 pairs of foramina that communicate with anterior sacral foramina; passages for blood vessels and nerves.
    f. *sacral canal*—continuation of vertebral canal.
 5. *coccyx* (Figure 11–5g).

## CONCLUSIONS

1.  How does the adult vertebral column differ from that of a child?


2.  Define the function of each of the following:

    a.  body

    b.  vertebral arch

    c.  vertebral foramen

    d.  intervertebral foramen

    e.  spinous process

    f.  superior articular process

    g.  inferior articular process

    h.  transverse process

    i.  transverse foramen

    j.  dens

    k.  facet

    l.  demifacet

3.  Summarize the distinguishing features of each of the following regions:

    a.  cervical



    b.  thoracic



    c.  lumbar

d. sacral

e. coccygeal

## PART VI   THORAX: STERNUM AND RIBS

The skeleton of the **thorax** consists of the sternum, costal cartilage, ribs, and bodies of the thoracic vertebrae.

Examine the articulated skeleton and disarticulated bones and identify the following. After your identification, label Figure 11–6a.

1. *sternum*—breastbone, flat bone in midline of anterior thorax.
   a. *manubrium*—superior triangular portion.
   b. *body*—middle, largest portion.

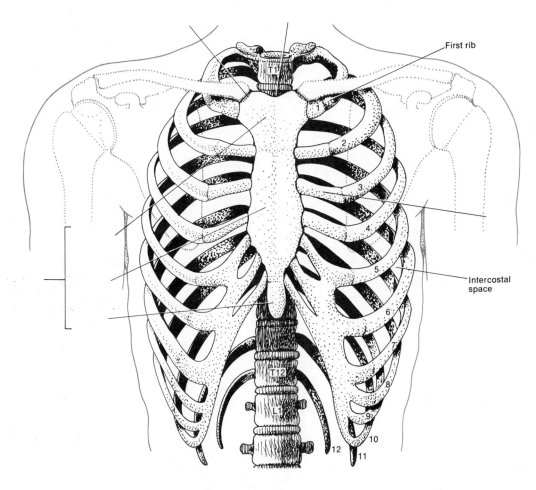

(a) Anterior view of thoracic cage

**FIGURE 11–6.** Bones of the thorax.

**85**

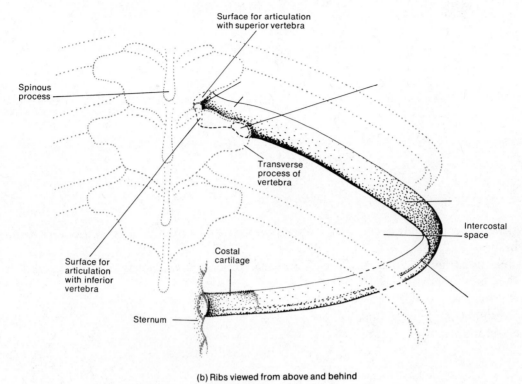

Surface for articulation
with superior vertebra

Spinous
process

Transverse
process of
vertebra

Intercostal
space

Surface for
articulation
with inferior
vertebra

Costal
cartilage

Sternum

(b) Ribs viewed from above and behind

**Figure 11–6 continued.**

    c. *xiphoid process*—inferior, smallest portion.
    d. *jugular notch*—depression on the superior surface of the manubrium that may be palpated.
    e. *clavicular notches*—articular surfaces for the clavicles.
    f. *costal notches*—points at which ribs articulate.
2. *costal cartilage*—strip of hyaline cartilage that attaches a rib to the sternum.
3. *ribs*—parts of a typical rib include:
    a. *body*—shaft, main part of rib.
    b. *head*—posterior projection.
       *neck*—constricted portion behind head.
    d. *tubercle*—knoblike elevation just below neck.
    e. *costal groove*—depression on the inner surface containing blood vessels and a nerve.
    Label the rib shown in Figure 11–6b as well.

## CONCLUSIONS

  1. Describe the shape of the thorax.

  2. What is the function of the thorax?

  3. What is a sternal puncture?

  4. Which ribs articulate with the manubrium and the body?

5. What is the function of the xiphoid process?

6. Describe the size of the ribs from 1-12.

7. Define a true rib.

8. What is a false rib?

9. Define a floating rib.

10. Examine ribs 2-9. How does each articulate with thoracic vertebrae?

11. What is an intercostal space?

12. Describe the shape, location, and function of the hyoid bone.

## PART VII  SHOULDER GIRDLES: CLAVICLES AND SCAPULAE

Each **shoulder girdle** consists of 2 bones—the clavicle (collar bone) and scapula (shoulder blade). Its purpose is to attach the bones of the upper extremity to the axial skeleton (see Figures 11-6a and 11-7).

Examine the articulated skeleton and disarticulated bones and identify the following. After your identification, label Figure 11-7.

1. *clavicle*—slender bone with a double curvature; lies horizontally in superior and anterior part of thorax.
   a. *sternal extremity*—rounded, medial end that articulates with manubrium of sternum.
   b. *acromial extremity*—broad, flat, lateral end that articulates with the acromion of the scapula.
   c. *conoid tubercle*—projection on the inferior, lateral surface for attachment of ligaments.
2. *scapula*—large, flat triangular bone in dorsal thorax between the levels of ribs 2-7.
   a. *body*—flattened, triangular portion.
   b. *spine*—ridge across posterior surface.
   c. *acromion*—flattened, expanded process of spine.
   d. *vertebral (medial) border*—edge of body near vertebral column.
   e. *axillary (lateral) border*—edge of body near arm.
   f. *inferior angle*—bottom of body where vertebral and axillary borders join.
   g. *glenoid cavity*—depression below acromion which articulates with head of humerus to form the shoulder joint.
   h. *coracoid process*—projection at lateral end of superior border.

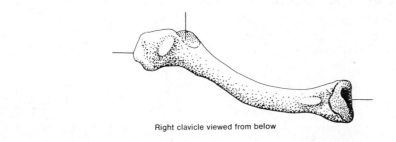

Right clavicle viewed from below

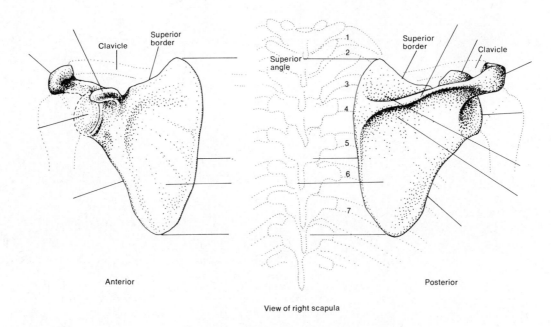

Clavicle    Superior
            border

Superior
angle

1
2
3
4
5
6
7

Superior
border       Clavicle

Anterior                                    Posterior

View of right scapula

**Figure 11–7.** Bones of the shoulder girdle.

    i. *supraspinatous fossa*—surface for muscle attachment above spine.
    j. *infraspinatous fossa*—surface for muscle attachment below spine.

**CONCLUSIONS**

1. Name all bones with which the clavicle articulates.

2. Name all bones with which the scapula articulates.

## PART VIII   UPPER EXTREMITIES

The skeleton of the **upper extremities** consists of a humerus in each arm, an ulna and radius in each forearm, carpals in each wrist, metacarpals in each palm, and phalanges in the fingers.

Examine the articulated skeleton and disarticulated bones and identify the following. After your identification, label Figure 11-8.

**88**

1. *humerus*—arm bone; longest and largest bone of the upper extremity.
   a. *head*—articulates with glenoid cavity of scapula.
   b. *anatomical neck*—oblique groove below head.
   c. *greater tubercle*—lateral projection below anatomical neck.
   d. *lesser tubercle*—anterior projection.
   e. *intertubercular groove*—between tubercles.
   f. *surgical neck*—constricted portion below tubercles.
   g. *body*—shaft.
   h. *deltoid tuberosity*—V-shaped area about midway down shaft.
   i. *capitulum*—rounded knob that articulates with radius.
   j. *radial fossa*—anterior depression which receives head of radius when forearm is flexed.
   k. *trochlea*—projection that articulates with the ulna.
   l. *coronoid fossa*—anterior depression that receives part of the ulna when the forearm is flexed.

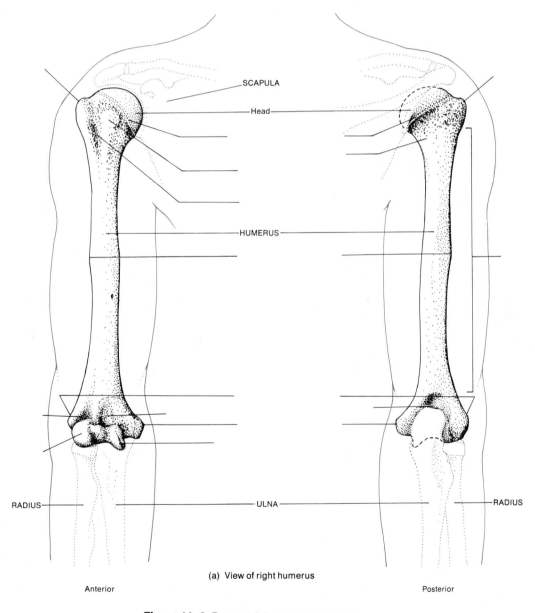

(a) View of right humerus

Anterior                                        Posterior

**Figure 11–8.** Bones of the upper extremity.

m. *olecranon fossa*—posterior depression which receives the olecranon of the ulna when the forearm is extended.

n. *medial epicondyle*—projection on medial side of distal end.

o. *lateral epicondyle*—projection on lateral side of distal end.

2. *ulna*—medial bone of forearm.

a. *olecranon*—prominence of elbow at proximal end.

b. *coronoid process*—anterior projection which, with olecranon, receives trochlea of humerus.

c. *trochlear notch*—curved area between olecranon and coronoid process into which trochlea of humerus fits.

d. *radial notch*—depression lateral and inferior to trochlear notch which receives the head of the radius.

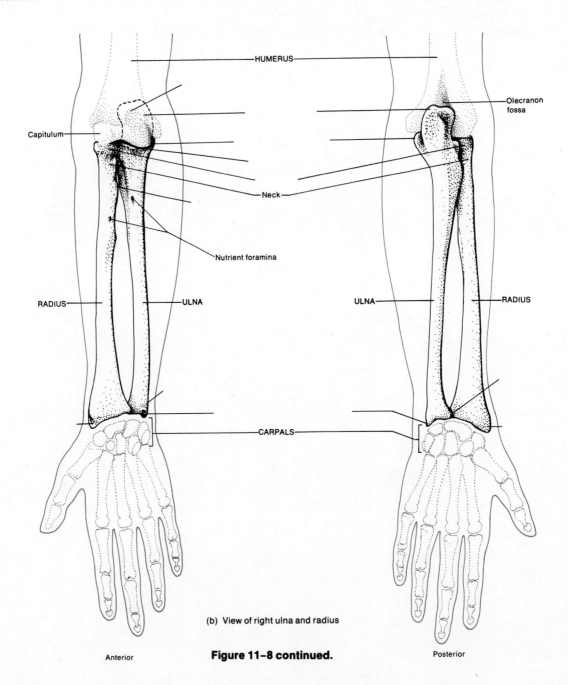

(b) View of right ulna and radius

**Figure 11–8 continued.**

Anterior                                              Posterior

e. *head*—rounded portion at distal end.

f. *styloid process*—projection on posterior side.

3. *radius*—lateral bone of forearm.

   a. *head*—disc-shaped process at proximal end.

   b. *radial tuberosity*—projection for insertion of the biceps muscle.

   c. *styloid process*—projection on lateral side of distal end.

   d. *ulnar notch*—medial, concave depression for articulation with ulna.

4. *carpus*—wrist, consists of 8 small bones, called carpals, united by ligaments.

   a. *proximal row*—from medial to lateral are called pisiform, triquetrum (triangular), lunate, and scaphoid (navicular).

   b. *distal row*—from medial to lateral are called hamate, capitate, trapezoid (lesser multangular), and trapezium (greater multangular).

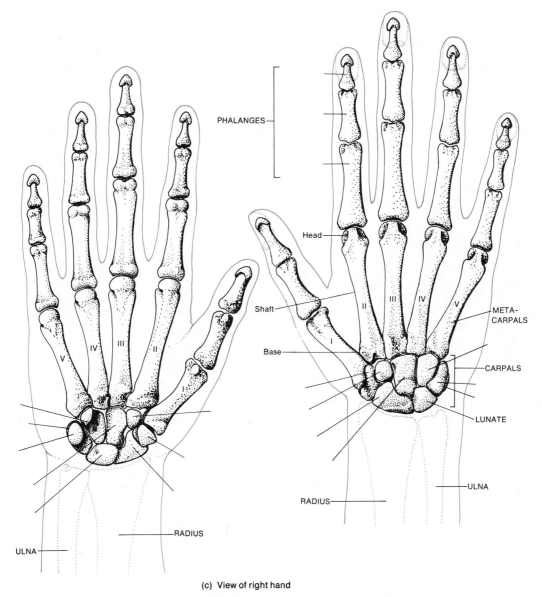

(c) View of right hand

**Figure 11–8 continued.**

Anterior                           Posterior

5. *metacarpus*—5 bones in the palm of the hand, numbered as follows beginning with thumb side: I, II, III, IV, and V metacarpals.
6. *phalanges*—bones of the fingers; 2 in each thumb (proximal and distal) and 3 in each finger (proximal, middle, and distal).

## CONCLUSIONS

1. List all bones or parts of bones with which the humerus articulates.

2. Why is the surgical neck so named?

3. List all bones or parts of bones with which the ulna articulates.

4. Where is the "funny bone"?

5. List all bones or parts of bones with which the radius articulates.

6. What is a Colle's fracture?

7. Which carpal bone is most commonly fractured and why?

8. Describe the points of articulation between the bases of the metacarpal bones and the carpals.

9. Describe the points of articulations between the heads of the metacarpal bones and the phalanges.

## PART IX   PELVIC GIRDLES

The **pelvic girdles** consist of the 2 coxal or hip bones. They provide a strong and stable support for the lower extremities on which the weight of the body is carried and attach the lower extremities to the axial skeleton.

Examine the articulated skeleton and disarticulated bones and identify the following. After you have completed the identification, label Figure 11-9.

1. *ilium*—superior flattened portion.
   a. *iliac crest*—superior border of ilium.
   b. *anterior superior iliac spine*—anterior projection of iliac crest.
   c. *anterior inferior iliac spine*—projection under anterior superior iliac spine.
   d. *posterior superior iliac spine*—posterior projection of iliac crest.
   e. *posterior inferior iliac spine*—projection below posterior superior iliac spine.
   f. *greater sciatic notch*—concavity under posterior inferior iliac spine.
   g. *iliac fossa*—medial concavity for attachment of iliacus muscle.
   h. *iliac tuberosity*—point of attachment for sacroiliac ligament posterior to iliac fossa.
   i. *auricular surface*—point of articulation with sacrum.
   j. *posterior, anterior,* and *inferior gluteal lines*—between these lines the gluteal muscles are attached on the lateral surface.
2. *ischium*—lower, posterior portion.
   a. *ischial spine*—posterior projection of ischium.
   b. *lesser sciatic notch*—concavity under the ischial spine.
   c. *ischial tuberosity*—roughened projection.
   d. *ramus*—portion of ischium that joins the pubis and surrounds the *obturator foramen.*

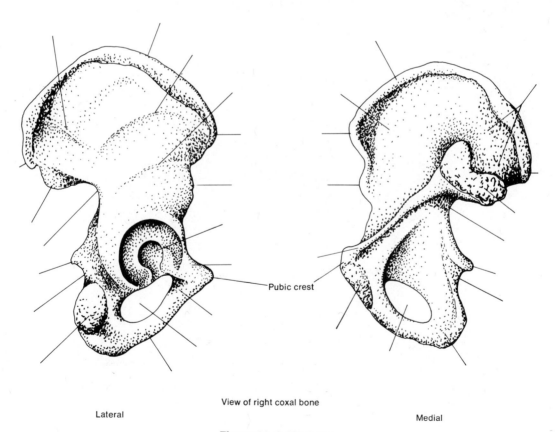

Pubic crest

View of right coxal bone

Lateral                                                                 Medial

**Figure 11-9.** Hip bones.

3. *pubis*—anterior, inferior portion.
   a. *superior ramus*—upper portion of pubis.
   b. *inferior ramus*—lower portion of pubis.
   c. *pubic symphysis*—joint between left and right hip bones.
4. *acetabulum*—socket that receives the head of the femur to form the hip joint; formed by ilium (2/5), ischium (2/5), and pubis (1/5).

Again, examine the articulated skeleton. This time compare the male and female pelvis. The **pelvis** consists of the 2 hip bones, sacrum, and coccyx. Identify the following and label Figure 11-10.

1. *greater (false) pelvis*—expanded portion situated above the brim of the pelvis; bounded laterally by the ilia and posteriorly by the upper sacrum.

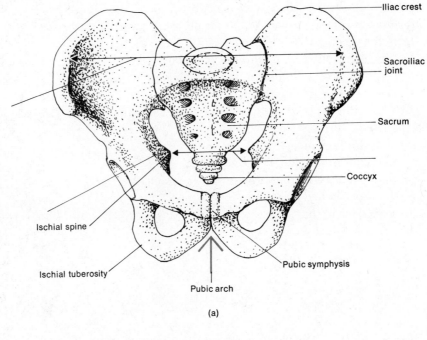

(a)

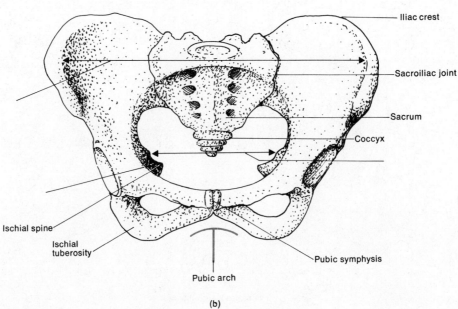

(b)

**Figure 11-10.** Comparison of (a) male and (b) female pelvis in anterior view.

2. *lesser (true) pelvis*—below and behind the brim of the pelvis; constructed of parts of the ilium, pubis, sacrum, and coccyx; contains an opening above, the *pelvic inlet,* and an opening below, the *pelvic outlet.*

**CONCLUSIONS**

1. Describe all articulations of the hip bones.


2. Compare the male and female pelvis with regard to the following:
   a. pelvis

   b. ilia

   c. inlet of true pelvis

   d. sacrum

   e. coccyx

   f. ischial spines

   g. pubic arch

3. Why is pelvimetry important to physicians prior to childbirth?


## PART X   LOWER EXTREMITIES

The bones of the **lower extremities** consist of a femur in each thigh, a patella in front of each knee joint, a tibia and fibula in each leg, tarsals in each ankle, metatarsals in each foot, and phalanges in the toes.

Examine the articulated skeleton and disarticulated bones and identify the following. After identification, label Figure 11-11.

1. *femur*—thigh bone; longest and heaviest bone in the body.
   a. *head*—rounded projection at proximal end that articulates with acetabulum of hip bone.
   b. *neck*—constricted portion below head.
   c. *greater trochanter*—prominence on lateral side.
   d. *lesser trochanter*—prominence on dorso-medial side.
   e. *intertrochanteric line*—ridge on anterior surface.
   f. *intertrochanteric crest*—ridge of posterior surface.
   g. *linea aspera*—vertical ridge on posterior surface.
   h. *medial condyle*—medial posterior projection on distal end that articulates with tibia.
   i. *lateral condyle*—lateral posterior projection on distal end that articulates with tibia.
   j. *intercondylar fossa*—depressed area between condyles on posterior surface.
   k. *medial epicondyle*—projection above medial condyle.
   l. *lateral epicondyle*—projection above lateral condyle.
2. *patella*—kneecap in front of knee joint; develops in tendon of quadriceps femoris muscle.
   a. *base*—broad superior portion.
   b. *apex*—pointed inferior portion.
   c. *articular facets*—articulating surfaces on posterior surface for medial and lateral epicondyles of femur.

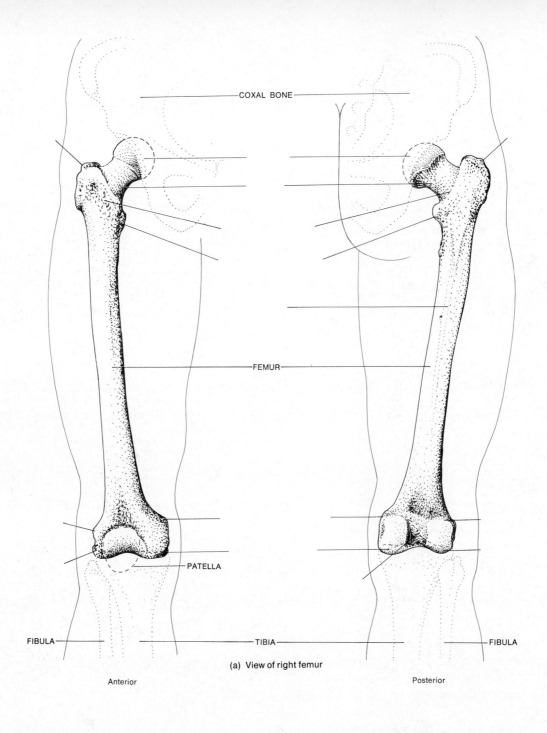

COXAL BONE

FEMUR

PATELLA

FIBULA          TIBIA         FIBULA

(a) View of right femur

Anterior          Posterior

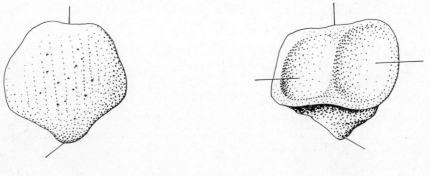

Anterior    (b) View of right patella    Posterior

**Figure 11–11.** Bones of the lower extremity.

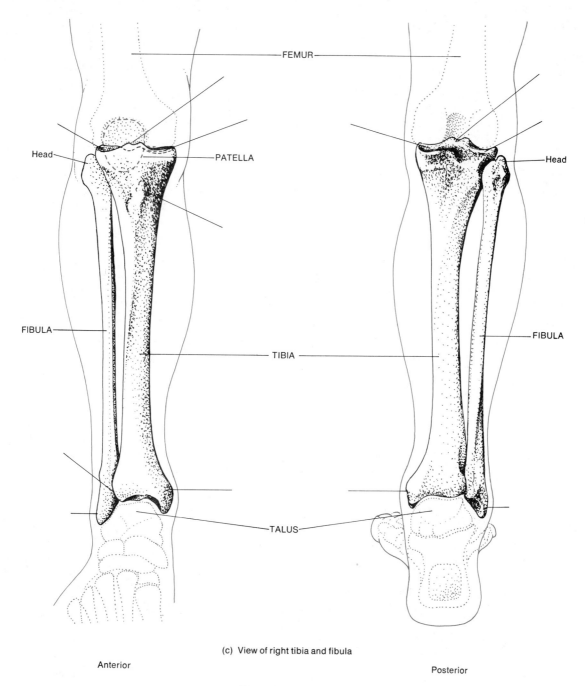

FEMUR

Head

PATELLA

FIBULA

TIBIA

Head

FIBULA

TALUS

(c) View of right tibia and fibula

Anterior

Posterior

**Figure 11–11 continued.**

3. *tibia*—shinbone; medial bone of leg.
   a. *lateral condyle*—articulates with condyle of femur.
   b. *medial condyle*—articulates with condyle of femur.
   c. *intercondylar eminence*—upward projection between condyles.
   d. *tibial tuberosity*—anterior projection for attachment of patellar ligament.
   e. *medial malleolus*—distal projection that articulates with talus bone of ankle.
   f. *fibular notch*—distal depression that articulates with the fibula.
4. *fibula*—lateral bone of leg.
   a. *head*—proximal projection that articulates with tibia.
   b. *lateral malleolus*—projection at distal end that articulates with the talus bone of ankle.

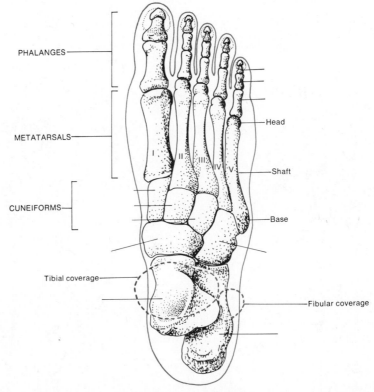

(d) Superior view of right foot

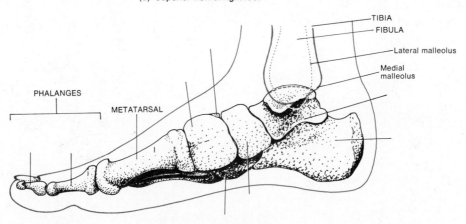

(e) Medial view of right foot

**Figure 11-11 continued.**

5. *tarsus*—7 bones of the ankle called tarsals.
   a. *posterior bones*—talus and calcaneus (heel bone).
   b. *anterior bones*—cuboid, navicular (scaphoid), and 3 cuneiforms called the first, second, and third cuneiforms.
6. *metatarsus*—consists of 5 bones of the foot called metatarsals, numbered as follows, beginning on the medial (large toe) side: I, II, III, IV, and V metatarsals.
7. *phalanges*—bones of the toes, comparable to phalanges of fingers; 2 in each large toe (proximal and distal) and 3 in each smaller toe (proximal, middle, and distal).

**98**

**CONCLUSIONS**

1. List all bones or parts of bones with which the femur articulates.

2. Why does the neck of the femur fracture frequently in older people?

3. Why is the patella a sesamoid bone?

4. List all bones or parts of bones with which the tibia articulates.

5. List all bones or parts of bones with which the fibula articulates.

6. What is a Pott's fracture?

7. Describe the points of articulations between the tibia, the fibula, and the ankle bones.

8. Describe the points of articulation between the ankle bones and metatarsus.


## PART XI    ARCHES OF THE FOOT

The bones of the foot are arranged in **2 arches** that enable the foot to support the weight of the body and provide leverage while walking. The bones comprising the arches are held in position by ligaments and tendons.

Refer to Figure 11-12. Consult your textbook and label the bones as well as the median part of the longitudinal arch, the lateral part of the longitudinal arch, and the transverse arch.

**CONCLUSIONS**

1. Define flatfoot.

2. What is a bunion?

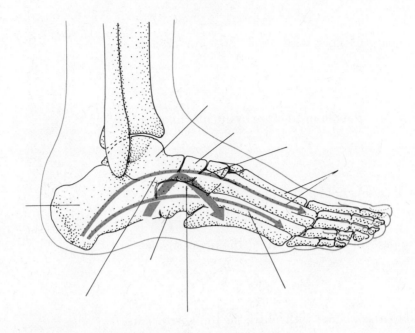

**Figure 11–12.** Arches of the foot.

## STUDENT ACTIVITIES

1. List each bone of the skull and determine all other bones with which they articulate. Be sure to include the direction of articulation (posterior, superior, lateral, etc.).
2. Select any disarticulated bone of the body. Have your partner name the bone, describe its articulations, and identify the markings studied in the module. Continue the procedure until you have learned all of the bones. Reverse roles with your partner.
3. Locate as many bones or parts of bones as you can on your partner's body. Reverse roles with your partner.
4. Label the skeleton in Figure 11–13.

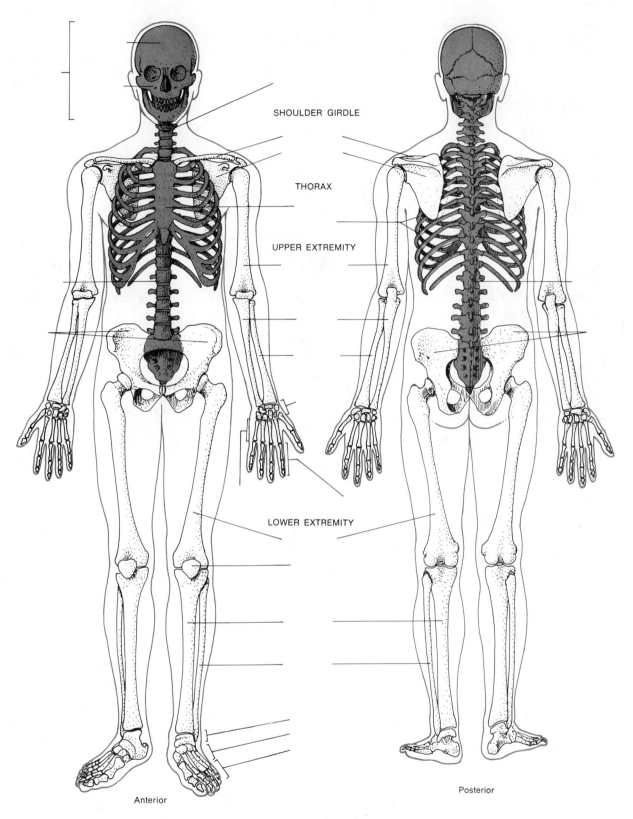

SHOULDER GIRDLE

THORAX

UPPER EXTREMITY

LOWER EXTREMITY

Anterior

Posterior

**Figure 11–13.** The skeleton.

# MODULE 12
# BONE FRACTURES

**OBJECTIVES**

1. To define and identify several kinds of common fractures

2. To explain the steps involved in fracture repair

A **fracture** is any break in a bone. Depending on the bone involved and the extent of injury, fractures may take several months to heal. In this module, you will identify several kinds of fractures and the sequence of events involved in fracture repair.

## PART I   KINDS OF FRACTURES

Fractures may be classified into the following types:

*partial*—the break across the bone is incomplete.
*complete*—the break occurs across the entire bone; the bone completely breaks in two.
*simple (closed)*—the fractured bone does not break through the skin.
*compound (open)*—the broken ends of the fractured bone protrude through the skin.
*comminuted*—the bone is splintered at the site of impact and smaller fragments of bone are found between the 2 main fragments.
*greenstick*—a partial fracture in which a portion of the bone bends.
*displaced*—the anatomical alignment of the bone fragments is not preserved.
*nondisplaced*—the anatomical alignment of the bone fragments has not been disrupted.

## CONCLUSIONS

Examine Figure 12-1 which contains illustrations of 6 of the fractures just described. Place the name of the fracture under the appropriate illustration.

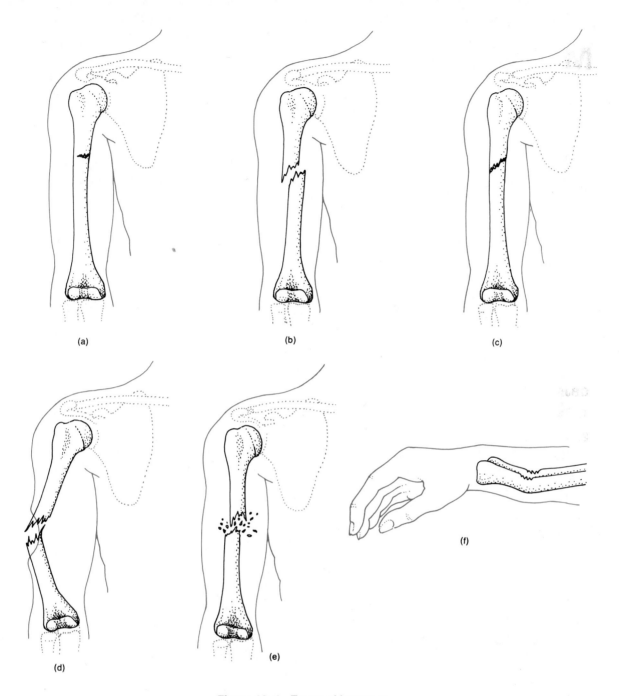

(a)

(b)

(c)

(d)

(e)

(f)

**Figure 12-1.** Types of fractures.

## PART II   FRACTURE REPAIR

Figure 12-2 illustrates the sequence of events involved in fracture repair. Using your textbook as a reference, study the sequence carefully and label the illustration, then answer the following conclusions:

**CONCLUSIONS**

1. What is a fracture hematoma?

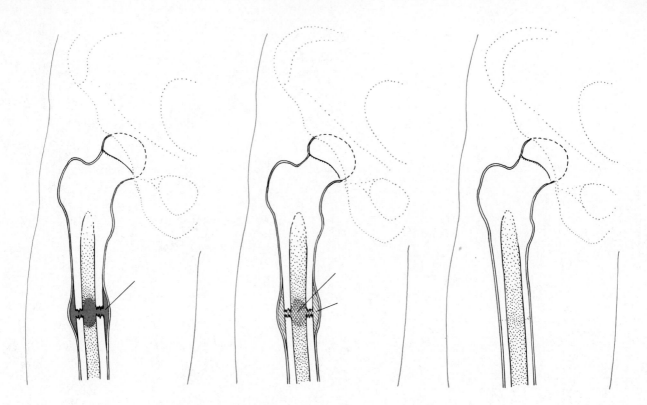

**Figure 12–2.** Fracture repair.

2. How does the fracture hematoma form?

3. Define a callus.

4. Distinguish between an external and internal callus.

   a. external

   b. internal

5. Which cells repair the fracture?

6. What occurs during the remodeling phase of fracture repair?

**STUDENT ACTIVITY**

Examine any roentgenograms that might be available. A **roentgenogram** is a film exposed to x-rays. See if you can determine which bone is fractured and the type of fracture involved.

# MODULE 13
# ARTICULATIONS: TYPES AND DISORDERS

**OBJECTIVES**

1. To compare the structure of joints

2. To contrast the movements possible at joints

3. To define selected disorders of joints

An **articulation** or **joint** is a point of contact between bones or between cartilage and bones. Some joints permit no movement, others permit a slight degree of movement, and still others afford free movement.

## PART I  KINDS OF JOINTS

The joints of the body may be classified into 3 principal kinds, on the basis of their structure and the type and degree of movement they allow.

1. *Fibrous*—allow little or no movement; do not contain a joint cavity; articulating bones are held together by fibrous connective tissue.
   a. *Sutures*—found between skull bones; bones are separated by a thin layer of fibrous tissue; immovable. Example: lambdoidal suture.
   b. *Syndesmosis*—articulating bones are united by dense fibrous tissue; slight movement. Example: distal ends of tibia and fibula.
2. *Cartilaginous*—allow little or no movement; do not contain a joint cavity; articulating bones are held together by cartilage.
   a. *Synchondrosis*—connecting material is hyaline cartilage; temporary joint; immovable. Example: epiphyseal plate (see Figure 9-2).
   b. *Symphysis*—connecting material is a broad, flat disc of fibrocartilage; slight movement. Example: between vertebrae and between anterior surfaces of hip bones.
3. *Synovial*—allow varying degrees of free movement; contain a joint (synovial) cavity between artic-

ulating bones; synovial membrane lines joint cavity; articulating bones are covered by hyaline cartilage; are held together by ligaments.

a. *gliding*—articulating surfaces usually flat; permits movement in 2 planes (biaxial—side to side, and back and forth. Example: between carpals, tarsals, sacrum and ilium, sternum and clavicle, scapula and clavicle.

b. *hinge*—spool-like surface of one bone fits into the concave surface of another; movement in a single plane (monaxial), usually flexion and extension. Example: elbow, knee, ankle, interphalangeal joints.

c. *pivot*—rounded, pointed, or conical surface of one bone articulates with a shallow depression of another bone; primary movement is rotation; joint is monaxial. Example: between atlas and axis and between proximal ends of radius and ulna.

d. *ellipsoidal*—oval-shaped condyle of one bone fits into an elliptical cavity of another bone; movement is biaxial—side to side and back and forth. Example: between radius and carpals.

e. *saddle*—surfaces of articulating bones are saddle-shaped—concave in one direction and convex in the other; movement similar to that of an ellipsoidal joint. Example: between trapezium and metacarpal of thumb.

f. *ball-and-socket*—ball-like surface of one bone fits into a cuplike depression of another bone; movement is in 3 planes (triaxial)—flexion-extension, abduction-adduction, and rotation. Example: shoulder and hip joint.

Examine the articulated skeleton and find as many examples as you can of the joints just described. As part of your examination, be sure to note the shapes of the articular surfaces and the movements possible at each joint.

## CONCLUSIONS

1. What determines the degree of movement at a joint?

2. Match the following:

| | | |
|---|---|---|
| _____ a. hinge | 1. | joint between skull bones |
| _____ b. ellipsoidal | 2. | joint between vertebrae |
| _____ c. symphysis | 3. | joint between radius and carpals |
| _____ d. synchondrosis | 4. | shoulder joint |
| _____ e. saddle | 5. | elbow |
| _____ f. ball-and-socket | 6. | joint between trapezium and metacarpal of thumb |
| _____ g. suture | 7. | epiphyseal plate |
| _____ h. pivot | 8. | joint at distal ends of tibia and fibula |
| _____ i. syndesmosis | 9. | joint between atlas and axis |

3. What are bursae?

## PART II   DISORDERS OF JOINTS

Consult your textbook and briefly define the following disorders. Be sure to include the cause, symptoms, and treatment wherever possible.

1. rheumatoid arthritis

2. osteoarthritis

3. gouty arthritis

4. bursitis

5. tendinitis

**CONCLUSIONS**

1. Define a pannus.

2. Why do joints become less mobile in arthritis?

3. How do rheumatoid arthritis and osteoarthritis differ?

4. What is the importance of uric acid to gout?

5. What is a "slipped disc"?

6. Define the following terms:
   a. ankylosis
   b. arthralgia
   c. arthrosis
   d. bursectomy
   e. chondritis
   f. dislocation
   g. sprain
   h. synovitis

## STUDENT ACTIVITIES

1. Have your partner assume the anatomical position. Point to a joint and ask your partner to name the type of joint and movements possible at the joint. Reverse roles.
2. Label the diagram of the stages of rheumatoid arthritis in Figure 13-1 and indicate below the diagram the changes that occur at each stage.

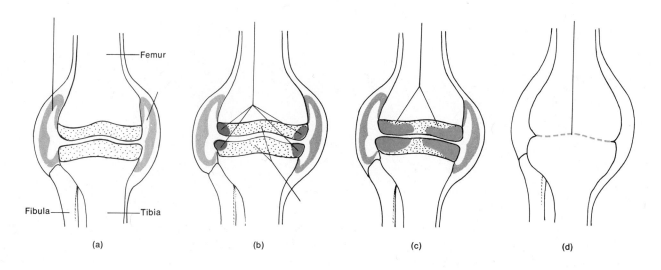

**Figure 13-1.** Stages in the development of rheumatoid arthritis.

3. Identify the following movements in Figure 13-2. Place your response next to the arrows and label any bones and joints indicated.

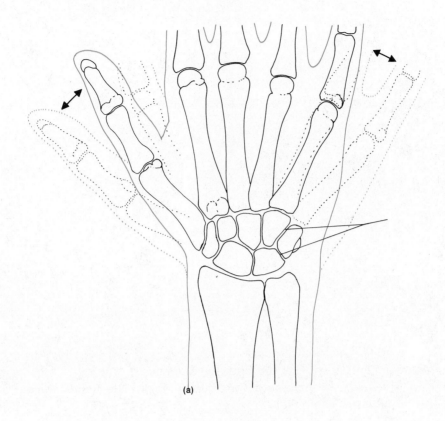

(a)

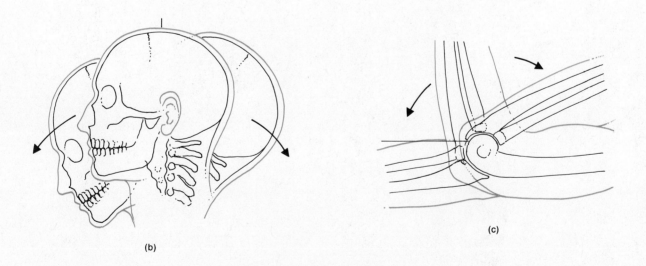

(b)

(c)

**Figure 13-2.** Movements at joints.

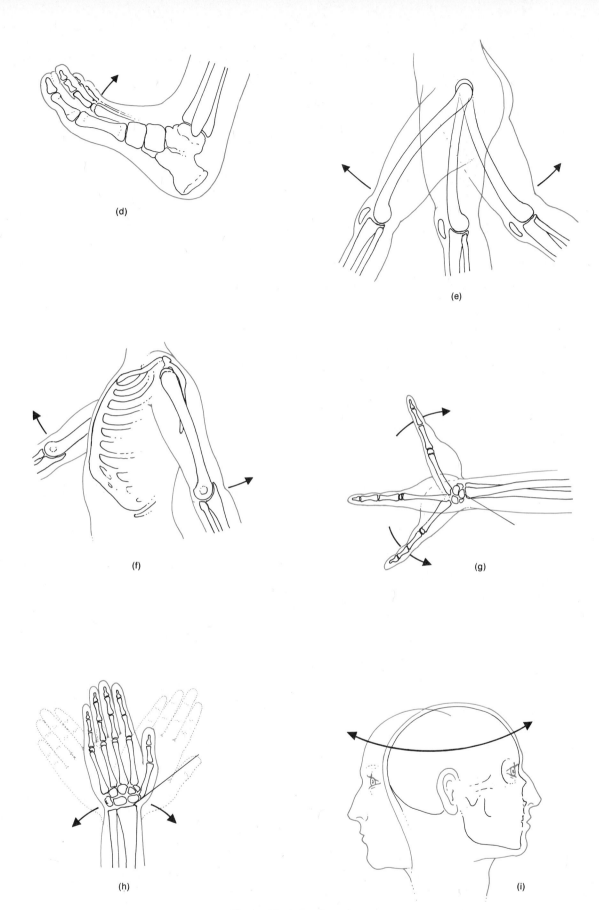

(d)

(e)

(f)

(g)

(h)

(i)

**Figure 13–2 continued.**

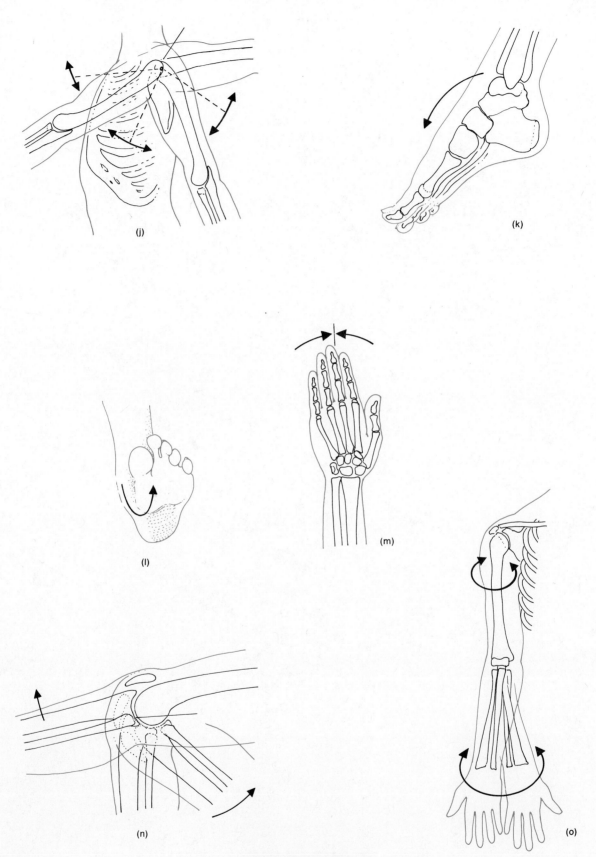

**Figure 13–2 continued.**

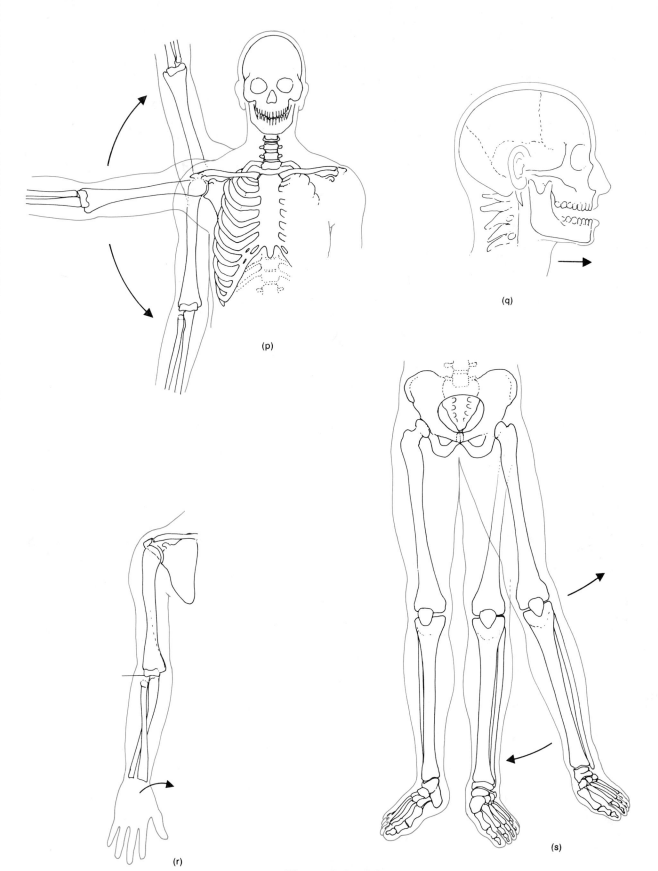

(p)

(q)

(r)

(s)

**Figure 13–2 continued.**

4. Label the following parts of a synovial joint in Figure 13-3.

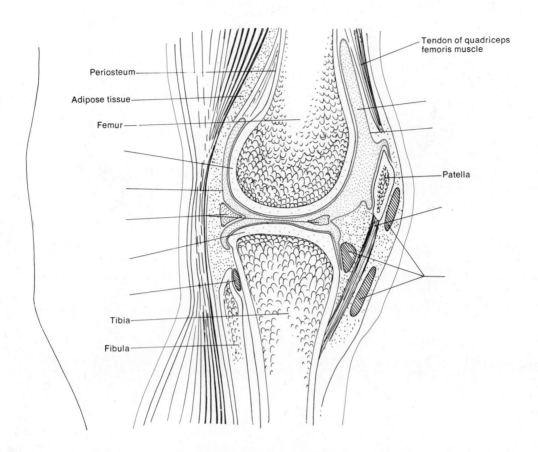

Periosteum

Adipose tissue

Femur

Tibia

Fibula

Tendon of quadriceps
femoris muscle

Patella

**Figure 13-3.** Structure of a synovial joint.

# MODULE 14
# HISTOLOGY OF MUSCLE TISSUE

**OBJECTIVES**

1. To compare the histology of skeletal, smooth, and cardiac muscle tissue

2. To examine skeletal muscle tissue based on electron micrographic studies

**Muscle tissue** comprises 40-50 percent of the total body weight and is composed of highly specialized cells that have 4 striking characteristics. *Irritability* is the ability of muscle tissue to receive and respond to stimuli. *Contractility* is its ability to contract (shorten and thicken). *Extensibility* is its ability to stretch when it is pulled. Finally, *elasticity* is the ability of muscle tissue to return to its original shape following contraction or extension.

Histologically, 3 kinds of muscle tissue are recognized:

1. **skeletal muscle**—usually attached to bones; contains conspicuous striations when viewed microscopically; voluntary because it contracts with conscious control.
2. **smooth muscle**—located in walls of viscera and blood vessels; does not contain striations, thus the name smooth; involuntary because it contracts without conscious control.
3. **cardiac muscle**—found only in the wall of the heart; striated; involuntary.

## PART I  SKELETAL MUSCLE

Obtain a prepared slide of skeletal muscle and examine it under high power. Look for the following:

*muscle fibers*—cylindrical, elongated muscle cells
*sarcolemma*—plasma membrane of the fiber
*sarcoplasm*—cytoplasm of the fiber
*nuclei*—several in each fiber lying close to sarcolemma
*striations*—cross stripes in each fiber

Refer to Figure 14-1a and label the structures indicated.

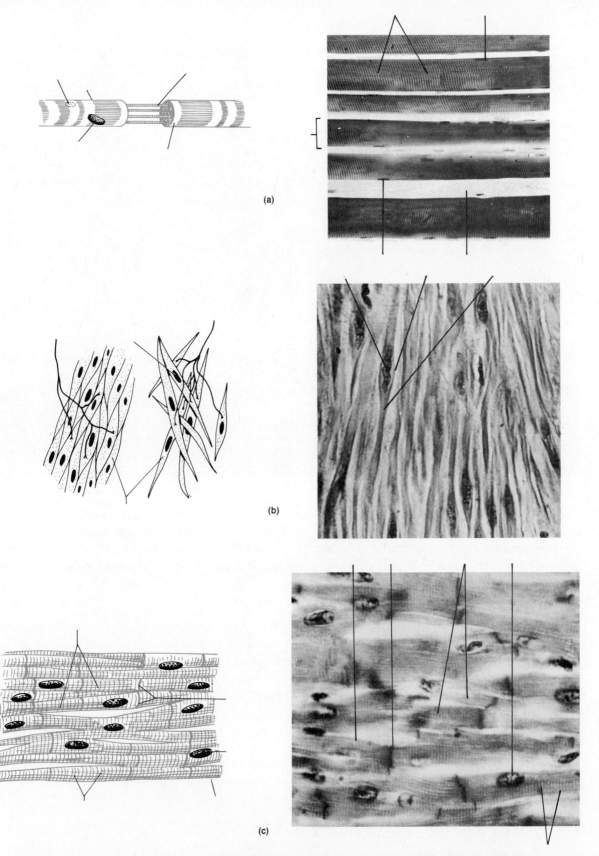

**Figure 14-1.** Muscle tissue. (a) Skeletal. (b) Smooth. (c) Cardiac. Can you identify the parts you labeled in the photos shown? (Photomicrographs courtesy of Edward J. Reith, from Atlas of Descriptive Histology, Edward J. Reith and Michael H. Ross, Harper & Row, Publishers, Inc., New York, N.Y., 1970).

With the use of an electron microscope, additional details of skeletal muscle may be noted. Among these are the following:

*mitochondria*
*sarcoplasmic reticulum*—tubelike network of roughly parallel vesicles.
*T tubules*—run perpendicular to and connect with the reticulum and open to the outside of the fiber.
*triad*—a T tubule and the segments of sarcoplasmic reticulum on both sides of it.
*myofibrils*—threadlike structures that run longitudinally through a fiber and consist of thin actin filaments and thick myosin filaments.
*sarcomere*—compartments within a myofibril formed by separations of zones of dense material called "Z lines."
*A band*—dark region in a sarcomere where actin and myosin filaments overlap.
*I band*—light region in a sarcomere composed of actin filaments only.
*striations*—combination of alternating dark and light regions in a sarcomere.
*H band*—region in a sarcomere consisting of myosin filaments only.

Figure 14-2 is a diagram of skeletal muscle tissue based upon electron micrographic studies. In conjunction with your textbook, label the structures shown.

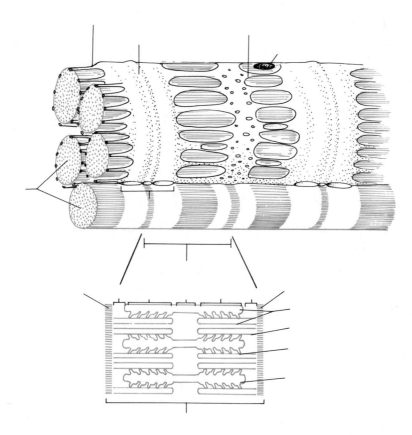

**Figure 14-2.** Diagram of skeletal muscle based on electron micrographic studies.

## PART II  SMOOTH MUSCLE

Obtain a prepared slide of smooth muscle tissue and examine under high power. Locate and label the following structures: *sarcolemma*, *sarcoplasm*, and *nucleus*, referring to Figure 14-1b.

**117**

## PART III  CARDIAC MUSCLE

Obtain a prepared slide of cardiac muscle tissue, examine it under high power, and locate the following structures: *sarcolemma, sarcoplasm, nuclei, striations,* and *intercalated discs* (transverse thickenings of the sarcolemma that separate individual fibers). Refer to Figure 14-c and label the diagram.

### CONCLUSIONS

1. Compare skeletal, smooth, and cardiac muscle cells with regard to the following:

   a. shape

   b. size

   c. position of nucleus

   d. presence of striations

   e. degree of branching

   f. location in body

2. Define a myofibril.

3. What chemical comprises actin and myosin filaments?

4. Define the following:

   a. sarcoplasmic reticulum
   b. T tubule
   c. triad
   d. sarcomere
   e. striation
   f. intercalated disc

### STUDENT ACTIVITY

If available, examine electron micrographs of skeletal muscle tissue. Compare the micrographs to Figure 14-2 and see how many structures you can identify.

# MODULE 15
# PHYSIOLOGY AND DISORDERS OF MUSCLES

**OBJECTIVES**

1. To describe the source of energy for muscular contraction, the motor unit, and the physiology of contraction

2. To define the all-or-none principle of muscular contraction

3. To contrast the kinds of normal contractions performed by skeletal muscles

4. To define common muscular disorders such as fatigue, fibrosis, fibrositis, muscular dystrophy, and myasthenia gravis

5. To define and compare abnormal contractions such as spasm, cramp, convulsion, tetany, and fibrillation

In this module you will study the mechanisms involved in the contraction of muscle tissue. You will also be introduced to selected disorders of muscles.

## PART I   ENERGY FOR CONTRACTION

In order for a muscle to contract, it must have a supply of energy. As far as we know, *ATP* is always the immediate source of energy for muscle contraction. Muscle cells build up ATP in 2 ways. A resting muscle needs little ATP and the surplus is stored. Also, when "storage space" runs out the excess ATP combines with creatine.

**CONCLUSIONS**

1. What is the immediate source of energy for muscle contraction?

2. Where is the energy source located in a muscle cell?

3. Complete the following equation by which a muscle cell synthesizes ATP.

$$+ \qquad\qquad + \qquad\qquad \longrightarrow \text{ATP}$$

4. How is the energy for ATP synthesis derived?

5. Define creatine phosphate.

6. When is creatine phosphate produced?

7. Complete the following equation which occurs during strenuous contractions.

$$+ \qquad\qquad \longrightarrow ATP + Creatine$$

## PART II   MOTOR UNIT

Contracting muscles not only need energy but must be stimulated by a nerve cell. The stimulus is carried by a nerve cell called a *motor neuron*. Upon entering a muscle, the axon of the neuron branches and the branches make contact with individual muscle cells. The area of contact is called a *motor end plate* or *neuromuscular junction*. A motor neuron together with the muscle cells it stimulates is referred to as a **motor unit**.

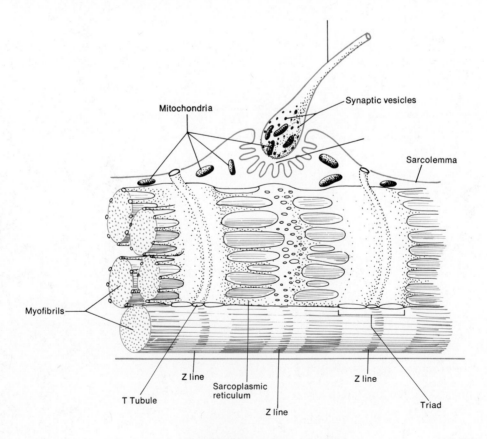

**Figure 15-1.** Motor end plate.

Obtain a prepared slide of a motor end plate and identify the following components: skeletal muscle fibers, axon of the motor neuron, and motor end plate. Now refer to Figure 15-1 and label the parts indicated.

**CONCLUSIONS**

1. Define a motor end plate.

2. How is the nerve impulse transmitted across the motor end plate?

3. Distinguish between a motor unit involved in precise movements and one involved in gross movements.

## PART III   PHYSIOLOGY OF CONTRACTION

The most commonly accepted explanation of muscle contraction is the **sliding filament hypothesis**. According to the hypothesis, the actin filaments slide toward each other and the muscle shortens. Now refer to Figure 15-2 and label the parts indicated.

**CONCLUSIONS**

1. How does the nerve impulse enter a muscle fiber?

2. What is the significance of the calcium ions in muscle contraction?

3. Describe the movement of actin during contraction.

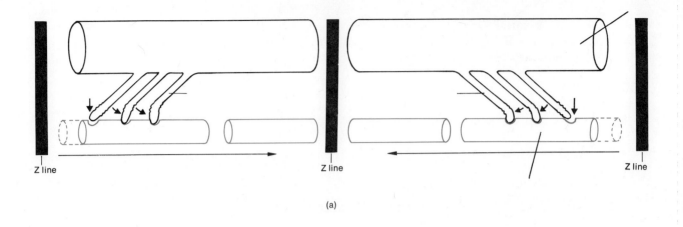

(a)

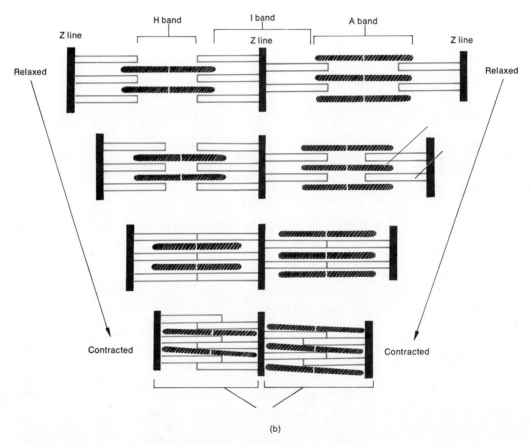

H band

I band

A band

Z line

Z line

Z line

Relaxed

Relaxed

Contracted

Contracted

(b)

**Figure 15–2.** Sliding filament hypothesis. (a) Direction of actin movement during contraction. (b) Position of various zones in relaxed and contracted sarcomeres.

4. How does the muscle return to its resting length?

## PART IV  ALL-OR-NONE PRINCIPLE

The **all-or-none principle** states that the muscle fibers of a motor unit will contract to their fullest extent, or will not contract at all. In other words, muscle fibers do not partially contract.

**CONCLUSIONS**

1. Define the following:

   a. liminal stimulus

   b. subliminal stimulus

2. How can the strength of a muscle fiber be altered?

## PART V  KINDS OF CONTRACTIONS

The various skeletal muscles are capable of producing different kinds of contractions depending upon the kind of stimulus applied. Among these contractions are isotonic, isometric, tonic, twitch, tetanic, and treppe. Although twitch contractions do not occur in the body, they are worth demonstrating because they show the different phases of muscle contraction quite clearly. Any record of a muscle contraction is called a *myogram*.

The observation of a twitch contraction may be done in groups, or as a demonstration by the instructor. The preparation of the animal for study is outlined in Appendix A. Refer to Appendix B for a description of the use of the kymograph, the instrument that will be used to record the twitch contraction.

Secure a living frog and double pith it according to the directions in Appendix A. Your instructor can demonstrate this technique. Depending on the size of the class and the equipment available, one or 2 students should be assigned to assist the instructor to set up the physiologic apparatus as described in Appendix B. One or two other students should be assigned as "surgeons" in isolating, removing, and suspending the frog muscle (Appendix A), and another student should act as the recorder and coordinate the experiment.

1. *Single skeletal muscle twitch.* With the kymograph running at the fastest speed available and marking the point of stimulus, record the single muscle twitch resulting from the single stimulus. Observe and calculate the time of the (1) latent period, (2) period of contraction, and (3) period of relaxation. Using the same muscle, repeat the experiment several times (use a new piece of paper for each run) and calculate average figures for the above 3 periods.

   Draw a typical twitch contraction labeling the 3 phases in Figure 15–3.

**Figure 15–3.** Myogram of a twitch contraction.

2. *Muscle response to increasing strength of stimulus.* With the stimulus intensity knob turned down to its minimum setting, and using a single shock stimulus, apply a subthreshold stimulus to the muscle by quickly hitting the single stimulus key. The recorder must note and write down the intensity of the stimulus applied. Increase the intensity and again stimulate the muscle. In a stepwise manner, elevate the stimulus intensity, and determine when a stimulus of threshold intensity has been reached.

   Once the minimum threshold stimulus has been determined, gradually increase the stimulus intensity above the threshold level, thus recording the height of contraction resulting from each stimulus. Observe and explain your results in terms of the "all-or-none" response.

3. *Muscle response to increasing load.* The equipment is set up exactly as in Sections 1 and 2 of this experiment except that a weight pan is added to the muscle lever. With the drum moving at a medium speed, and using a single threshold stimulus, determine the work performed by the muscle when individual 5-gram weights are added to the muscle in a stepwise manner. Put the additional weight on the pan before each stimulus, and keep repeating the experiment (adding a weight and then stimulating) until the muscle is no longer able to lift the pan containing all of the weights. The recorder should note the total number of grams that the muscle lifts at each contraction. Remind the students that once they choose a stimulus, they are *not* to change either its frequency or intensity throughout this experiment.

4. *Calculation of work performed by muscle (optional).* If time permits, the gram-centimeters of work performed by the muscle at each contraction may be calculated by using the following formula: L: l = H:h

   when  L = length of long arm of lever, in centimeters
   
         l = length of short arm of lever, in centimeters
   
         H = height of *recorded* contraction, in centimeters
   
         h = calculated *actual* distance weight was lifted at its point of attachment in centimeters

   Solve for h for *each* load carried.

   To calculate the *gram-centimeters of work* performed by the muscle for *each load* carried, multiply *each calculated* h by the grams of load lifted for each particular h (work = distance moved X load lifted). Record these and other data in Table 15-1.

Table 15–1. WORK RECORD OF MUSCLE

| CONTRACTION NUMBER | LOAD LIFTED (IN GRAMS) | H (HEIGHT OF THE RECORDED CONTRACTION, IN CENTIMETERS) | h (ACTUAL CALCULATED HEIGHT THE LOAD WAS LIFTED, IN CENTIMETERS) | WORK IN GRAM-CENTIMETERS (h X LOAD IN GRAMS) |
|---|---|---|---|---|
| 1 | | | | |
| 2 | | | | |
| 3 | | | | |
| 4 | | | | |
| 5 | | | | |
| 6 | | | | |
| 7 | | | | |
| 8 | | | | |
| 9 | | | | |

5. *Muscle response to variation of stimulus frequency.* In this experiment we will demonstrate a phenomenon known as *tetanus.* If we repeatedly apply many stimuli to a skeletal muscle without allowing the muscle to relax, there will be a fusion of twitches. This state of sustained contraction is called tetanus. The equipment is the same as in Sections 1 and 2.

Stimulate the muscle slowly at first and gradually increase the rate until the muscle remains in a completely contracted state. If little time is given for relaxation between contractions, a record of what is called *incomplete tetanus* may be demonstrated first. By increasing the rate of stimulation still more, the curve written by the muscle lever is almost a smooth line, and the muscle appears to be in a state of sustained contraction or *complete tetanus.* As the muscle becomes fatigued by being held in a contracted state, it may relax even though receiving stimuli.

Draw in Figure 15-4 a myogram of incomplete and complete tetanus that you just observed.

**Figure 15-4.** Incomplete and complete tetanus.

6. *Optional muscle experiments may include:* Effect of temperature on muscle contraction by contractions of a chilled versus a warmed muscle.

Observation of the "staircase phenomenon" (treppe) in skeletal muscle by rapidly stimulating the muscle with constant intensity, observing that each twitch will be a little greater than the preceding one. It is believed that this phenomenon is caused by the increase in temperature or warming up of the muscle as it continues to contract, which increases its irritibility.

Applying multiple (tetanizing) *above threshold* stimuli to a muscle until complete fatigue occurs, determine the time interval from the beginning of stimulation until tetanus and complete fatigue. After allowing the fatigued muscle to rest for 5 minutes, repeat the multiple stimuli. This time note just how long the muscle remains in tetanus, and how long it takes before it exhibits complete fatigue. Again, using the same muscle, allow it to remain quiet for 20 minutes. Repeat the above experiment. Determine if this longer "resting" period makes any difference. Ask the students to explain their results in chemical-physiological terms.

## CONCLUSIONS

1. Distinguish between an isotonic and an isometric contraction.

2. What is the importance of a tonic contraction?

3. Define flaccid.

4. What conditions could cause muscle atrophy?

5. What is a myogram?

6. Define refractory period and how does it compare for cardiac and skeletal muscle?

7. Define summation.

8. Compare an incomplete and complete tetanic contraction.

9. Explain a treppe (staircase) contraction and why it is important for athletes.

10. In what respects do you think myograms are important in the diagnosis of muscle diseases?

## PART VI   MUSCLE DISORDERS

Consult your textbook and briefly describe the following disorders. Be sure to indicate the cause, symptoms, and treatment wherever possible.

1. fatigue

2. fibrosis

3. fibrositis

4. muscular dystrophy

5. myasthenia gravis

**CONCLUSIONS**

1. What factors contribute to muscle fatigue?

2. Contrast atrophy and myopathy.

3. What structural changes occur in muscle fibers in muscular dystrophy?

4. Define the following abnormal contractions.
   a. spasm

   b. cramp

   c. convulsion

   d. fibrillation

5. Define each of the following:

  a. gangrene

  b. myology

  c. myomalacia

  d. myopathy

  e. myoscelerosis

  f. myospasm

  g. myotonia

  h. sprain

  i. trichinosis

  j. Volkmann's contracture

  k. wryneck

**STUDENT ACTIVITY**

The experiments outlined in the module involved skeletal muscle. Write a short paragraph explaining the main differences between contractions of skeletal, smooth, and cardiac muscle, and why skeletal muscle was chosen for these particular experiments.

**128**

# MODULE 16
# THE MUSCULAR SYSTEM: GROSS FEATURES AND MOVEMENTS

## OBJECTIVES

1. To identify the connective tissue coverings and attachments of skeletal muscles

2. To define the origin, insertion, and belly of a skeletal muscle

3. To identify the relationship between bones and skeletal muscles in producing movements

4. To explain the relationship between skeletal muscles, blood vessels, and nerves

5. To explain skeletal muscle movements as activities of agonists, antagonists, and synergists

In this module you will study the connective tissue components of muscles, blood, and nerve supplies, their relationships to bones, and the action of muscles. The skeletal muscles you will learn are described in Module 17.

## PART I   GENERAL CONSIDERATIONS

Skeletal muscles are protected, strengthened, and attached to other structures by several connective tissue components. The entire muscle is usually wrapped in fibrous connective tissue called the *epimysium*. Extensions of the epimysium divide the muscle into fasciculi, or bundles. These extensions are referred to as the *perimysium*. Extensions of the perimysium, the *endomysium*, penetrate into each fasciculus and surround each fiber. The epimysium is continuous with a *tendon*, a connective tissue that attaches muscle to bone, or an *aponeurosis*, a broad flat tendon that attaches muscles to bone or to each other. Skeletal muscles are also well supplied with nerves and blood vessels. Generally, an artery and one or two veins accompany a nerve into a skeletal muscle.

Using your textbook as a reference, label Figure 16-1.

Skeletal muscles produce movements by pulling on tendons which, in turn, pull on bones. Ordinarily, the tendon attachment to the more stationary bone is called the *origin*, while the tendon attachment to the movable bone is referred to as the *insertion*. The portion of the muscle between the origin and insertion is the *belly*. Refer to Figure 16-2a and label the origin, insertion, and belly.

**129**

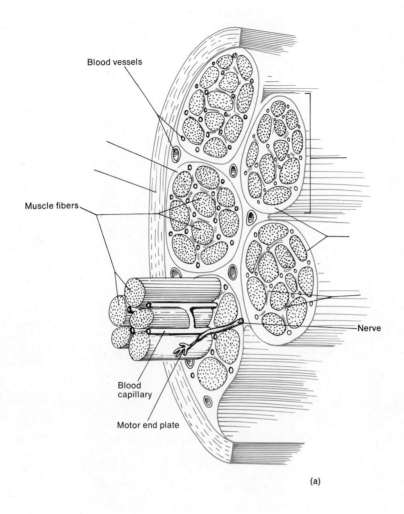

Blood vessels

Muscle fibers

Blood
capillary

Motor end plate

Nerve

(a)

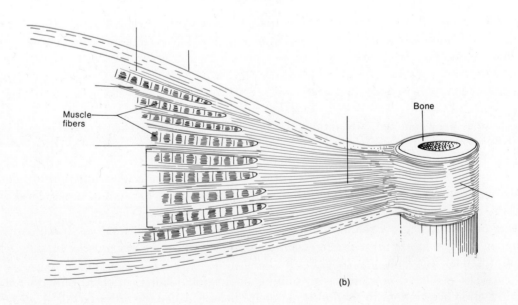

Muscle
fibers

Bone

(b)

**Figure 16–1.** Gross features of a skeletal muscle. (a) Cross section. (b) Longitudinal section.

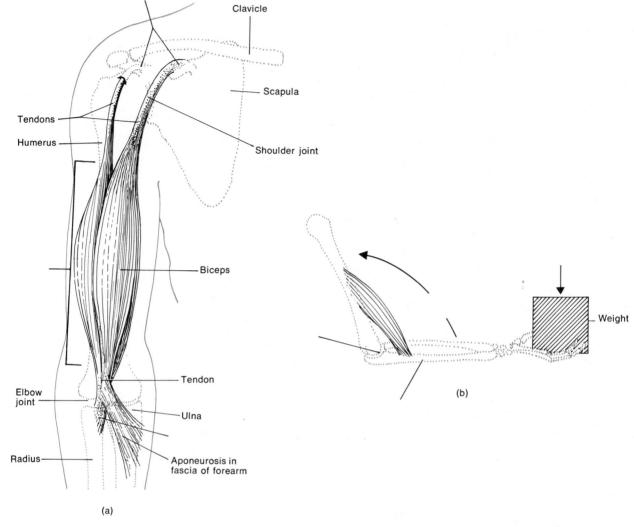

Figure 16-2. Muscle attachments.

Labels in figure (a): Clavicle, Scapula, Shoulder joint, Tendons, Humerus, Biceps, Tendon, Elbow joint, Ulna, Radius, Aponeurosis in fascia of forearm

Labels in figure (b): Weight

    When skeletal muscles produce movements, bones act as levers and joints function as fulcrums of the levers. A *lever* is a rigid rod free to move about a fixed point or support called a *fulcrum*. A lever is acted upon at 2 different points by 2 different forces:

1.  resistance (load)—something to be overcome or balanced, such as the weight of the body to be moved or some object to be lifted.
2.  effort (force)—contraction of the muscle that overcomes the resistance.

    Refer to Figure 16-2b and label the lever, fulcrum, resistance, and effort. Assume that the forearm is being flexed by the biceps muscle.

    Most body movements are coordinated by the activity of several skeletal muscles acting together. Consider flexing the forearm at the elbow again. The biceps muscle is the *agonist* (prime mover) since its contraction produces the desired action. The triceps is the *antagonist* because its action (relaxation) is opposite that of the agonist. The deltoid (abductor of humerus) and pectoralis major (adductor and rotator of humerus) are *synergists* because they assist the agonist by reducing undesired action or unnecessary actions. In this case, they hold the arm and shoulder in a suitable position for the flexing action.

**131**

**CONCLUSIONS**

1. Describe the 3 connective tissues that surround a skeletal muscle.

2. Distinguish between a tendon and an aponeurosis.

3. Give an example of a first, second, and third class lever in the body. Be sure to indicate the positions of the fulcrum, effort, and resistance.

4. What is the importance of each of the following in producing movement?

   a. agonist

   b. antagonist

   c. synergist

5. Describe how a skeletal muscle receives its blood and nerve supply.

6. Define origin and insertion.

7. Distinguish between superficial and deep fascia.

**STUDENT ACTIVITY**

   Write a brief paragraph describing the relationships between skeletal muscles and skin, blood vessels, nerves, fascia, tendons, aponeurosis, and bones.

# MODULE 17
# MUSCLES OF THE BODY

**OBJECTIVES**

1. To define several distinctive criteria employed in naming skeletal muscles

2. To identify the origins, insertions, and actions of selected skeletal muscles

In this module you will learn the names, locations, and actions of the principal skeletal muscles of the body.

## PART I   NAMING SKELETAL MUSCLES

Most of the almost 700 skeletal muscles of the body are named on the basis of one or more distinctive criteria. If you understand these criteria, you will find it much easier to learn and remember the names of individual muscles. Some muscles are named on the basis of the *direction of the muscle fibers*, e.g., rectus (meaning straight), transverse, and oblique. Rectus fibers run parallel to some imaginary line, usually the midline of the body, transverse fibers run perpendicular to the line, and oblique fibers run diagonal to the line. Muscles named according to this criterion include the rectus abdominis, transversus abdominis, and external oblique, respectively. Another criterion employed is *location*. For example, the temporalis is so named because of its proximity to the temporal bone, while the tibialis anterior is located near the tibia. *Size* is also commonly employed. For instance, "maximus" means largest, "minimus" means smallest, "longus" refers to long, and "brevis" refers to short. Examples include the gluteus maximus, gluteus minimus, adductor longus, and peroneus brevis.

Some muscles such as the biceps, triceps, and quadriceps are named on the basis of the *number of origins* they have. For instance, the biceps has 2 origins, the triceps 3, and the quadriceps 4. Other muscles are named on the basis of *shape*. Common examples include the deltoid (meaning triangular) and trapezius (meaning trapezoid). Muscles may also be named after their *insertion* and *origin*. Two such examples are the sternocleidomastoid (originates on sternum and clavicle, and inserts at mastoid process of temporal bone), and the stylohyoideus (originates on styloid process of temporal bone, and inserts at the hyoid bone).

Still another criterion used for naming muscles is *action*. Listed here are the principal actions of muscles, their definitions, and examples of muscles that perform the actions. For convenience, the actions are grouped as antagonistic pairs where possible.

*Flexor:*            Decreases the angle at a joint
                     (flexor carpi radialis)
*Extensor:*          Increases the angle at a joint
                     (extensor carpi ulnaris)
*Abductor:*          Moves a bone away from the midline
                     (abductor hallucis)
*Adductor:*          Moves a bone closer to the midline
                     (adductor longus)
*Levator:*           Produces an upward or superiorly directed movement
                     (levator scapulae)
*Depressor:*         Produces a downward or inferiorly directed movement
                     (depressor labii inferioris)
*Supinator:*         Turns the palm upward or to the anterior
                     (supinator)
*Pronator:*          Turns the palm downward or to the posterior
                     (pronator teres)
*Dorsiflexor:*       Flexes the ankle joint
                     (entensor digitorum longus)
*Plantar flexor:*    Extends the ankle joint
                     (plantaris)
*Invertor:*          Turns the sole of the foot inward
                     (tibialis anterior)
*Evertor:*           Turns the sole of the foot upward
                     (peroneus tertius)
*Sphincter:*         Decreases the size of an opening
                     (pyloric sphincter between stomach and duodenum). Many sphincters are composed
                     of smooth muscle.
*Tensor:*            Makes a body part more rigid
                     (tensor fasciae latae)
*Rotator:*           Moves a bone around its longitudinal axis
                     (obturator)

## PART II  PRINCIPAL MUSCLES OF THE BODY

In the pages that follow, a series of Tables has been provided for you to learn the principal skeletal muscles by region, diagrams of the skeleton already studied in Module 11, spaces for listing the origins, insertions, and actions of the muscles; and a learning key. The learning key is a listing of prefixes, roots, suffixes, and definitions that explain the derivation of the muscles' names.

Do the following with each Table:

1. Take each muscle, in sequence, and study the learning key. This will help you to understand the reason for giving a muscle its name.
2. As you learn the name of each muscle, consult your textbook to determine its origin, insertion, and action. Write in the origins, insertions, and actions in the spaces provided in the Table.
3. Again, using your textbook as a guide, label the diagram shown in the Table.
4. Try to visualize what happens when the muscle contracts so that you will understand its action.
5. Do the above for each muscle in the Table. Before moving to the next Table, examine a torso or

**134**

chart of the skeletal system, so that you can compare and approximate the positions of the muscles that you have sketched.

6. When possible, try to feel each muscle on your own body.

Table 17–1. MUSCLES OF FACIAL EXPRESSION.

| MUSCLE | LEARNING KEY | ORIGIN    INSERTION    ACTION |
|---|---|---|
| Epicranius | *epi* = over; *crani* = skull | This muscle is divisible into 2 portions: the frontalis, over the frontal bone, and the occipitalis, over the occipital bone. The 2 muscles are united by a strong aponeurosis, the galea aponeurotica, which covers the top and sides of the skull. |

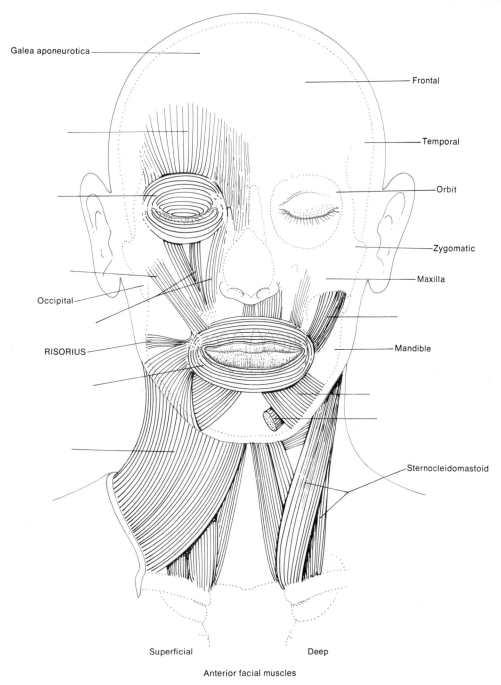

Anterior facial muscles

TABLE 17– 1 continued.

| | LEARNING KEY | ORIGIN | INSERTION | ACTION |
|---|---|---|---|---|
| Frontalis | *front* = forehead | | | |
| Occipitalis | *occipito* = base of skull | | | |
| Orbicularis oris | *orb* = circular; *or* = mouth | | | |
| Zygomaticus major | *zygomatic* = cheek bone; *major* = greater | | | |
| Levator labii superioris | *levator* = raises or elevates; *labii* = lip; *superioris* = upper | | | |
| Depressor labii inferioris | *depressor* = depresses or lowers; *inferioris* = lower | | | |
| Buccinator | *bucc* = cheek | | | |
| Mentalis | *mentum* = chin | | | |
| Platysma | *platy* = flat, broad | | | |
| Risorius | *risor* = laughter | | | |
| Orbicularis oculi | *ocul* = eye | | | |

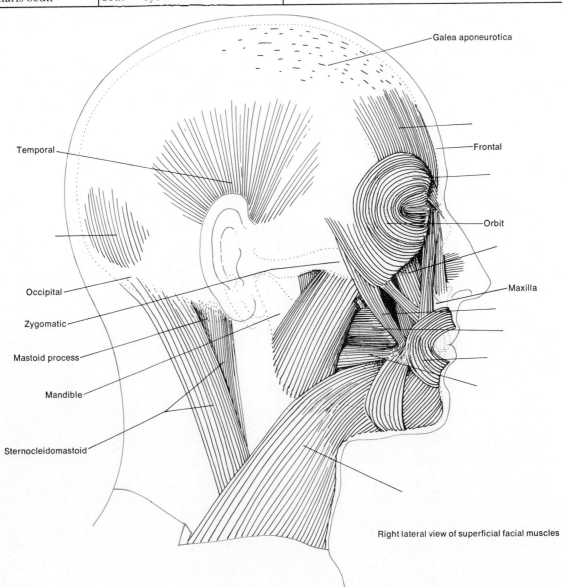

Right lateral view of superficial facial muscles

Table 17–2.  MUSCLES THAT MOVE THE MANDIBLE.

| MUSCLE | LEARNING KEY | ORIGIN | INSERTION | ACTION |
|---|---|---|---|---|
| Masseter | *maseter* = chewer | | | |
| Temporalis | *tempora* = temples | | | |
| Medial pterygoid | *medial* = closer to midline; *pterygoid* = pterygoid plate of sphenoid | | | |
| Lateral pterygoid | *lateral* = farther from midline | | | |

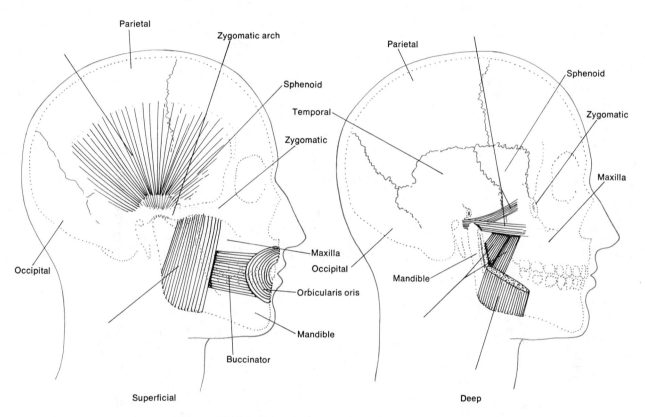

Right lateral view of muscles that move the mandible

**137**

Table 17–3. MUSCLES THAT MOVE THE EYEBALLS.

| MUSCLE | LEARNING KEY | ORIGIN | INSERTION | ACTION |
|---|---|---|---|---|
| Superior rectus | *superior* = above; *rectus* = muscle fibers running parallel to midline | | | |
| Inferior rectus | *inferior* = below | | | |
| Lateral rectus | | | | |
| Medial rectus | | | | |
| Superior oblique | *oblique* = muscle fibers running diagonal to midline of eyeball | | | |
| Inferior oblique | | | | |

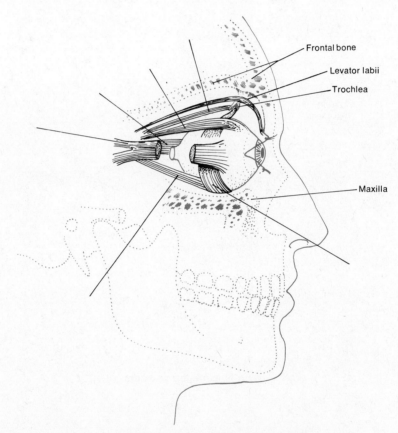

Frontal bone

Levator labii

Trochlea

Maxilla

Right lateral view of muscles of the right eyeball

### Table 17–4. MUSCLES THAT MOVE THE TONGUE.

| MUSCLE | LEARNING KEY | ORIGIN | INSERTION | ACTION |
|---|---|---|---|---|
| Genioglossus | *geneion* = chin; *glossus* = tongue | | | |
| Styloglossus | *stylo* = styloid process of temporal | | | |
| Stylohyoid | *hyoeides* = hyoid bone | | | |
| Hyoglossus | *hyo* = hyoid bone | | | |

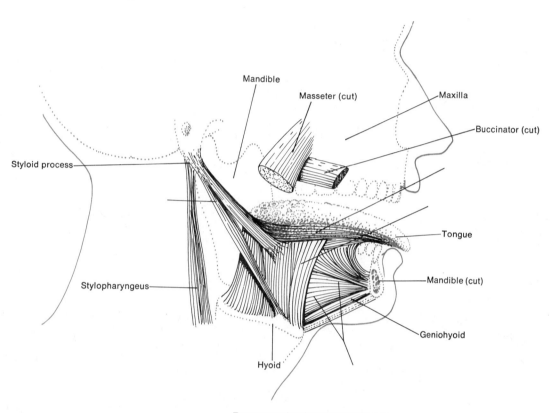

Tongue muscles viewed from right side

### Table 17–5. MUSCLES THAT MOVE THE HEAD.

| MUSCLE | LEARNING KEY | ORIGIN | INSERTION | ACTION |
|---|---|---|---|---|
| Sternocleidomastoid (see Table 17-8) | *sternum* = breastbone; *cleido* = clavicle; *mastoid* = mastoid process of temporal | | | |
| Semispinalis capitis (see Table 17-12) | *semi* = half; *spine* = vertebral spinous processes; *caput* = head | | | |
| Splenius capitis (see Table 17-12) | *splen* = bandage | | | |
| Longissimus capitus (see Table 17-12) | *longissimus* = longest | | | |

## Table 17–6. MUSCLES THAT ACT ON THE ANTERIOR ABDOMINAL WALL.

| MUSCLE | LEARNING KEY | ORIGIN | INSERTION | ACTION |
|---|---|---|---|---|
| Rectus abdominis | *abdominus* = abdomen | | | |
| External oblique | *external* = closer to the surface | | | |
| Internal oblique | *internal* = farther from the surface | | | |
| Transversus abdominis | *transverse* = muscle fibers run transversely to midline | | | |

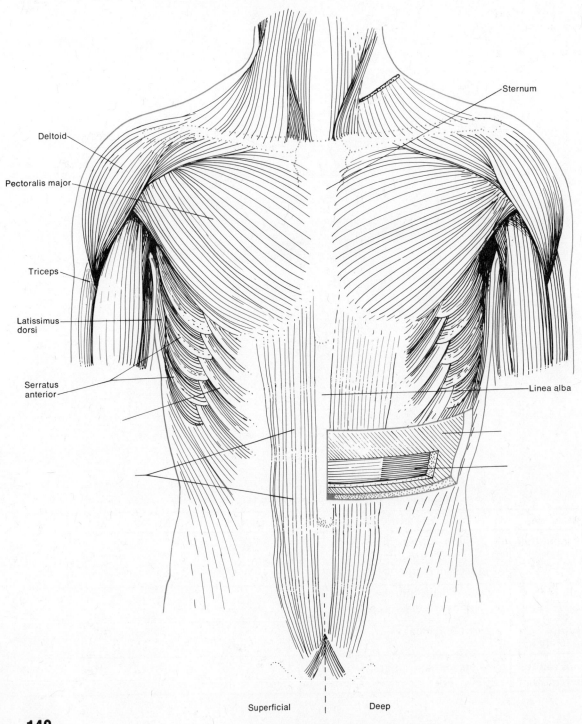

Deltoid

Pectoralis major

Triceps

Latissimus dorsi

Serratus anterior

Sternum

Linea alba

Superficial       Deep

Anterior abdominal muscles

**140**

Table 17–7.  MUSCLES USED IN BREATHING.

| MUSCLE | LEARNING KEY | ORIGIN | INSERTION | ACTION |
|---|---|---|---|---|
| Diaphragm | *diaphragm* = a partition wall | | | |
| External intercostals | *inter* = between; *costa* = rib | | | |
| Internal intercostals | | | | |

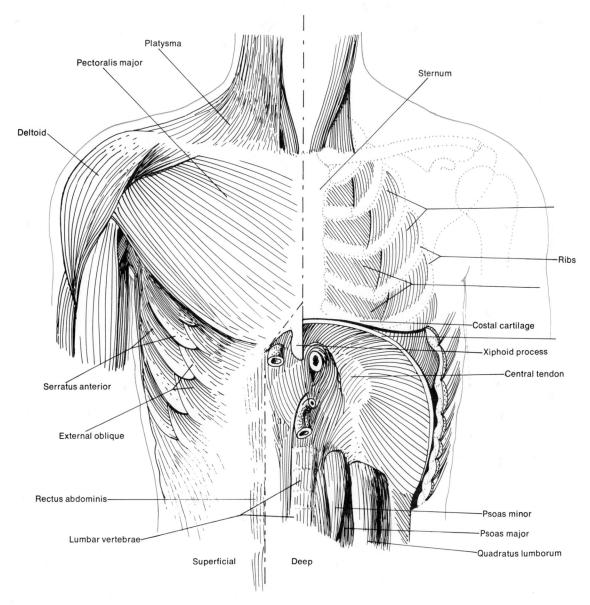

Anterior muscles used in breathing

**141**

### Table 17-8. MUSCLES THAT MOVE THE SHOULDER GIRDLE.

| MUSCLE | LEARNING KEY | ORIGIN | INSERTION | ACTION |
|---|---|---|---|---|
| Subclavius | *sub* = under; *clavius* = clavicle | | | |
| Pectoralis minor | *pectus* = breast, chest, thorax; *minor* = lesser | | | |
| Serratus anterior | *serratus* = serrated; *anterior* = front | | | |
| Trapezius | *trapezoeides* = trapezoid-shaped | | | |
| Levator scapulae | *levator* = raises; *scapulae* = scapula | | | |
| Rhomboideus major | *rhomboides* = rhomboid-shaped | | | |
| Rhomboideus minor | | | | |

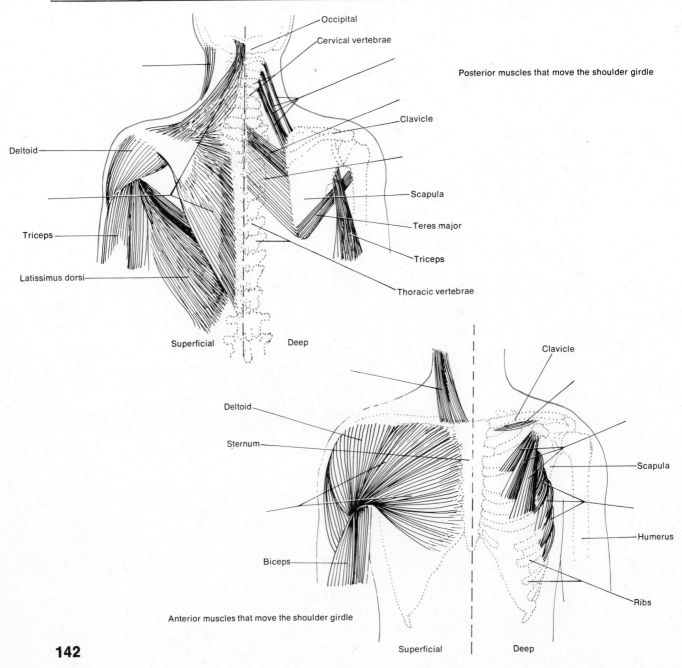

Posterior muscles that move the shoulder girdle

Anterior muscles that move the shoulder girdle

142

## Table 17-9. MUSCLES THAT MOVE THE HUMERUS.

| MUSCLE | LEARNING KEY | ORIGIN | INSERTION | ACTION |
|---|---|---|---|---|
| Pectoralis major (see Table 17-8) | | | | |
| Latissimus dorsi | *latissimus* = broadest; *dorsum* = back | | | |
| Deltoid | *delta* = triangular-shaped | | | |
| Supraspinatus | *supra* = above; *spinatus* = spine of scapula | | | |
| Infraspinatus | *infra* = below | | | |
| Teres major | *teres* = round | | | |
| Teres minor | | | | |

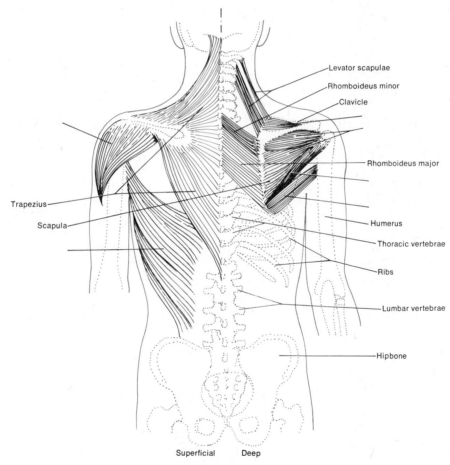

Posterior muscles that move the humerus

Table 17–10. MUSCLES THAT MOVE THE FOREARM

| MUSCLE | LEARNING KEY | ORIGIN | INSERTION | ACTION |
|---|---|---|---|---|
| Biceps brachii | *biceps* = 2 heads of origin; *brachi* = arm | | | |
| Brachialis | | | | |
| Brachioradialis | *radialis* = radius | | | |
| Triceps brachii | *triceps* = 3 heads of origin | | | |
| Supinator | *supination* = turning palm upward or anterior | | | |
| Pronator teres | *pronation* = turning palm downward or posterior | | | |

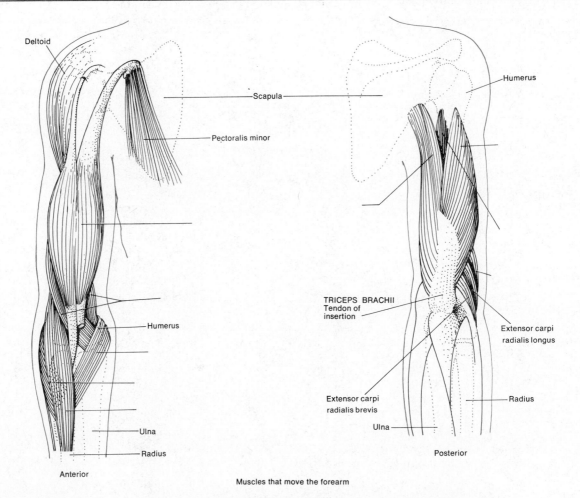

Muscles that move the forearm

Table 17–11. MUSCLES THAT MOVE THE WRIST AND FINGERS.

| MUSCLE | LEARNING KEY | ORIGIN | INSERTION | ACTION |
|---|---|---|---|---|
| Flexor carpi radialis | *flexor* = decreases angle; *carpus* = wrist | | | |
| Flexor carpi ulnaris | *ulnaris* = ulna | | | |
| Extensor carpi radialis longus | *extensor* = increases angle at a joint; *longus* = long | | | |

**144**

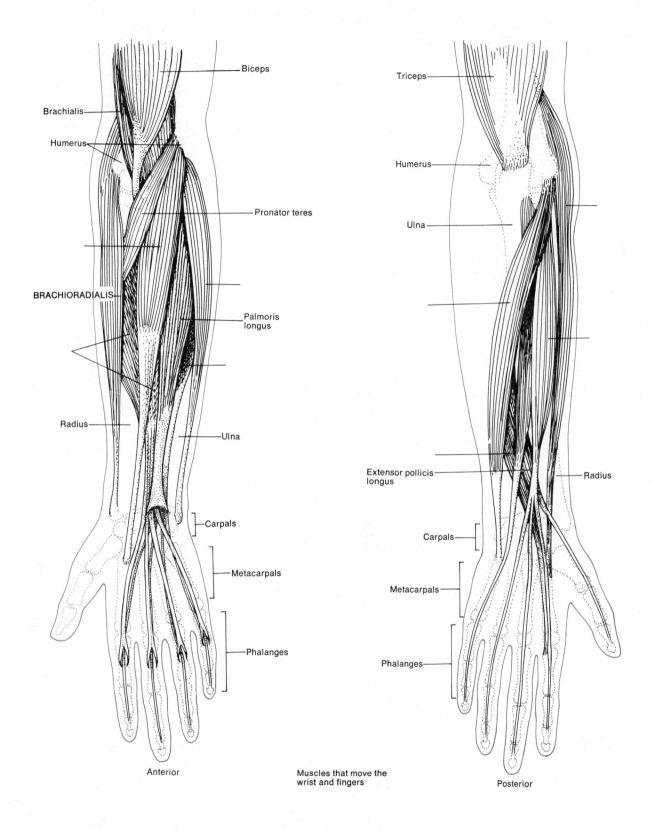

Biceps

Brachialis

Humerus

Pronator teres

BRACHIORADIALIS

Palmoris
longus

Radius

Ulna

Carpals

Metacarpals

Phalanges

Anterior

Triceps

Humerus

Ulna

Extensor pollicis
longus

Radius

Carpals

Metacarpals

Phalanges

Posterior

Muscles that move the
wrist and fingers

TABLE 17–11 continued.

| MUSCLE | LEARNING KEY | ORIGIN | INSERTION | ACTION |
|---|---|---|---|---|
| Extensor carpi ulnaris | | | | |
| Flexor digitorum profundus | *digit* = finger or toe; *profundus* = deep | | | |
| Flexor digitorum superficialis | *superficialis* = superficial | | | |
| Extensor digitorum | | | | |
| Extensor indicis | *indicis* = of index | | | |

Table 17–12. MUSCLES THAT MOVE THE VERTEBRAL COLUMN.

| MUSCLE | LEARNING KEY | ORIGIN | INSERTION | ACTION |
|---|---|---|---|---|
| Rectus abdominis (see Table 17–6) | | | | |
| Quadratus lumborum | *quad* = quadrilateral; *lumb* = of lumbar region | | | |
| Sacrospinalis (erector spinae) | This muscle consists of 3 posterior groupings: iliocostalis, longissimus, and spinalis and these groups, in turn, consist of a series of overlapping muscles. The iliocostalis group is laterally placed, the longissimus group is intermediate in placement, and the spinalis is medially placed. | | | |
| *Lateral* Iliocostalis lumborum | *ili* = flank; *lumb* = loin | | | |
| Iliocostalis thoracis | *thorax* = chest | | | |
| Iliocostalis cervicis | *cervix* = neck | | | |
| *Intermediate* Longissimus thoracis | | | | |
| Longissimus cervicis | | | | |
| Longissimus capitis | | | | |
| *Medial* Spinalis thoracis | | | | |

**146**

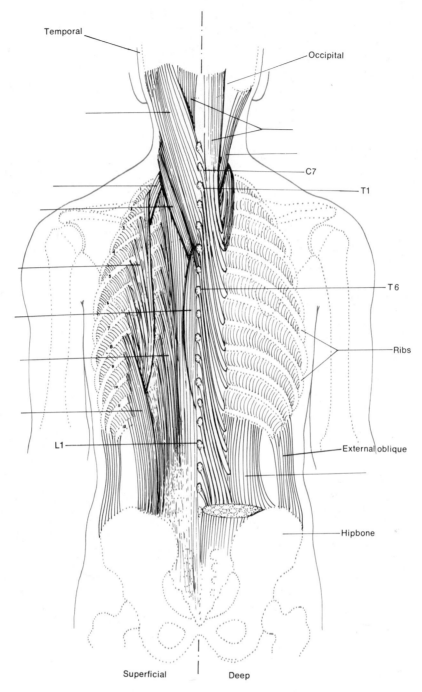

Temporal

Occipital

C7

T1

T6

Ribs

External oblique

L1

Hipbone

Superficial          Deep

Muscles that move the vertebral column

### Table 17-13. MUSCLES THAT MOVE THE FEMUR.

| MUSCLE | LEARNING KEY | ORIGIN | INSERTION | ACTION |
|---|---|---|---|---|
| Psoas major | *psoas* = muscle of loin | | | |
| Iliacus | *iliacus* = ilium | | | |
| Gluteus maximus | *gloutos* = buttock; *maximus* = largest | | | |
| Gluteus medius | *medi* = middle | | | |

TABLE 17–13 continued.

| MUSCLE | LEARNING KEY | ORIGIN | INSERTION | ACTION |
|---|---|---|---|---|
| Gluteus minimus | *minimus* = smallest | | | |
| Tensor fasciae latae | *tensor* = makes rigid; *fasci* = band; *lat* = broad, wide | | | |
| Adductor longus | *adductor* = moves a part closer to the midline | | | |
| Adductor brevis | *brevis* = short | | | |
| Adductor magnus | *magnus* = large | | | |
| Piriformis | *piri* = pear; *forma* = form, shape | | | |
| Quadratus femoris | *femoris* = femur | | | |
| Obturator externus | *obturator* = closed because it arises over the obturator foramen which is closed by a membrane; *external* = outside | | | |

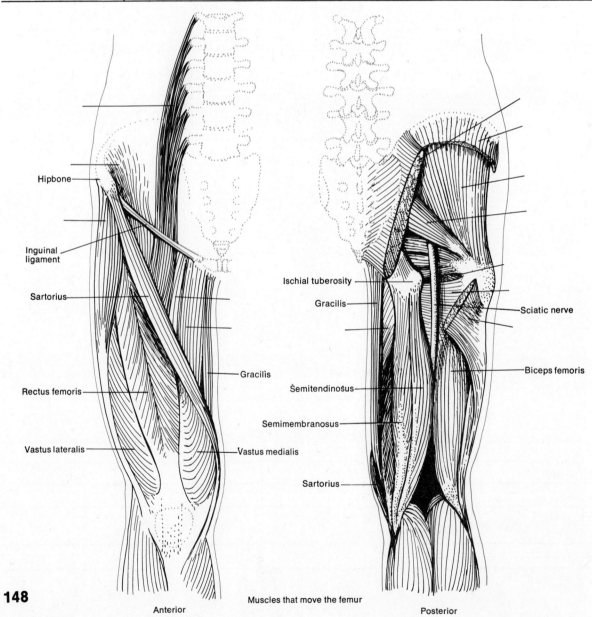

Hipbone

Inguinal ligament

Sartorius

Rectus femoris

Vastus lateralis

Gracilis

Vastus medialis

Ischial tuberosity

Gracilis

Semitendinosus

Semimembranosus

Sartorius

Sciatic nerve

Biceps femoris

Muscles that move the femur

Anterior

Posterior

### Table 17–14. MUSCLES THAT ACT ON THE KNEE JOINT.

| MUSCLE | LEARNING KEY | ORIGIN | INSERTION | ACTION |
|---|---|---|---|---|
| Quadriceps femoris | A composite muscle which includes 4 distinct parts, usually described as 4 separate muscles. The common tendon from the patella to the tibial tuberosity is known as the patellar ligament. | | | |
| Rectus femoris | | | | |
| Vastus lateralis | *vastus* = vast, large; *lateralis* = lateral | | | |
| Vastus medialis | *medialis* = medial | | | |
| Vastus intermedius | *intermedius* = middle | | | |
| Hamstrings | A collective designation for 3 separate muscles. | | | |
| Biceps femoris | | | | |
| Semitendinosus | *semi* = half; *tendo* = tendon | | | |
| Semimembranosus | *membran* = membrane | | | |
| Gracilis | *gracilis* = slender | | | |
| Sartorius | *sartor* = tailor (refers to cross-legged position in which tailors sit) | | | |

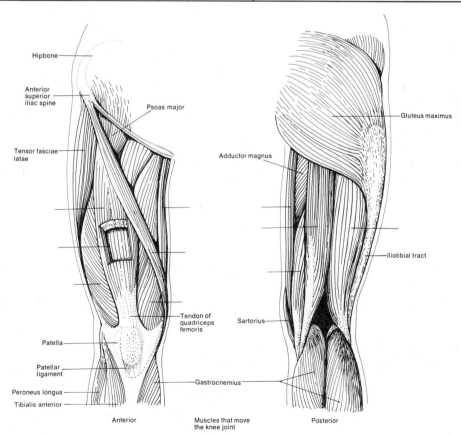

Anterior      Muscles that move the knee joint     Posterior

**149**

## Table 17–15.  MUSCLES THAT MOVE FOOT AND TOES.

| MUSCLE | LEARNING KEY | ORIGIN | INSERTION | ACTION |
|---|---|---|---|---|
| Gastrocnemius | *gaster* = belly; *kneme* = leg | | | |
| Soleus | *solea* = sole of foot | | | |
| Peroneus longus | *perone* = brooch, or fibula; *longus* = long | | | |

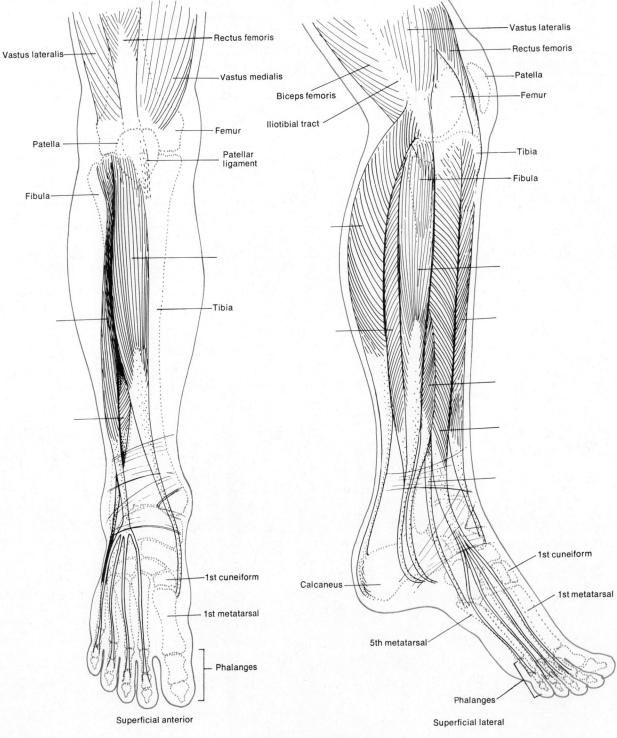

Rectus femoris

Vastus lateralis

Vastus medialis

Biceps femoris

Iliotibial tract

Femur

Patella

Patellar ligament

Fibula

Tibia

1st cuneiform

1st metatarsal

Phalanges

Superficial anterior

Vastus lateralis

Rectus femoris

Patella

Femur

Tibia

Fibula

1st cuneiform

Calcaneus

1st metatarsal

5th metatarsal

Phalanges

Superficial lateral

Muscles that move the foot and toes

**150**

TABLE 17–15 continued.

| MUSCLE | LEARNING KEY | ORIGIN | INSERTION | ACTION |
|---|---|---|---|---|
| Peroneus brevis | *brevis* = short | | | |
| Peroneus tertius | *tertius* = third | | | |
| Tibialis anterior | *tibialis* = tibia | | | |
| Tibialis posterior | *posterior* = back | | | |
| Flexor digitorum longus | *flexor* = flexor; *digitorum* = of the digits | | | |
| Extensor digitorum longus | *extensor* = extensor | | | |

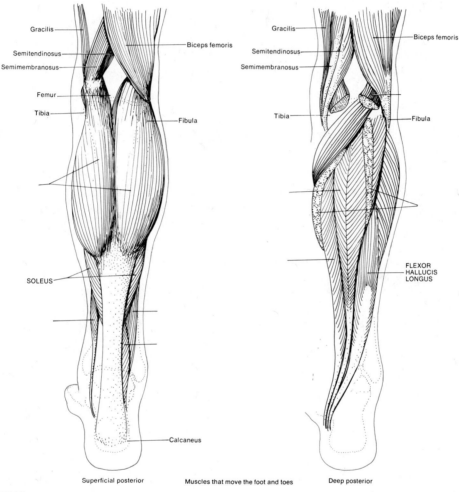

Superficial posterior          Muscles that move the foot and toes          Deep posterior

## CONCLUSIONS

1. What criterion (or criteria) is used in naming the following muscles?

   a. epicranius

   b. levator labii superioris

   c. mentalis

   d. medial pterygoid

   e. superior oblique

   f. styloglossus

**151**

g. sternocleidomastoid

h. rectus abdominus

i. internal intercostals

j. serratus anterior

k. levator scapulae

l. teres minor

m. biceps brachii

n. supinator

o. extensor digitorum

p. quadratus lumborum

q. gluteus maximus

r. adductor brevis

s. quadriceps femoris

t. gracilis

u. gastrocnemius

v. peroneus longus

2. List the muscle or muscles you would use to do the following:
   a. squint

   b. pucker your lips

   c. frown

   d. open your mouth

   e. look to the left

   f. stick your tongue out

   g. move your head as in signifying "yes"

   h. raise your shoulders

   i. abduct the humerus

   j. pronate the forearm

   k. make a fist

   l. bend at the waist

   m. adduct the femur

   n. crossing your legs

   o. curling your toes

3. Under what circumstances are intramuscular injections preferred?

4. Describe the landmarks you would use before giving an intramuscular injection in each of the following:
   a. buttock

b. lateral side of thigh

c. deltoid

**STUDENT ACTIVITY**

In order to have a total picture of skeletal muscles, label the superficial muscles indicated in Figure 17-1.

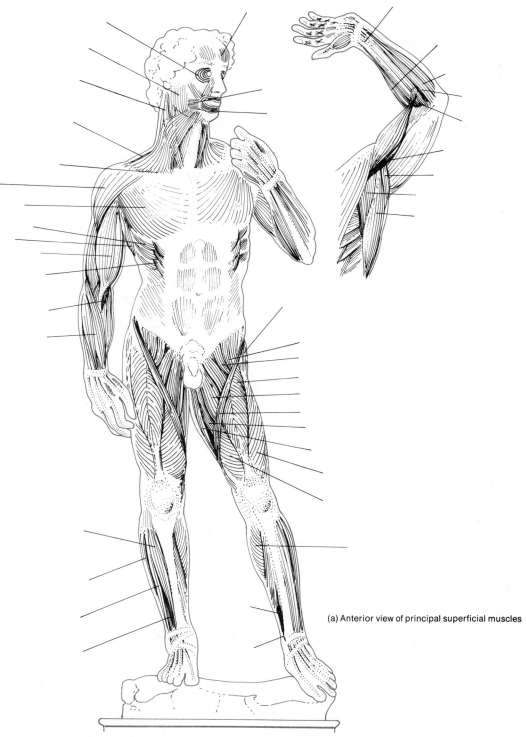

(a) Anterior view of principal superficial muscles

**Figure 17-1.** Anterior view of principal superficial muscles.

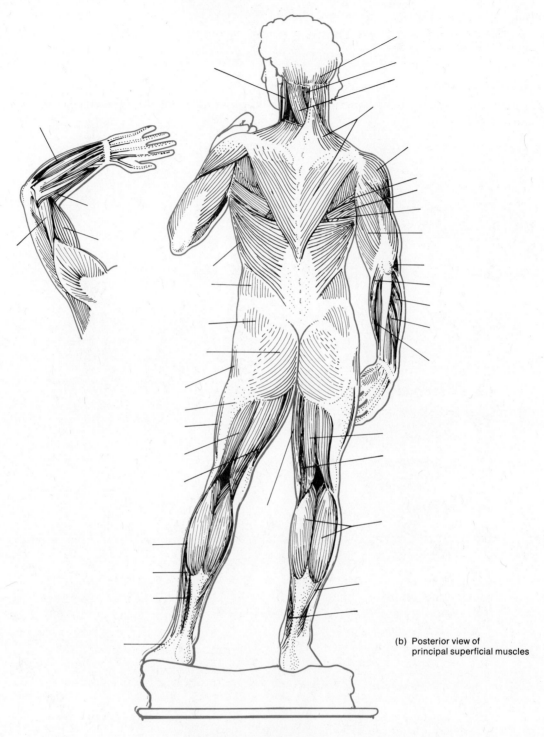

(b) Posterior view of
principal superficial muscles

**Figure 17–1 continued.**

# MODULE 18
## STRUCTURE AND PHYSIOLOGY OF NERVOUS TISSUE

**OBJECTIVES**

1. To identify the outstanding structural features of a neuron

2. To define the factors involved in the initiation of a nerve impulse

3. To explain the properties of nervous tissue that determine the speed of impulse conduction

4. To identify the components of a reflex arc

5. To list the steps involved in the transmission of a nerve impulse across a synapse

6. To define a reflex

7. To describe several clinically important reflexes

Nervous tissue, unlike any other tissue of the body, is specialized to conduct impulses. It is through these impulses and the effects of hormones that all body reactions are controlled and coordinated to maintain homeostasis. In this module, you will examine some of the structural features of nervous tissue and some of the functional aspects of nerve impulse conduction.

## PART I   HISTOLOGY OF NERVOUS TISSUE

Despite the complexity of the nervous system, it consists of only 2 principal kinds of cells. The first of these, the **neurons** or **nerve cells**, make up the nervous tissue, which forms the structural and functional portion of the system. Neurons are highly specialized for impulse conduction and are responsible for all the special attributes associated with the nervous system, such as thinking, controlling muscle activity, and regulating glands. The second type of cell, the **neuroglia**, forms a special kind of connective tissue component of the nervous system. Neuroglia generally perform the less specialized activities of binding together nervous tissue, forming myelin sheaths around axons within the central nervous system, and serving as a selective barrier through which substances must pass before entering nervous tissue. They do not transmit impulses.

Using your textbook as a reference, label Figure 18-1, the neuron. You should include the

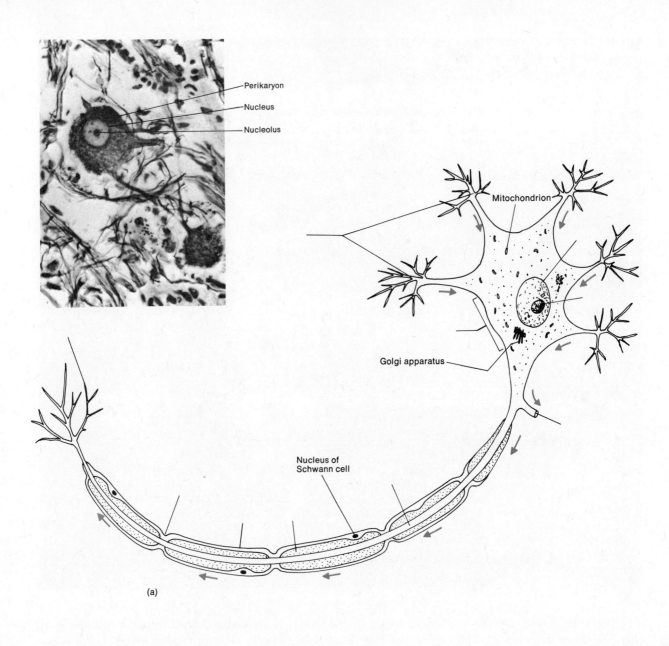

Perikaryon

Nucleus

Nucleolus

Mitochondrion

Golgi apparatus

Nucleus of
Schwann cell

(a)

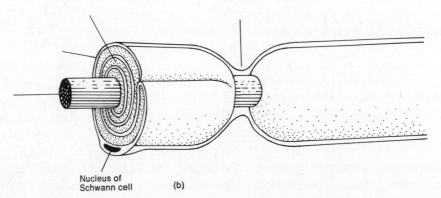

Nucleus of
Schwann cell

(b)

**Figure 18–1.** Structure of a neuron. (a) Shown in an entire neuron. (b) Cross section through a myelinated axon.

following parts of a typical neuron: cell body, dendrites, axon, axon collateral, myelin sheath, Schwann cells, nodes of Ranvier, neurilemma, and cell body.

Examine a prepared slide of ox spinal cord under low, high, and oil-immersion objectives, and identify as many parts of the neuron as you can.

Examine prepared slides of human spinal cord (cross and longitudinal section), nerve endings in skeletal muscle, and nerve trunk (cross and longitudinal section).

## CONCLUSIONS

1. Distinguish between neurons and neuroglia.

2. Consult your textbook and define the following:

  a. cell body

  b. dendrites

  c. axon

  d. axon collaterals

  e. myelin sheath

  f. Schwann cells

  g. nodes of Ranvier

  h. neurilemma

## PART II   THE NERVE IMPULSE

Nerve cells have the ability to respond to stimuli and convert them into nerve impulses. This property is called **irritability.** Nerve cells also have **conductivity,** which is the ability to transmit the impulse to another neuron or to another tissue. Nerve impulses revolve around a simple phenomenon, i.e., the passive and active transport of sodium and potassium ions. In the neuron, sodium ions are actively transported out of the cell where they are found in great numbers, producing a positive charge outside of the cell membrane, a phenomenon called the "sodium pump." However, there is also a "potassium pump" produced because potassium ions are actively transported inward so that more of these ions lie inside the cell than out, producing a negative charge. This ionic difference between the inside and outside of the cell membrane is called the **resting potential**, and such a membrane is said to be **polarized**.

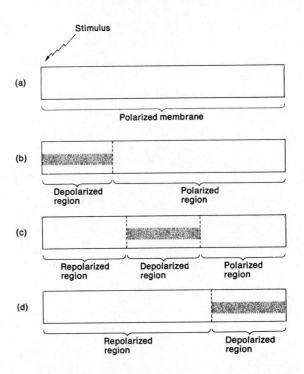

**Figure 18-2.** Initiation and conduction of a nerve impulse.

Consult your textbook and complete Figure 18-2 showing details of how a neuron becomes depolarized and then repolarized back to its resting potential after stimulation. Place the positive and negative charges in the proper positions on either side of the membrane.

## CONCLUSIONS

1. Compare irritability and conductivity.

2. Define "sodium pump."

3. Define "potassium pump."

4. Define resting potential.

5. Contrast depolarization and repolarization.

## PART III   SPEED OF A NERVE IMPULSE

The speed of a nerve impulse is independent of the strength of the stimulus and is normally determined by the size, type, and physiologic condition of the fiber. Fibers with the greatest diameter are called *A fibers*, and they transmit impulses at speeds up to 394 to 410 feet/second. These fibers generally connect the brain and spinal cord with sensors that detect potentially dangerous situations in the outside environment. *B fibers* have middle-sized diameters transmitting impulses at speeds of 10 to 46 feet/second. *C fibers* have the smallest diameters and transmit impulses at the rate of about 2 to 8 feet/second.

### CONCLUSIONS

1. What factors influence the speed of a nerve impulse?

2. Differentiate between *A, B* and *C fibers* comparing their speeds of impulse transmission.

3. Consult your textbook and list some specific functions for each of the 3 types of fibers.

   a. A fiber

   b. B fiber

   c. C fiber

## PART IV   REFLEX ARCS

The route that an impulse takes from its origin in the dendrites or cell body of a stimulated neuron, to its termination somewhere else in the body, is called its "conduction pathway." The most basic conduction pathway is the **reflex arc**, the functional unit of the nervous system. A reflex arc consists of a receptor, a sensory neuron, an integrating center, a motor neuron, and an effector. There are different types of reflex arcs and the simplest is the *two-neuron reflex arc* (also called monosynaptic). An example is the patellar reflex. The withdrawal of the hand from a hot object is typical of a *three-neuron reflex arc* (bisynaptic).

Completely label Figure 18–3 using your textbook as a reference.

### CONCLUSIONS

1. Define a reflex arc.

2. What is the difference between a two-neuron and three-neuron reflex arc?

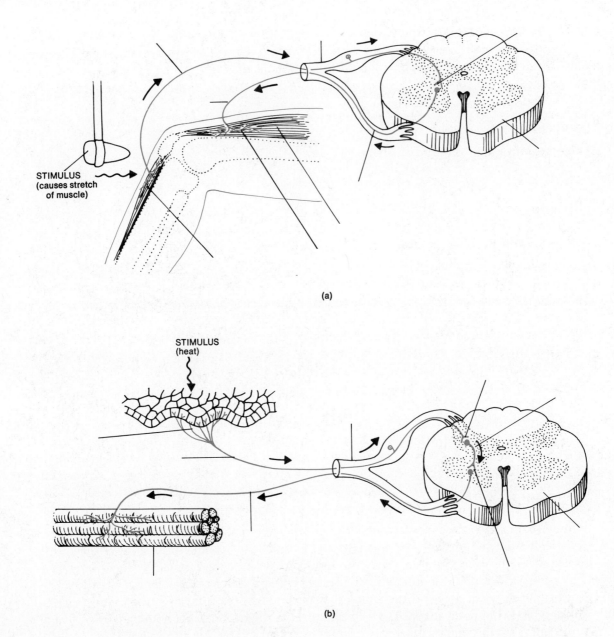

STIMULUS
(causes stretch
of muscle)

(a)

STIMULUS
(heat)

(b)

**Figure 18–3.** Reflex arcs. (a) Two-neuron reflex arc. (b) Three-neuron reflex arc.

3. Define the following:

a. receptor

b. sensory neuron

**160**

c. center

d. association neuron

e. motor neuron

f. effector

## PART V  CONDUCTION ACROSS SYNAPSES

The axons of a neuron never quite touch the dendrites of the next neuron in a conduction pathway. In order for an impulse to move from one neuron to the next, it must bridge a small gap called a **synapse**. The terminal end of an axon contains round or oval expansions called *synaptic knobs* or *end feet*. These end feet contain granular structures called *synaptic vesicles* which, in turn, contain chemical transmitter substances. The impulse arriving at the end of the axon causes the rupture of the synaptic vesicles with the release of these chemicals. The chemicals diffuse from the end feet, into the synapse making contact with the dendrites of the next neuron, stimulating its membrane to become more permeable to sodium ions. If enough sodium ions flow inward, the impulse is transmitted along the second neuron, to the next, and so on.

The best known chemical transmitter at synapses is *acetylcholine*, which is the chemical that also stimulates skeletal muscle contraction. This stimulation would be never-ending if it were not for another chemical called *cholinesterase*, which is also released at a synapse. Cholinesterase destroys acetylcholine. Normally cholinesterase works quickly enough to inactivate all of the acetylcholine by the time the second neuron is returned to its resting state and ready to transmit a second impulse. Consult your textbook for other types of chemical transmitters beside acetylcholine, and other factors that affect impulse transmission. Label Figure 18-4.

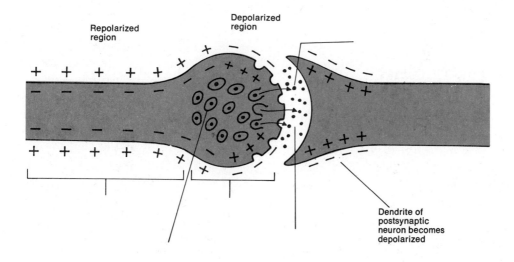

**Figure 18-4.** Impulse conduction at a synapse.

**CONCLUSIONS**

1. Define the following:

   a. synapse

   b. end feet

   c. synaptic vesicles

2. Describe the relationship between acetylcholine and cholinesterase.

3. What other chemical transmitters are there?

4. Explain several factors that affect impulse transmission.

## PART VI    REFLEXES

A **reflex** is a quick, involuntary response that occurs when a nerve passes over a reflex arc. Reflexes may be classified as visceral or somatic, depending upon the kind of organ stimulated. *Visceral reflexes* pass over visceral sensory fibers and then pass back over the autonomic motor fibers to adjust activities such as heartbeat, breathing, and blood pressure. *Somatic reflexes* are conducted over somatic neurons and stimulate skeletal muscle to contract. Reflexes may also be classified on the basis of location of the receptors that pick up the impulse. In an *exteroceptive reflex*, the receptor is located at or near the external body surface. In a *visceroceptive reflex*, the receptor is within a visceral organ or blood vessel. In a *proprioceptive reflex*, the receptor is within a muscle or a tendon.

Reflexes are important tools for diagnosing certain disorders of the nervous system and for locating the area of an injured tissue. Usually, somatic reflexes are tested for such purposes. If a reflex ceases to function or functions abnormally, the physician may suspect that the damage lies somewhere along the conduction pathway.

Among the more important somatic reflexes that have clinical significance are the following:

1. *Patellar reflex*. Extension of the lower leg by contraction of the quadriceps muscle in response to tapping the tendon over the patella.

2. *Achilles reflex.* Extension of the foot in response to tapping the Achilles tendon. Also called the ankle jerk.
3. *Babinski reflex.* Stimulation of the outer margin of the sole of the foot resulting in extension of the great toe, with or without fanning of the other toes. This reflex is considered normal in children under one and a half years of age due to incomplete development of the nervous system.
4. *Plantar reflex.* Curling under of all of the toes accompanied by a slight turning in and flexion of the anterior part of the foot. This is a normal response to stimulation of the outer edge of the sole after one and a half years of age.
5. *Abdominal reflex.* Pulling in the abdominal wall in response to stroking the side of the abdomen.

## CONCLUSIONS

1. Define a reflex.

2. What is the clinical importance of a reflex?

3. How are reflexes classified by organ stimulated and location of receptors?

4. Indicate the probable site of injury or disease if each of the following reflexes is abolished or exaggerated:

   a. patellar

   b. Achilles

   c. Babinski

   d. abdominal

## STUDENT ACTIVITY

Since only axons liberate transmitter substances, there is only one way impulse conduction at a synapse. The direction is from axon to synapse to dendrite. Explain why this is important in maintaining the homeostasis of the body.

# MODULE 19
# STRUCTURE AND FUNCTION OF THE CENTRAL NERVOUS SYSTEM

**OBJECTIVES**

1. To define the structural groupings of nervous tissue

2. To identify the parts of the brain

3. To explain the functions of the brain

4. To identify the parts of the spinal cord

5. To explain the structure and functions of the tracts of the spinal cord

6. To describe the formation and circulation of cerebrospinal fluid

In this module you will be concerned with the anatomy and physiology of the brain and spinal cord, that is, the **central nervous system**. In addition, you will study how the central nervous system is maintained and protected.

## PART I   ORGANIZATION OF NERVOUS TISSUE

Before undertaking your study of the central nervous system (CNS), be sure that you can define each of the following terms:

a. white matter

b. gray matter

c. nerve

d. ganglion

e. tract

f. funiculus

g. nucleus

h. horn

## PART II  THE BRAIN: STRUCTURE AND PHYSIOLOGY

The **brain** is divided into 3 principal areas: **cerebrum**, **cerebellum**, and **brain stem**. The brain stem consists of the medulla oblongata, pons varolii, midbrain, thalamus, and hypothalamus. Label the main parts of the brain shown in Figure 19-1.

Generally, examine a model of a human brain or a preserved mammalian brain and locate the following parts of the cerebrum: cerebral cortex, convolutions, fissures, sulci, hemispheres, and corpus callosum.

Now, carefully examine the brain and identify each of the following parts of the cerebrum: longitudinal fissure, central sulcus, lateral cerebral sulcus, parieto-occipital sulcus, transverse fissure, frontal lobe, parietal lobe, temporal lobe, occipital lobe, and insula. Label the parts of the brain in Figure 19-2.

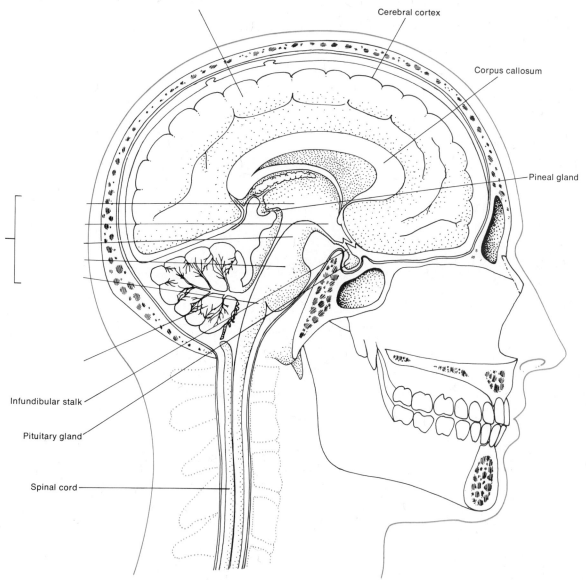

Cerebral cortex

Corpus callosum

Pineal gland

Infundibular stalk

Pituitary gland

Spinal cord

**Figure 19–1.** Parts of the brain as seen in sagittal section.

**165**

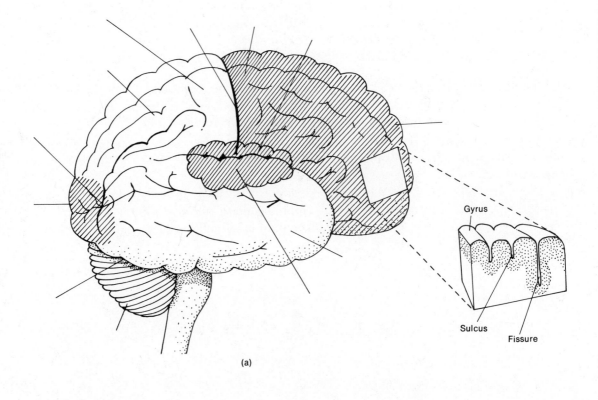

Gyrus

Sulcus

Fissure

(a)

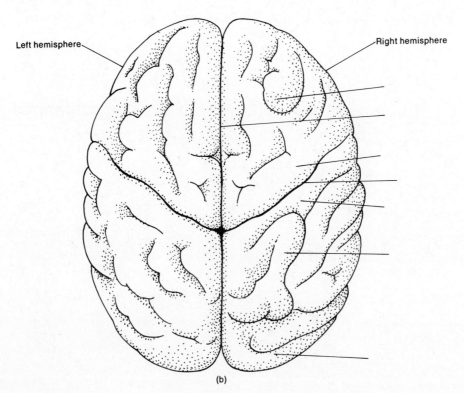

Left hemisphere

Right hemisphere

(b)

**Figure 19–2.** Sulci, fissures, and lobes of the brain (a) Right lateral view. (b) Superior view.

In a very general way, the cerebral cortex is divided into motor, sensory, and association areas. The *motor areas* are regions that govern muscular activities; the *sensory areas* are concerned with the interpretation of sensory impulses; and the *association areas* are concerned with emotional and intellectual processes. Using a red colored pencil, locate motor areas 4, 6, 8, and 44 on Figure 19-3. Where possible indicate the name of each area and its function. Now, using a blue colored pencil, locate the sensory areas 3, 1, 2, 41, 42, 17, and 43 in Figure 19-3. Where possible, indicate the name of each area and its function. Be sure to label the indicated parts.

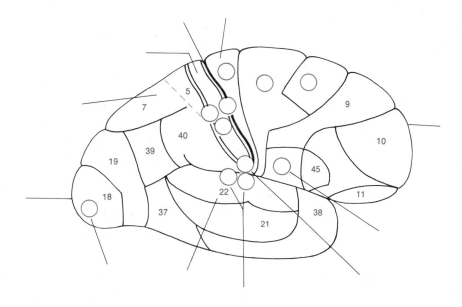

**Figure 19-3.** Motor and sensory areas of the cerebrum.

Examine your model of the brain or preserved specimen and identify the cerebellum. Locate the transverse fissure, vermis, hemispheres, sulci, arbor vitae, and cerebellar nuclei. Label these parts in Figure 19-4.

## CONCLUSIONS

1. Distinguish between the following:

   a. central nervous system

   b. peripheral nervous system

2. Describe the location and composition of the cerebral cortex.

3. Why do convolutions form?

4. Distinguish between fissures and sulci.

5. How are the cerebral hemispheres held together?

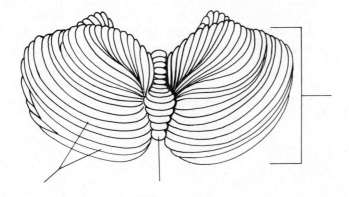

(a)

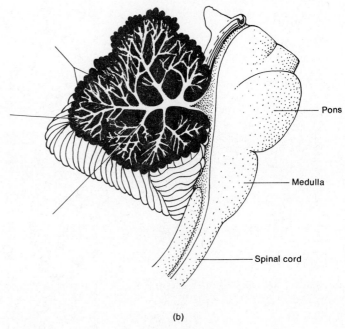

— Pons

— Medulla

— Spinal cord

(b)

**Figure 19–4.** Cerebellum. (a) Viewed from below. (b) Sagittal section.

6. Explain the location and function of each of the following:

    a. association fibers

    b. commissural fibers

    c. projection fibers

7. What is a cerebral nucleus?

8. Describe the function of the caudate and lentiform nuclei.

9. Define the following terms:

    a. aphasia

    b. agraphia

c. word deafness

d. word blindness

10. What is an electroencephalogram?

    Why is it important?


11. What are cerebellar peduncles?

12. Name and locate 3 cerebellar peduncles.

  a.

  b.

  c.

13. Explain the role of the cerebellum in coordinating the movement of skeletal muscles.


14. Describe some signs associated with ataxia.


15. Where is the medulla located?

16. Identify the importance of each of the following parts of the medulla:

  a. decussation of pyramids


  b. nucleus gracilis and cuneatus


  c. cardiac center

  d. respiratory center

  e. vasoconstrictor center

17. Describe the location of the pons.

18. What are the functions of the pons?


19. Define the midbrain.

20. Describe the function of the basilar and tegmental portions of the midbrain.


21. Where is the thalamus located?

    What are its functions?


22. Describe the homeostatic function of the hypothalamus with regard to the following:

  a. control of the autonomic nervous system

b. intermediary between the nervous and endocrine systems

c. center for mind-over-body phenomena

d. control of body temperature

e. regulation of food intake and thirst

## PART III   SPINAL CORD:  STRUCTURE AND PHYSIOLOGY

Examine a model, diagram, or preserved specimen of a **spinal cord** and identify the following parts: cervical enlargement, lumbar enlargement, conus medullaris, filum, cauda equina, and the cervical thoracic, lumbar, sacral, and coccygeal nerves. Label these parts in Figure 19-5a.

Now examine a model, diagram, specimen, or microscope slide of a spinal cord in cross section. Identify the posterior median sulcus, dorsal root, dorsal root ganglion, ventral root, central canal, gray commissure, anterior horn, lateral horn, posterior horn, anterior median fissure, anterior funiculus, lateral funiculus, posterior funiculus, and spinal nerve. Label each of these parts in Figure 19-5b.

Each funiculus of the cord is divided into tracts that convey impulses between spinal nerves and the brain. Label the following tracts in Figure 19-6 and next to each name, indicate its function: anterior spinocerebellar, anterior spinothalamic, lateral spinothalamic, rubrospinal, fasciculus gracilis and cuneatus, lateral corticospinal, posterior spinocerebellar, anterior corticospinal, and tectospinal. Color the ascending tracts red and the descending tracts blue.

### CONCLUSIONS

1. Where does the spinal cord begin and end?


2. What is the significance of the enlargements of the spinal cord?


3. Distinguish between a horn and a funiculus.


4. How many spinal nerves are there? _____ How are they connected to the spinal cord?


5. Compare the function of an ascending and descending tract.



6. On what basis are tracts named?


7. Define transection.

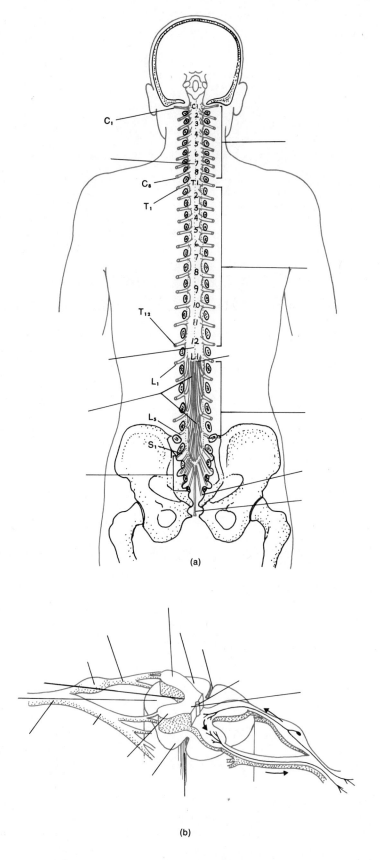

$C_1$

$C_8$

$T_1$

$T_{12}$

$L_1$

$L_5$

$S_1$

(a)

(b)

**Figure 19–5.** The spinal cord. (a) Dorsal view. (b) Cross section.

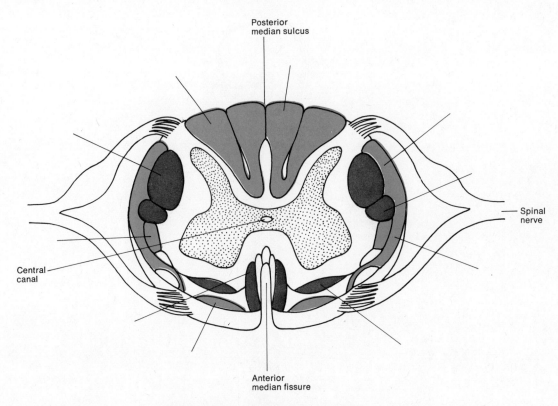

Posterior
median sulcus

Spinal
nerve

Central
canal

Anterior
median fissure

**Figure 19-6.** Tracts of the spinal cord.

8. How does transection interfere with body functions?

9. Distinguish between a quadriplegic and paraplegic.

10. Summarize the functions of the spinal cord.

## PART IV   MAINTENANCE AND PROTECTION OF THE CNS

Examine a diagram or preserved specimen of a brain and spinal cord and locate the dura mater, arachnoid, and pia mater. Label each of these parts shown in Figure 19-7.

Refer to Figure 19-7 again and label the lateral ventricle, interventricular foramen of Monroe, third ventricle, cerebral aqueduct of Sylvius; fourth ventricle, median aperture, lateral apertures, and choroid plexuses. Also label the parts of the brain indicated.

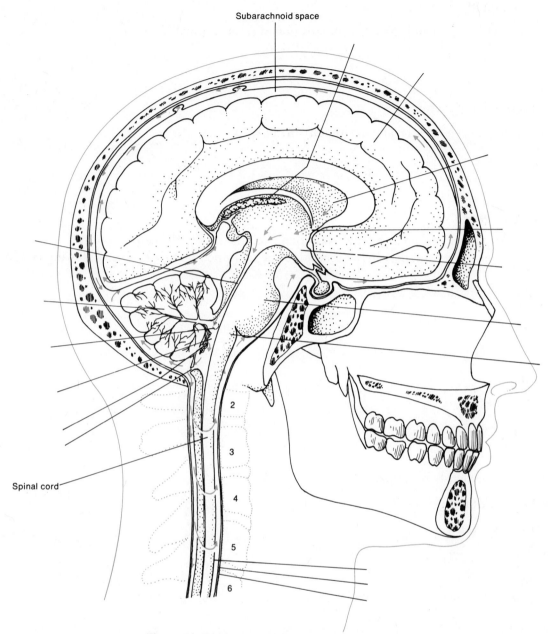

Subarachnoid space

Spinal cord

2

3

4

5

6

**Figure 19-7.** Meninges and ventricles of the brain.

## CONCLUSIONS

1. Explain the importance of oxygen and glucose to the brain.

2. What are meninges?

3. Describe the location and structure of each of the meninges.

4. Define meningitis.

5. Describe the formation and chemical composition of cerebrospinal fluid.



6. What is a lumbar puncture?

7. Distinguish between internal and external hydrocephalus.



**STUDENT ACTIVITY**

1. Describe the pathway taken by a touch sensation from the skin to the cerebral cortex.
2. Which parts of the central nervous system might be malfunctioning when the following conditions occur?

tremors

aphasia

improper vision

loss of memory

inability to taste

poor proprioception

abnormal eye movements

inability to consciously recognize pain

loss of equilibrium

uncontrollable thirst

transection

meningitis

external hydrocephalus

chewing problems

inability to swallow

abnormal body temperature

# MODULE 20
# STRUCTURE AND FUNCTION OF THE SOMATIC NERVOUS SYSTEM

**OBJECTIVES**

1. To identify by number and name the 12 pairs of cranial nerves

2. To describe the distribution, function, and effect of injury or disease on the cranial nerves

3. To distinguish between sensory, motor, and mixed nerves

4. To identify the regional names of the 31 pairs of spinal nerves

5. To identify the components of a spinal nerve

6. To identify the principal plexuses, branches, distributions, and functions of spinal nerves

7. To explain peripheral nerve regeneration

Your investigation in this module concerns the structure and physiology of the **somatic** portion of the peripheral nervous system. The *peripheral nervous system* refers to all the neural tissue outside the brain and spinal cord. The somatic portion of the peripheral nervous system includes the cranial nerves and spinal nerves.

## PART I   CRANIAL NERVES

Of the 12 pairs of **cranial nerves**, 10 pairs originate from the brain stem and all 12 pairs pass through foramina in the base of the skull. The cranial nerves are designated by roman numerals that indicate the order in which the nerves originate from front to back in the brain, and names that indicate their distribution or function.

Label the cranial nerves indicated in Figure 20–1 by roman numeral and name.

**CONCLUSIONS**

1. Define the following and indicate which cranial nerves fall into each category:

   a. mixed

b. motor

c. sensory

2. Describe the distribution, function, and effect of injury or disease for the following cranial nerves:

a. olfactory

b. optic

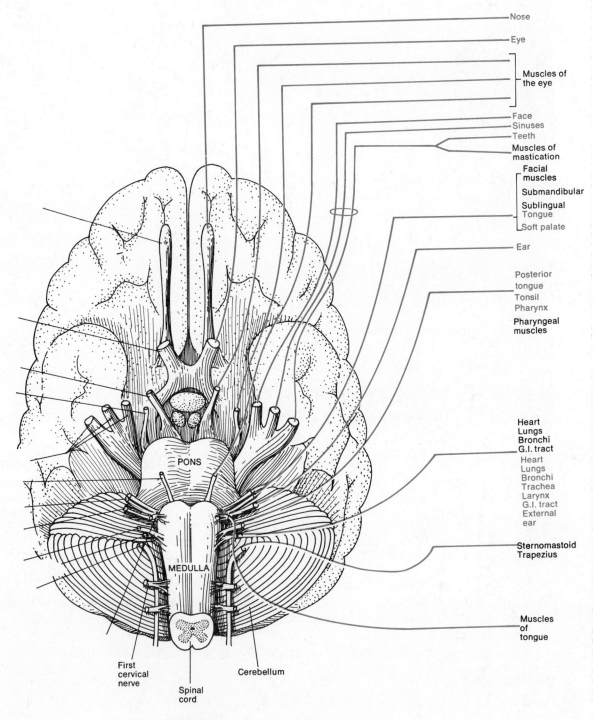

**Figure 20–1.** Cranial nerves.

c. oculomotor

d. trochlear

e. trigeminal

f. abducens

g. facial

h. vestibulocochlear

i. glossopharyngeal

j. vagus

k. accessory

l. hypoglossal

## PART II   SPINAL NERVES

Thirty-one pairs of **spinal nerves** originate from the spinal cord. All are mixed nerves and they are named for the region of the vertebral column from which they emerge.

Refer to Figure 20-2a and label the spinal nerves indicated.

Now examine the cross section of the spinal cord in Figure 20-2b and on a slide of the cross section of the spinal cord. Label the following parts in Figure 20-2b: dorsal root, dorsal root ganglion, ventral root, spinal nerve, dorsal branch, ventral branch, and visceral branch.

### CONCLUSIONS

1. Describe the contents of the:

   a. dorsal root

   b. dorsal root ganglion

   c. ventral root

2. Explain the distribution of the following branches of a spinal nerve:

   a. ventral

   b. dorsal

   c. visceral

   d. meningeal

3. Define a plexus.

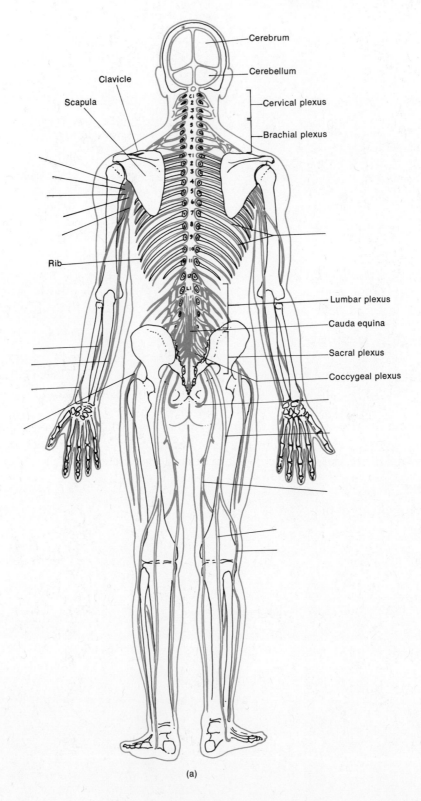

Cerebrum

Cerebellum

Cervical plexus

Brachial plexus

Clavicle

Scapula

Rib

Lumbar plexus

Cauda equina

Sacral plexus

Coccygeal plexus

(a)

**Figure 20–2.** Spinal nerves. (a) Plexuses and their distributions. (b) Structure of a typical spinal nerve.

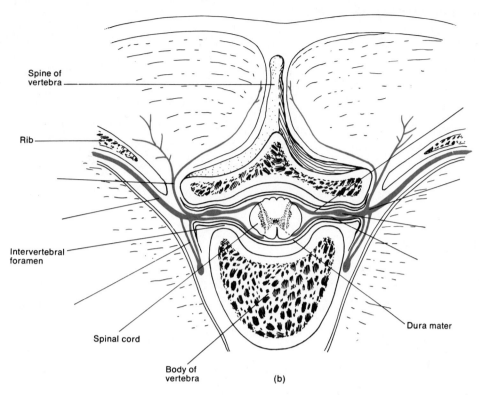

Spine of
vertebra

Rib

Intervertebral
foramen

Spinal cord

Body of
vertebra

Dura mater

(b)

**Figure 20–2 continued.**

4. For each of the following plexuses, indicate the spinal nerves involved, at least 2 principal branches, and the function of the branches:

   a. cervical

   b. brachial

   c. lumbar

   d. sacral

5. What are the intercostal nerves?

6. Describe how a peripheral nerve regenerates.

7. What is necessary in order for a fiber to regenerate?

**STUDENT ACTIVITY**

1. Given the following conditions, which cranial nerve(s) would be directly involved?

   difficulty in chewing and speaking

   reduced production of saliva

   anosmia

   deafness

   blindness

   Bell's palsy

   turning of the eyeball inward

   ptosis of the eyelid

   paralysis of chewing muscles

   loss of balance

   inability to rotate head

   loss of touch in face

   diplopia

2. Which spinal nerve(s) would innervate the following parts of the body:

   external genitals

   head, neck, and shoulders

   diaphragm

   posterior thigh and calf muscles

   posterior arm, forearm, and hand

   anterior and lateral aspects of thigh and leg

   abdominal wall

   dorsum of foot

   most anterior muscles of forearm

# MODULE 21
# STRUCTURE AND FUNCTION OF THE AUTONOMIC NERVOUS SYSTEM

**OBJECTIVES**

1. To describe the components of an autonomic pathway

2. To distinguish the location of autonomic ganglia

3. To contrast the structural differences between the sympathetic and parasympathetic divisions of the ANS

4. To compare the types and actions of cholinergic and adrenergic fibers of the ANS

5. To explain the general roles of the sympathetic and parasympathetic divisions of the ANS

6. To describe several examples of antagonistic effects of the ANS

This module is concerned with the anatomy and physiology of the autonomic nervous system. The **autonomic nervous system (ANS)** is that portion of the peripheral nervous system that controls the activities of smooth muscle, cardiac muscle, and glands. It differs from the somatic nervous system in that it is involuntary, entirely motor, and generally produces antagonistic effects between its two divisions.

## PART I   STRUCTURE

### AUTONOMIC PATHWAY

Refer to Figure 21-1 which shows an autonomic pathway from the central nervous system to a visceral effector. Label the following parts of the pathway: preganglionic neuron, ganglion, synapse, and postganglionic neuron.

### CONCLUSIONS

1. Define the following terms:

   a. preganglionic neuron

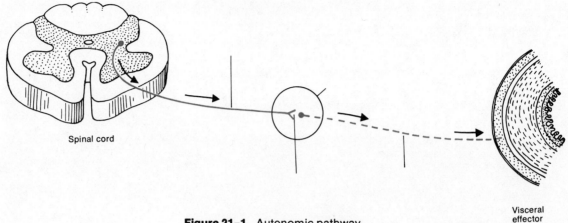

**Figure 21–1.** Autonomic pathway.

Spinal cord

Visceral effector

  b. autonomic ganglion

  c. postganglionic neuron

  d. visceral effector

2. Distinguish the following autonomic ganglia on the basis of location and function:

  a. sympathetic (lateral or vertebral) trunk


  b. prevertebral (collateral)


  c. terminal


3. What is a gray ramus?


4. What is a white ramus?


## AUTONOMIC NERVOUS SYSTEM

    Now examine Figure 21-2, the sympathetic and parasymphathetic divisions of the autonomic nervous system. Label the indicated ganglia and nerves.

### CONCLUSIONS

1. Why is the sympathetic division also called the thoracolumbar division?


2. Describe the sympathetic pathway from the spinal cord to a visceral effector.


3. Why do sympathetic responses generally have widespread effects on the body?

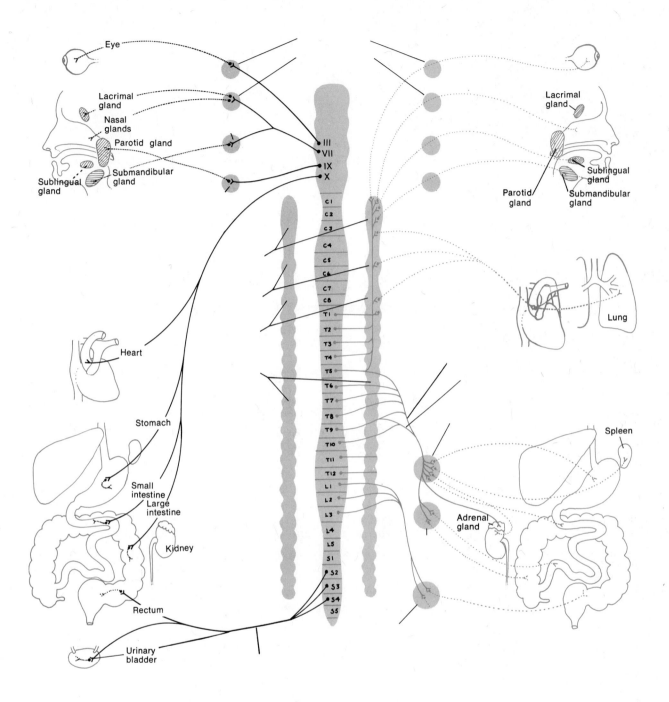

**Figure 21-2.** Structure of the autonomic nervous system.

4. Why is the parasympathetic division also called the craniosacral division?

5. Describe the parasympathetic pathway from the brain or cord to a visceral effector.

6. Why do parasympathetic responses usually produce specific rather than widespread effects?

## PART II  PHYSIOLOGY

Most visceral effectors receive fibers from both the sympathetic and parasympathetic divisions of the ANS. In these cases, the effects are antagonistic.

**CONCLUSIONS**

1. What is a neuroeffector junction?

2. Distinguish between cholinergic and adrenergic fibers and give examples of each.

   a. cholinergic

   b. adrenergic

3. Why are sympathetic responses longer lasting than parasympathetic responses?

4. Under normal circumstances, what are the primary functions of the parasympathetic division?

5. What is the primary responsibility of the sympathetic division? Give examples.

6. Give several examples of antagonistic effects of the autonomic nervous system and indicate which division mediates which response.

**184**

**STUDENT ACTIVITY**

Assume that the body is under severe stress due to a fear situation. Describe the sympathetic responses of the body to fear based upon the following diagram (Figure 21-3). Indicate your response next to each label.

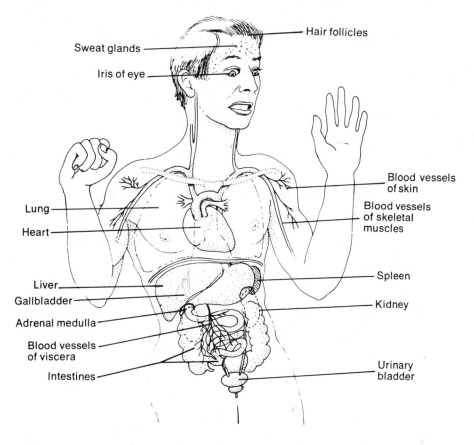

**Figure 21–3.** Sympathetic responses to fear.

# MODULE 22
# DISORDERS OF
# THE NERVOUS SYSTEM

**OBJECTIVES**

1. To describe the causes, symptoms, and treatment of selected disorders of the nervous system

2. To define medical terms associated with the nervous system

In this module you will be concerned with a study of common disorders of the nervous system and medical terminology associated with the nervous system.

## DISORDERS

For each of the followng disorders, describe the cause, symptoms, and treatment (where applicable):

1. poliomyelitis

2. syphilis

3. cerebral palsy

4. Parkinsonism

5. epilepsy

6. multiple sclerosis

7. stroke

**CONCLUSIONS**

1. How does the polio virus spread in the body?

2. What part of the nervous system is affected by the polio virus?

3. Define neurosyphilis.

4. What is tabes dorsalis?

5. How is syphilis treated?

6. What parts of the nervous system are affected by cerebral palsy?

7. What causes the tremor and rigidity associated with Parkinsonism?

8. Describe the use of levadopa in the treatment of Parkinsonism.

9. What is the diagnostic value of an EEG in epilepsy?

10. Distinguish between grand mal and petit mal epileptic seizures.

11. What is psychomotor epilepsy?

12. Distinguish between a glioma and meningioma.

13. How are tumors of the nervous system diagnosed?

## MEDICAL TERMINOLOGY

Define each of the following terms:

a. analgesia

b. anesthesia

c. coma

d. neuralgia

e. neuritis

f. paralysis

g. sciatica

h. shingles

i. spastic

j. torpor

## STUDENT ACTIVITY

Write a short paragraph explaining the widespread homeostatic imbalances that occur when there is a disorder of the nervous system.

# MODULE 23
# CUTANEOUS, PAIN, AND PROPRIOCEPTIVE SENSATIONS

**OBJECTIVES**

1. To define a sensation, and list the 4 prerequisites necessary for its transmission

2. To define projection, adaptation, after images, and modality as characteristics of sensations

3. To compare the location and function of exteroceptors, visceroceptors, and proprioceptors

4. To describe the distribution of cutaneous receptors by interpreting the results of the two-point discrimination test

5. To find the location and function of the receptors for touch, pressure, cold, heat, pain, and proprioception

6. To distinguish between somatic, visceral, referred, and phantom pain

7. To define acupuncture and discuss its possible use as a pain reliever

Now that we have finished the nervous system, we will look at sensations, which are closely related and integrated with this system. If you could not "sense" your environment, and make the necessary homeostatic adjustments, you could not survive on your own. In its broadest context, the term *sensation* refers to the state of awareness of external or internal conditions of the body.

## PART I   CHARACTERISTICS OF SENSATIONS

In order for a sensation to occur, 4 prerequisites must be met: (1) A *stimulus*, or change in the environment, capable of initiating a response by the nervous system, must be present. (2) a *receptor* or sense organ must pick up the stimulus and convert it to a nervous impulse. (3) The impulse must be *conducted* along a nervous pathway from the receptor or sense organ to the brain. (4) A *region* of the brain must *translate* the impulse into a sensation. A *sense receptor* or *sense organ* may be viewed as specialized nervous tissue that exhibits a high degree of sensitivity to internal or external conditions.

A receptor might be very simple, like the dendrites of a single neuron that detect pain in the skin, or it may be a complex organ, such as the eye, that contains highly specialized neurons, epithelium, and connective tissues. All sense receptors contain the dendrites of sensory neurons, exhibit a very high degree of excitability, and possess a low threshold stimulus. Furthermore, the majority of sensory impulses are conducted to the sensory areas of the cerebral cortex, for it is only in this region of the body that a stimulus can produce conscious feeling.

We see with our eyes, hear with our ears, and feel pain in an injured part of our body only because the cortex interprets the sensation as coming from the stimulated sense receptor. One of the characteristics of sensations, that of *projection*, describes the process by which the brain refers sensations to their point of stimulation. A second characteristic of many sensations is *adaptation*. According to this phenomenon, a sensation may disappear even though a stimulus is still being applied. For example, when you first get into a tub of hot water, you might feel an intense burning sensation. But, after a brief period of time, the sensation decreases to one of comfortable warmth, even though the stimulus (hot water) is still present. Another characteristic is *after images*: some sensations tend to persist even though the stimulus has been removed. One common example of after image occurs when you look at a bright light and then look away. You will still see the light for several seconds afterward.

The fourth characteristic of sensations is *modality*, which refers to the specific kind of sensation felt. Examples are pain, pressure, touch, body position, equilibrium, hearing, vision, smell, or taste. In other words, the distinct properties by which one sensation may be distinguished from another is its modality.

## CONCLUSIONS

1. Define sensation.

2. List the 4 prerequisites that must be met before a sensation occurs.

3. Define a sense receptor or sense organ.

4. Give an example of a simple and a complex sense receptor.

5. List 3 characteristics of sense receptors.

6. What part of the brain receives the majority of sensory impulses?
   Why?

7. Define the 4 characteristics of sensations and give an example where applicable.

## PART II    CLASSIFICATION OF SENSATION

One convenient method of classifying sensations is to categorize them according to the location of the receptor. On this basis, receptors may be classified as exteroceptors, visceroceptors, and proprioceptors.

*Exteroceptors* provide information about the external environment, and are located near the surface of the body. They pick up stimuli outside the body and transmit sensations of hearing, sight, touch, pressure, temperature, and pain on the skin.

*Visceroceptors* pick up information about the internal environment and are located in blood vessels and viscera. This information arises from within the body and may be felt as pain, taste, fatigue, hunger, thirst, and nausea.

*Proprioceptors* allow us to feel position and movement and are located in muscles, tendons, and joints. Sensations of position, movement, and equilibrium are caused by the stretching or movement of parts where these receptors are located. They give us information about muscle tension, the position and tension of our joints, and equilibrium.

### CONCLUSIONS

1. What is one way of classifying sensations?


2. Classify and define the 3 basic types of receptors and locate each of them in the human body.




3. Which type of receptor is involved in the following actions?

   a. not eating for 6 hours

   b. leaning over a railing

   c. being poked in the side

   d. seeing an accident

   e. using a trampoline

   f. getting seasick

   g. playing too much tennis

   h. walking in the Sahara Desert

## PART III    CUTANEOUS SENSATIONS

Touch, pressure, cold, heat, and pain are known as the *cutaneous sensations*. The receptors for these sensations are located in the skin, connective tissue, and the ends of the gastrointestinal tract. Inasmuch as the sensation of pain is not limited to cutaneous receptors, we will consider pain under a separate heading.

The cutaneous receptors are randomly distributed over the body surface, so that some parts of the skin are densely populated with receptors and other parts contain only a few, widely separated ones. Areas of the body that have few cutaneous receptors are relatively insensitive, whereas those regions containing large numbers of cutaneous receptors are very sensitive. This can be demonstrated by using the *two-point discrimination test* for touch (Figure 23-1). In the following tests, students will work in pairs, one acting as subject, the other as experimenter. The subject will keep his eyes closed during the experiments.

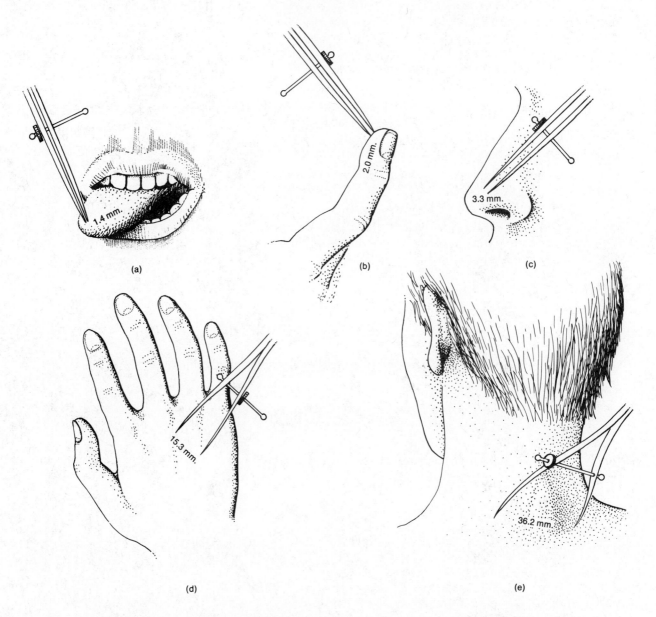

**Figure 23–1.** Two-point discrimination test for touch. (a) Tip of tongue. (b) Tip of finger. (c) Side of nose. (d) Back of hand. (e) Back of neck.

In this test, a compass is applied to the skin and the distance in millimeters between the 2 points of the compass is varied. The subject then indicates when he feels 2 points and when he feels only one. The following order, from greatest sensitivity to least, has been established from the test: tip of tongue, tip of finger, side of nose, back of hand, and back of neck.

The compass may be placed on the tip of the tongue, an area where receptors are very densely packed. The distance between the 2 points can then be narrowed to 1.4 mm. At this distance, the points are able to stimulate 2 different receptors, and the subject feels that he is being touched by 2 objects. However, if the distance is decreased below 1.4 mm., he feels only one point, even though both points are touching him. This is because the points are so close together that they reach only one receptor. The compass can then be placed on the back of the neck, where receptors are relatively few and far between. In this case, the subject feels 2 distinctly different points only if the distance between them is 36.2 mm. or more.

The result of this test indicates that the more sensitive the area, the closer the compass points may be placed and still be felt separately. Record your results, testing each of the areas of the body mentioned previously. Compare the results obtained with Figure 23-1.

Results:

| PART OF BODY | DISTANCE |
|---|---|
| Tip of tongue | _____ |
| Tip of finger | _____ |
| Side of nose | _____ |
| Back of head | _____ |
| Back of neck | _____ |

A. Cutaneous receptors for *touch* include Meissner's corpuscles, Merkel's discs, and root hair plexuses (Figure 23-2). *Meissner's corpuscles* and *Merkel's discs* are most numerous in the fingertips, palms of the hand, and soles of the feet. They are also abundant in the eyelids, tip of the tongue, lips, nipples, clitoris, and tip of penis. *Root hair plexuses* are dendrites arranged in networks around the roots of hairs.

Draw a one-inch square on the back of the forearm, and divide it into 16 smaller squares. In your laboratory manual mark a similar area. With the subject's eyes closed, press the Von Frey hair or a bristle against the skin, just enugh to cause it to bend, once in each of the 16 squares. The pressure should be applied in the same manner each time. The subject should indicate when the sensation of touch is experienced, and the experimenter should record on the paper with the square the corresponding points at which the sensations are felt.

The subject and the experimenter switch places and repeat the test.

## CONCLUSIONS

1. Name 5 cutaneous sensations.

2. What does the two-point discrimination test prove?

3. What parts of the body are most sensitive?

4. What parts of the body are insensitive?

5. Name the 3 cutanous receptors for touch.

6. Name 6 parts of the body that contain many touch receptors.

7. Are the touch receptors uniformly distributed in the 16 squares investigated?

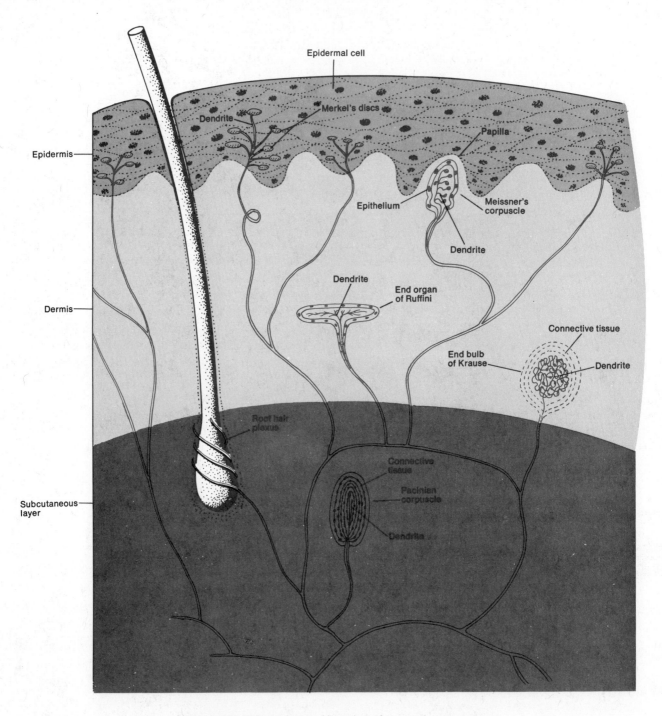

**Figure 23–2.** Structure and location of cutaneous receptors.

B. Sensations of *pressure* are longer lasting and have less variation in intensity than do sensations of touch. Moreover, whereas touch is felt in a small, "pinprick" area, pressure is felt over a much larger area. The pressure receptors are called *Pacinian corpuscles* and are found in the deep, subcutaneous tissues that lie under mucous membranes, in serous membranes of the abdominal cavity, around joints and tendons, and in some viscera.

The experimenter touches the skin of the subject (who has his eyes closed) with a point of a piece of colored chalk. The subject then tries to touch the same spot with a piece of differently

**194**

colored chalk. The distance between the 2 points is then measured and recorded. Proceed using various parts of the body, such as palm of hand, upper arm, forearm, and back of neck.

**CONCLUSIONS**

1. What are pressure receptors called?

2. What does the chalk test prove?

3. Where are these pressure receptors located?

C. The cutaneous receptors for the sensation of cold are called *end bulbs of Krause*. They are widely distributed in the dermis and subcutaneous connective tissue, and are also located in the cornea of the eye, the tip of the tongue, and external genitals. The cutaneous receptors for heat are referred to as *end organs of Ruffini*. They are deeply embedded in the dermis and are less abundant than cold receptors.

On the back of the wrist draw a one-inch square. Place a forceps or other metal probe in ice cold water for a minute, dry it quickly, and, with the dull point, explore the area in the square for the presence of cold spots. Keep the probe cold and mark the position of each spot found with ink. Immerse the forceps in hot water so that it will give a sensation of warmth when removed and applied to the skin, but avoid having it too hot. Proceeding as before, locate the position of the warm spots in the same area. Mark these spots with ink of a different color. Expose the back of the hand and the palm to the heated forceps.

**CONCLUSIONS**

1. Name the cutaneous receptors for the sensation of cold and their locations.

2. What are the cutaneous receptors for heat called?

3. Which of these 2 types of sensations is more numerous within the one-inch square?

4. Which of the 2 areas of your hand is more sensitive to heat?

Why?


**PART IV   PAIN SENSATIONS**

The receptors for *pain* are simply the branching ends of the dendrites of certain sensory neurons. Pain receptors are found in practically every tissue of the body and adapt only slightly or not at all. They may be stimulated by any type of stimulus. Excessive stimulation of any sense organ causes pain. For example, when stimuli for other sensations such as touch, pressure, heat, and cold reach a certain threshold, they stimulate pain receptors as well. Pain receptors, because of their sensitivity to all stimuli, have a general protective function of informing us of changes that could be potentially dangerous to health or life. Adaptation to pain does not readily occur. This is important, since pain indicates disorder or disease. If we became used to it and ignored it, irreparable damage could result.

Sensory impulses for pain are conducted through spinal and cranial nerves. Recognition of the kind and intensity of pain is localized in the cerebral cortex.

In general, pain may be divided into 2 types: somatic and visceral. *Somatic pain* arises from stimulation of the skin receptors, and is called superficial somatic pain. It may also arise from stimulation of receptors in skeletal muscles, joints, tendons, and fascia, in which case it is called deep somatic pain. *Visceral pain* results from stimulation of receptors in the viscera. In most instances of visceral pain, the sensation of pain is not projected back to the point of stimulation. Rather, the pain may be felt in or just under the skin that overlies the stimulated organ. In certain cases, the pain may be felt in a surface area of the body that is quite far removed from the stimulated organ. This phenomenon is called *referred pain*. In general, the area to which pain is referred and the visceral organ involved receive their innervation from the same segment of the spinal cord. Consider the following example: Afferent fibers from the heart enter segments $T_1$ to $T_4$ of the spinal cord, as do afferent fibers from the skin over the heart, and the skin over the medial surface of the left arm. Thus, the pain of a heart attack is typically felt in the skin over the heart and down the left arm. The source of pain is misinterpreted. Figure 23-3 illustrates cutaneous regions to which visceral pain may be referred.

A kind of pain frequently experienced by amputees is called *phantom pain*. In this instance, the person experiences pain in a limb or part of a limb after it has been amputated. Here is how phantom pain occurs: let us say that a foot has been amputated. A sensory nerve that originally terminated in the foot is severed during the operation but repairs itself and returns to function within the remaining leg. From past experience the brain has always projected stimulation of the neuron back to the foot. So when the distal end of this neuron is now stimulated the brain continues to project the sensation back to the missing part. Thus, even though the foot has been amputated, the patient still "feels" pain in his toes.

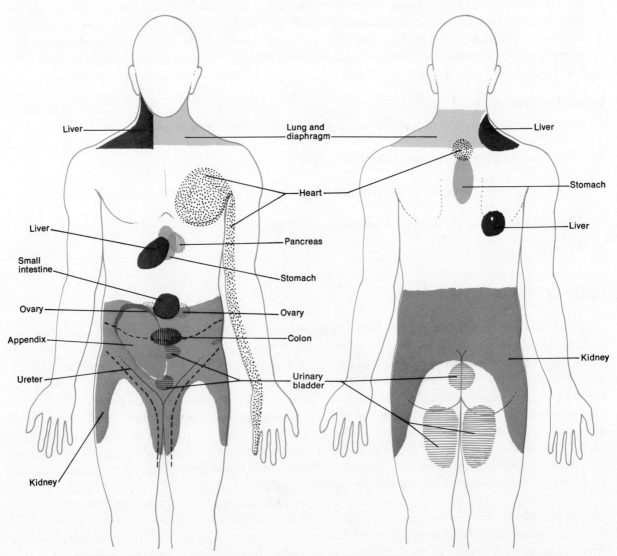

**Figure 23–3.** Referred pain. The shaded parts of the diagram indicate cutaneous areas to which visceral pain is referred.

Most pain sensations respond to pain-killing drugs or surgery. However, another method of inhibiting pain impulses is called *acupuncture*. The word comes from the terms *acus*, meaning needle, and *pungere*, meaning to sting. Here is how the procedure is performed: Needles are inserted through selected areas of the skin. The needles are then twirled by the acupuncturist or by a battery-operated device. In about 20 to 30 minutes after the twirling starts, pain is deadened for 6 to 8 hours. The location of needle insertion varies with the part of the body the acupuncturist desires to anesthetize. To pull a tooth, one needle is inserted in the web between the thumb and the index finger. For a tonsillectomy, one needle is inserted about 2 inches above the wrist. For removal of the lung, one needle is placed in the forearm, midway between the wrist and the elbow.

There is no complete or totally satisfactory explanation of why acupuncture works. According to the "gate control" theory, the twirling of the acupuncture needle stimulates 2 sets of nerves that eventually enter the spinal cord and synapse with the same association neurons. One very fine nerve is the nerve for pain and the other, a much thicker nerve, is the nerve for touch. The speed of the impulse passing along the touch nerve is faster than that passing along the pain nerve. Recall that fibers with larger diameters conduct impulses faster than those with smaller diameters. Because the touch impulse reaches the dorsal horn of the cord first, it has the right of way over the pain impulse. It thus "closes the gate" to the brain before the pain impulse reaches the cord. Since the pain impulse does not pass to the brain, no pain is felt. We should mention, however, that before acupuncture can be used as a routine procedure by American physicians, additional research and understanding of the process must take place.

Using the same one-inch square of the forearm that was previously used for the touch test in Part III, A, perform the following experiment. Mark a similar area in your laboratory manual below. Apply a piece of absorbent cotton soaked with water to the area of the wrist for 5 minutes to soften the skin. Use water as needed during the experiment. Place the point of a needle to the surface of the skin and press enough to produce a sensation of pain. Explore the marked area in a systematic manner, recording in the manual the locations of the points that give pain sensation when stimulated. Distinguish between sensations of pain and touch. Are the areas for touch and pain identical?

Perform the following test to demonstrate the "referred pain" phenomenon. Place your elbow in a large shallow pan of ice water, and note the progression of sensation which you experience. At first, you will feel some discomfort in the region of the elbow. Later, pain sensations will be felt elsewhere. Where do you feel the referred pain?

## CONCLUSIONS

1. Where are pain receptors found?

2. Why is it to our advantage to have pain receptors that are sensitive to all stimuli?

3. Do pain receptors adapt like other sensory receptors? Why is this to our advantage?

4. How are sensory impulses conducted?

5. What part of the brain recognizes the kind and intensity of pain?

6. Name the 2 main types of pain.

7. Where are the receptors located in each type?

8. Describe referred pain.

9. Explain referred pain as it relates to a heart attack.

10. Name the type of pain experienced by some amputees.

11. What is acupuncture?

12. How is it supposed to work?

13. Where is the needle inserted for the following: to pull a tooth, a tonsillectomy, lung removal?

## PART V  PROPRIOCEPTIVE SENSATIONS

An awareness of the activities of muscles, tendons, and jonts is provided by the *proprioceptive* or *kinesthetic sense.* It informs us of the degree to which tendons are tensed and muscles are contracted. The proprioceptive sense enables us to recognize the location and rate of movement of one part of the body in relation to other parts. It also allows us to estimate weight and to determine the muscular work necessary to perform a task. With the proprioceptive sense, we can judge the position and movements of our limbs without using our eyes when we walk, type, play a musical instrument, or dress in the dark.

Proprioceptive receptors are located in muscles, tendons, and joints, and in the connective tissue that surrounds muscle fibers. Proprioceptors adapt only slightly. This is beneficial since the brain must be appraised of the status of different parts of the body at all times so that adjustments can be made to ensure coordination.

A. Face a blackboard close enough so that you can easily reach to mark it. Mark a small "X" on the board in front of you. Close your eyes, raise your right hand above your head and then, with your eyes still closed, make a mark as near as possible to the X. Repeat the procedure by placing your chalk on the X, closing your eyes, and raising your arm above your head, then making another mark as close as possible to the X. Repeat the procedure a third time. Record your results by estimating or measuring how far you missed the X for each trial.

1st trial _____ 2nd trial _____ 3rd trial _____

B. Write the word "physiology" on the left line provided just below. Now, with your *eyes closed,* write the same word immediately to the right. How do the 2 samples of writing compare?

_____     _____

Explain your results:


The following experiments demonstrate that the kinesthetic sensations make it possible to repeat more easily certain acts involving muscular coordination.

Students will work in pairs for these experiments.

1. The experimenter asks the subject to carry out certain movements with his eyes closed, e.g., point to the middle finger of his left hand with the index finger of his right hand.

2. With his eyes closed, the subject extends his right arm as far as possible behind him, then brings his index finger quickly to the tip of his nose. How accurate is the subject in doing this?

3. Ask the subject, with eyes shut, to touch the named fingers of one hand with the index finger of the other hand. How well does the subject carry out the directions?

## CONCLUSIONS

1. What is proprioception or the kinesthetic sense and what does it enable us to do?

2. Where are the proprioceptive receptors located?

3. Explain why it is beneficial to us that the proprioceptive receptors adapt only slightly.

## STUDENT ACTIVITY

Write a short paragraph, clearly indicating the relationship between the different receptors involved in the various sensations discussed in this module and homeostasis of the human body.

# MODULE 24
# SPECIAL SENSES: PHYSIOLOGY AND DISORDERS

**OBJECTIVES**

1. To locate the receptors for olfaction and describe the neural pathway for smell

2. To identify the gustatory receptors and describe the neural pathway for taste

3. To describe the structure and physiology of the accessory visual organs

4. To list the structural divisions of the eye

5. To discuss retinal image formation by describing refraction, accommodation, constriction of the pupil, convergence, and inverted image formation

6. To define emmetropia, myopia, hypermetropia, and astigmatism

7. To describe the efferent pathway of light impulses to the brain

8. To define the anatomical subdivisions of the ear

9. To list or describe the principal events involved in the physiology of hearing

10. To identify the receptor organs for equilibrium

11. To discuss the receptor organs' roles in the maintenance of dynamic and static equilibrium

12. To contrast the causes and symptoms of cataracts, glaucoma, conjunctivitis, trachoma, Meniere's disease, and impacted cerumen

13. To define medical terminology associated with the sense organs

    In contrast to the general senses, the 4 special senses of smell, taste, sight, and hearing have receptor organs that are highly complex. The sense of smell is the least specialized, as opposed to the sense of sight, which is the most specialized. Like the general senses, however, the special senses allow us to detect changes in our environment.

## PART I   OLFACTORY SENSATIONS

The receptors for the *olfactory sense* are located in the nasal epithelium at the upper portion of each nasal cavity on either side of the nasal septum (Figure 24-1a). The nasal epithelium consists of 2 principal kinds of cells: *supporting cells*, which are columnar epithelium cells of the mucous membrane lining the nose, and *olfactory cells*, that are bipolar neurons whose cell bodies lie between the supporting cells (Figure 24-1b).

The distal (free) end of each olfactory cell contains 6 to 8 dendrites, called *olfactory hairs*. The axons of the olfactory cells unite to form several nerves, which pass through openings in the cribriform plate of the ethmoid bone. These olfactory nerves terminate in masses of gray matter called *olfactory bulbs*. The olfactory bulbs lie beneath the frontal lobes of the cerebrum, on either side of the crista galli of the ethmoid bone.

The first synapse of the olfactory neural pathway occurs in the olfactory bulbs between the axons of the olfactory nerves and the dendrites of neurons inside the olfactory bulbs. Axons of these

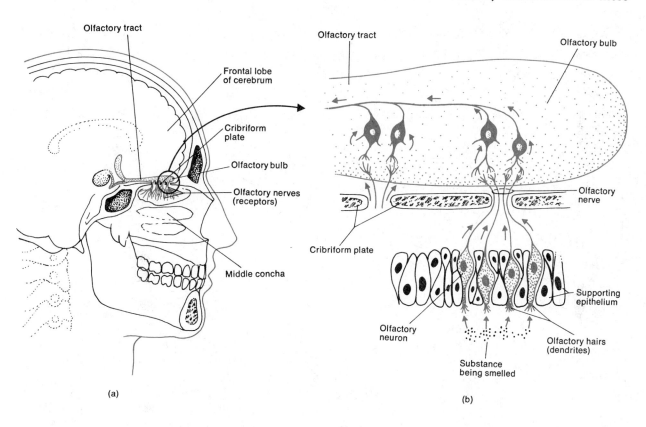

(a)

(b)

**Figure 24-1.** Receptors for olfaction. (a) Location of receptors in nasal cavity. (b) Enlarged aspect of olfactory receptors. *(Photograph courtesy of Donald I. Patt, from Comparative Vertebrate Histology, Donald I. Patt and Gail R. Patt, Harper & Row, Publishers, Inc., New York, N.Y., 1969. Magnification x 450.)*

neurons run posteriorly to form the *olfactory tract*. From here, impulses are conveyed to the olfactory portion of the cortex. In the cortex, the impulses are interpreted as odor and give rise to the sensation of smell.

This process happens quickly, especially the adaptation to odors. For this reason, we become accustomed to some odors and are also able to endure unpleasant ones. Rapid adaptation also accounts for the failure of a person to detect gas that accumulates slowly in a room.

The subject should close his eyes after plugging one of his nostrils with cotton. Hold a bottle of oil of cloves, or other substance having a distinct odor, under the open nostril. The subject breathes in through the open nostril, and exhales through the mouth. Note the time required for the odor to disappear, and repeat with the other nostril. As soon as *olfactory adaptation* has occurred, test an entirely different substance. Compare results for the various materials tested. Olfactory stimuli, such as pepper, onions, ammonia, ether, and chloroform, are irritating and may cause tearing because they stimulate the receptors of the trigeminal nerve as well as the olfactory neurons.

## CONCLUSIONS

1.  Where are the receptors for the olfactory sense located?

2.  What types of cells are involved in the sense of smell?

3.  The olfactory cell contains dendrites that are called

4.  How is the olfactory tract formed?

5.  What part of the brain is involved in smell?

6.  What is smell adaptation?

7.  What causes tearing?

## PART II  GUSTATORY SENSATIONS

The receptors for *gustatory sensations*, or sensations of taste, are located in the taste buds (Figure 24-2b). Taste buds are most numerous on the tongue, but they are also found on the soft palate and in the throat. The *taste buds* consist of 2 kinds of cells. The *supporting cells* are specialized epithelium that forms a capsule. Inside each capsule are 4 to 20 *gustatory cells*, which are the sensory neurons for taste. Each gustatory cell contains a dendrite that projects to the external surface through an opening in the taste bud called the *taste pore*. Gustatory cells make contact with taste stimuli through the taste pore.

Taste buds are found in some connective tissue elevations on the tongue called *papillae* (Figure 24-2a). The papillae give the upper surface of the tongue its characteristic rough appearance. *Circumvallate papillae*, the largest type, are circular and form an inverted V-shaped row at the posterior portion of the tongue. *Fungiform papillae* are knoblike elevations and are found primarily on the tip and sides of the tongue. All circumvallate and most fungiform papillae contain taste buds. *Filiform papillae* are threadlike structures that cover the anterior two-thirds of the tongue (Figure 24-3).

In order for gustatory cells to be stimulated, the substances we taste must be in solution in the saliva so that they can enter the taste pores in the taste buds. Despite the many substances we taste, there are basically only 4 taste sensations: sour, salt, bitter, and sweet. Each taste is due to a different

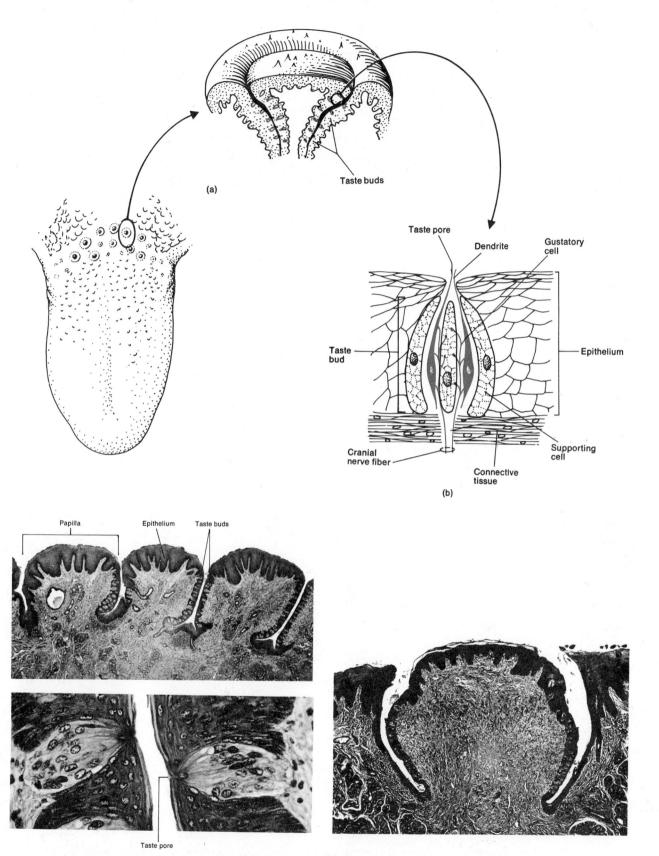

**Figure 24-2.** Taste receptors. (a) Location of taste buds relative to papillae. (b) Structure of a taste bud. *(Photographs courtesy of Edward J. Reith, from Atlas of Descriptive Histology, Edward J. Reith and Michael H. Ross, Harper & Row, Publishers, Inc., New York, N.Y., 1970.)*

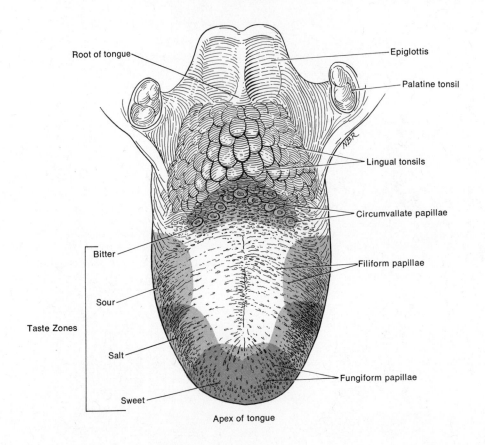

**Figure 24-3.** The tongue.

response to different chemicals. Certain regions of the tongue react more strongly than other regions to particular taste sensations. For example, although the tip of the tongue reacts to all 4 taste sensations, it is highly sensitive to sweet and salty substances. The posterior portion of the tongue is highly sensitive to bitter substances, and the lateral edges of the tongue are more sensitive to sour substances (Figure 24-3).

The nerves that supply afferent fibers to taste buds are: the facial nerve (VII, See Figure 20-1), which supplies the anterior two-thirds of the tongue; the glossopharyngeal (IX), which supplies the posterior one-third of the tongue; and the vagus (X), which supplies the epiglottis area of the throat. From these nerves the impulses enter the medulla, pass through the thalamus, and terminate in the parietal lobe of the cortex.

Students will work in pairs for the following experiments.

1. Subject thoroughly dries tongue (use absorbent cotton). The experimenter places some granulated sugar on the tip of the tongue. The subject indicates when the sugar is tasted by raising his hand. The experimenter records the time interval till tasted.

 The experiment is repeated using a drop of sugar solution. These results are also recorded. How do you explain the difference in time periods?

2. Have the subject rinse his mouth and extend his tongue. The experimenter places a drop of sugar solution on the back and on the tip of the tongue; the subject indicates where the sensation of sweetness is more strongly detected.

 The subject rinses his mouth again. The experiment is then repeated using the quinine solution, and the salt solution.

 After rerinsing, the experiment is repeated using the acetic acid solution or vinegar placed on the tip and *sides* of the tongue. Record the results in Table 24-1 by inserting + or − in the appropriate places.

**204**

Table 24–1. TONGUE AREAS INVOLVED IN BASIC TASTES

|  | SOUR | SALT | BITTER | SWEET |
|---|---|---|---|---|
| Tip of tongue |  |  |  |  |
| Base of tongue |  |  |  |  |
| Sides of tongue |  |  |  |  |

3. Taste for certain substances is inherited, and geneticists for many years have been using the chemical phenylthiocarbamide for testing taste. To some individuals this substance tastes bitter, to others it is sweet, and some cannot taste it at all. Place a few crystals of phenylthiocarbamide on the subject's tongue. Does he taste it? If so, describe the taste. Special paper that is flavored with this chemical may be chewed and mixed with saliva and tested in the same manner.

Record on the blackboard your response to the phenylthiocarbamide test. Usually about 70 percent of the people tested can taste this compound; 30 percent cannot. Compare this percentage with the class results.

4. The last test combines the effect of smell on the sense of taste. Obtain small cubes of carrot, onion, potato, and apple. The subject dries his tongue, closes his eyes, and pinches his nostrils shut. The experimenter places the materials, one by one, on the subject's tongue. The subject attempts to identify each (1) immediately, (2) after chewing (nostrils closed), and (3) after opening the nostrils. Record your results in Table 24-2.

Table 24–2. TASTE SENSATION

|  | SENSATIONS WHEN PLACED ON DRY TONGUE | SENSATIONS WHILE CHEWING (NOSTRILS CLOSED) | SENSATIONS WITH NOSTRILS OPENED |
|---|---|---|---|
| Carrot |  |  |  |
| Onion |  |  |  |
| Potato |  |  |  |
| Apple |  |  |  |

## CONCLUSIONS

1. Where are the receptors for the gustatory sense located?


2. Where are the taste buds found?


3. Which cells make up the taste buds?


4. What structures connect the special taste cells with the taste stimuli?


5. The upper surface of the tongue is rough because of what structures?

6. Name, describe, and locate the 3 kinds of structures in which taste buds are located on the tongue.

7. Name the 4 basic taste sensations.

8. Which nerves supply the taste buds?

9. Where do these nerves terminate in the brain?

## PART III  VISUAL SENSATIONS

The structures related to *vision* are the eyeball, which is the receptor organ for visual sensations, the optic nerve (II, see Figure 20-1), the brain, and a number of accessory structures.

A. Among the *accessory organs* are the eyebrows, eyelids, eyelashes, and the lacrimal apparatus which allows us to tear (Figure 24-4). The *eyebrows* protect the eyeball from falling objects, prevent perspiration from getting into the eye, and shade the eye from the direct rays of the sun. The *eyelids*, or *palpebrae*, consist primarily of smooth muscle that is covered externally by skin. The underside of the muscle is lined with a mucous membrane called the *conjunctiva*, which also covers the surface of the eyeball. The eyelids shade the eye during sleep, protect the eye from light rays and foreign objects, and spread lubricating secretions over the surface of the eyeball. Projecting from the border of each eyelid is a row of short, thick hairs, the *eyelashes*. Sebaceous glands at the base of the hair follicles of the eyelashes pour a lubricating fluid into the follicles. Infection of these glands is called a *sty*. The *lacrimal apparatus* is a term used for a group of structures that manufactures and drains away tears. These structures are the lacrimal glands, the lacrimal ducts, the lacrimal sacs, and the nasolacrimal ducts. The *lacrimal secretion* is a watery solution containing salts and some mucus. It cleans, lubricates, and moistens the external surface of the eyeball.

Using your textbook as a reference, label Figure 24-4.

B. The adult eye measures about one inch in diameter. Of its total surface area, only the anterior one-sixth is exposed. The remainder is recessed and protected by the orbit into which it fits. Anatomically, the eyeball can be divided into 3 principal layers: (1) fibrous tunic, (2) vascular tunic, and (3) retina.

The *fibrous tunic* is the outer coat of the eyeball, and is divided into 2 regions; the posterior sclera and the anterior cornea. The *vascular tunic* is the middle layer of the eyeball and is composed of 3 portions: the posterior choroid, the anterior ciliary body, and iris. The third and inner coat of the eye, the *retina*, lies in the posterior portion of the eye and its primary function is image formation. In addition to these principal layers, the eyeball itself contains the lens.

Using your textbook as a reference, label Figure 24-5.

The dendrites of the photoreceptor neurons are called rods and cones because of their respective shapes. They are visual receptors highly specialized for stimulation by light rays. *Rods*

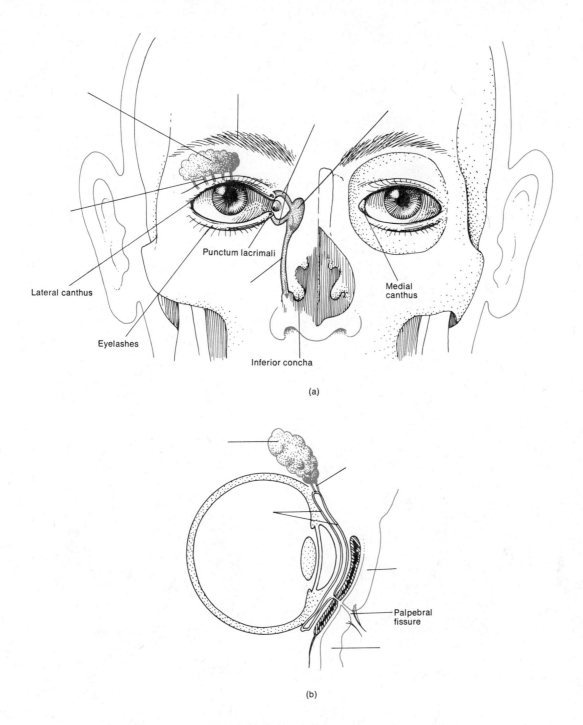

Lateral canthus

Punctum lacrimali

Medial canthus

Eyelashes

Inferior concha

(a)

Palpebral fissure

(b)

**Figure 24–4.** The lacrimal apparatus seen in front view and sectional view.

are specialized for vision in dim light (see Figure 24-6), and they also allow us to discriminate between different shades of dark and light and permit us to see shapes and movement. *Cones*, by contrast, are specialized for color vision and for sharpness of vision, called *visual acuity*. Cones are stimulated only by bright light. This is why we cannot see color by moonlight.

The acuteness of vision may be tested by means of a *Snellen Chart*. It consists of letters of different sizes which are read at a distance normally designated at 20 feet. If the subject reads to the line that is marked "50," he is said to possess 20/50 vision in that eye. If he reads to the line marked "20," he has 20/20 vision in that eye. The normal eye can sufficiently refract light rays from an object 20 feet away to focus a clear object on the retina. Therefore, if you have 20/20

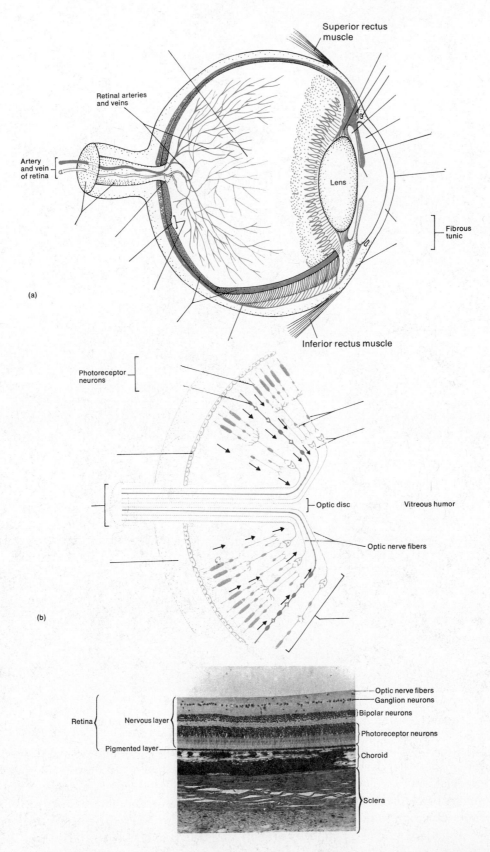

**Superior rectus muscle**

**Retinal arteries and veins**

**Artery and vein of retina**

**Lens**

**Fibrous tunic**

(a)

**Inferior rectus muscle**

**Photoreceptor neurons**

**Optic disc**

**Vitreous humor**

**Optic nerve fibers**

(b)

**Retina** {
**Nervous layer** {

**Pigmented layer**

**Optic nerve fibers**
**Ganglion neurons**
**Bipolar neurons**
**Photoreceptor neurons**
**Choroid**
**Sclera**

**Figure 24–5.** Structure of the eye. (a) Gross anatomy. (b) Microscopic structure of retina. *(b. Courtesy of Donald Patt, from Comparative Vertebrate Histology, Donald I. Patt and Gail R. Patt, Harper & Row, Publishers, Inc., New York, N.Y., 1969. Magnification x 100.)*

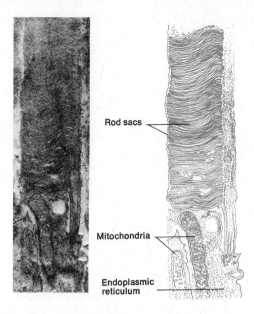

Figure 24-6. Rod cell sacs.

vision, your eyes are perfectly normal, and the higher the number, the larger the letter must be for you to see it clearly, and of course the worse or weaker your eyes are.

Have the subject stand 20 feet from the Snellen Chart and cover the right eye with a 3″ X 5″ card. Instruct him to slowly read down the chart until he can no longer focus the letters. Record the number of the last line that can be successfully read. Repeat this procedure covering the left eye. Now the subject should read the chart with both eyes. Record your results and change place.

The axons of the ganglion neurons extend posteriorly to a small area of the retina called the *optic disc* or *blind spot*. This region contains openings through which the fibers of the ganglion neurons emerge as the optic nerve. Since this area contains neither rods nor cones, and only nerve fibers, no image is formed on it. Thus it is called the blind spot.

Hold this page about 20 inches from your face with the cross in Figure 24-7 directly in front of your right eye. You should be able to see the cross and the circle when you close your left eye. Now, keeping the left eye closed, slowly bring the page closer to your face while fixing the right eye on the cross. At a certain distance the circle will disappear from your field of vision because its image falls upon the blind spot.

C. The formation of an image on the retina requires 4 basic processes, all concerned with focusing light rays. These basic processes are: (1) refraction of light rays, (2) accommodation of the lens, (3) constriction of the pupil, and (4) convergence of the eyes.

When light rays traveling through a transparent medium (such as air) pass into a second transparent medium with a different density (such as water), the rays bend at the surface of the 2 media. This is called *refraction* (Figure 24-8a). The eye has 4 such media of refraction—the cornea, aqueous humor, lens, and vitreous humor.

Figure 24-7. Demonstration of the blind spot.

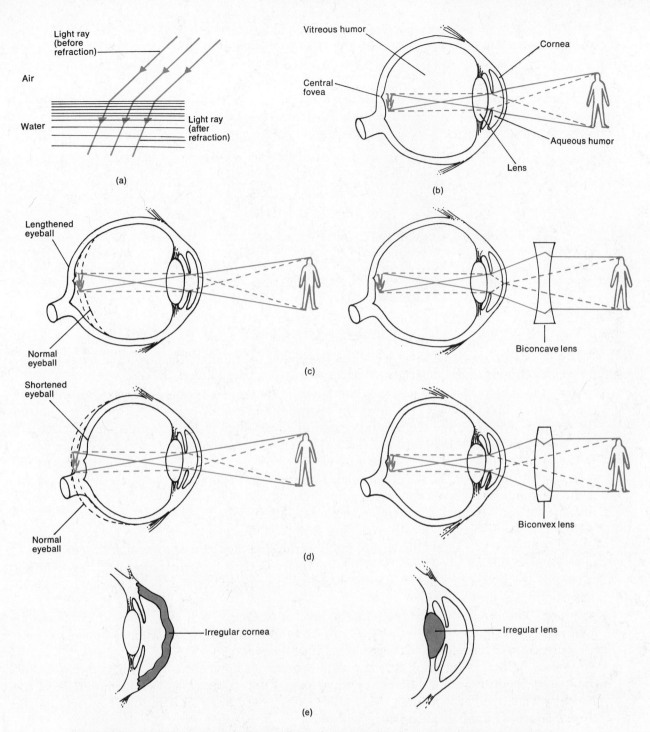

**Figure 24-8.** Refraction. (a) Refraction of light rays passing from air into water. Normal and abnormal refraction in the eye. (b) In the normal or emmetropic eye, light rays from an object are bent sufficiently by the 4 refracting media and converged on the central foveal. A clear image is formed. (c) In the myopic eye, the image is focused in front of the retina. The condition may result from an elongated eyeball or a thickened lens. Correction is by use of a concave lens. (d) In the hypermetropic eye, the image is focused behind the retina. The condition may be the result of the eyeball being too short or the lens being too thin. Correction is by a convex lens. (e) Astigmatism. This is a condition in which the curvature of the cornea or lens is uneven. As a result, horizontal and vertical rays are focused at 2 different points on the retina. Suitable glasses correct the refraction of an astigmatic eye. On the left, astigmatism resulting from an irregular cornea. On the right, astigmatism resulting from an irregular lens. The image is not focused on the area of sharpest vision of the retina. This results in blurred or distorted vision.

The lens of the eye has the unique ability to change the focusing power of the eye by becoming moderately curved at one moment and greatly curved the next. When the eye is focusing on a close object, the lens curves greatly in order to bend the rays toward the central fovea of the eye. This increase in the curvature of the lens is called *accommodation*.

The following test determines your *near point accommodation*. Using any card with a letter printed upon it, close one eye and focus on the letter. Measure the distance of the card from the eye with a ruler or meter stick. Now slowly bring the card as close to your open eye as possible, and stop when you no longer see a clear detailed letter. Measure and record this distance. This value is your near point accommodation. Repeat this procedure 3 times and then test your other eye. Check the table below to see whether the near point for your eyes corresponds with that recorded for your age group.

CORRELATION OF AGE AND NEAR POINT OF ACCOMMODATION
NEAR POINT MEASUREMENT

| AGE | INCHES | CM |
|---|---|---|
| 10 | 2.95 | 7.5 |
| 20 | 3.54 | 9.0 |
| 30 | 4.53 | 11.5 |
| 40 | 6.77 | 17.2 |
| 50 | 20.67 | 52.5 |
| 60 | 32.80 | 83.3 |

Place a 3″ X 5″ card on the side of the nose so that a light shining on one side of the face will not affect the eye on the other side. Place a lamp or shine a flashlight on one eye, 6 inches away (approx. 15 cm), for about 5 seconds. Note the change in the size of the pupil of this eye. Remove the light, wait about 2 minutes and repeat, but this time observe the pupil of the opposite eye. Wait a few minutes and this time repeat the test observing the pupils of both eyes.

The normal eye, referred to as an *emmetropic eye,* can sufficiently refract light rays from an object 20 feet away to focus a clear object on the retina. Many individuals, however, do not have this ability because of abnormalities related to improper refraction. Among these are *myopia* (near sightedness), *hypermetropia* (farsightedness), and *astigmatism* (irregularities in the surface of the lens or cornea). The conditions are illustrated and explained in Figure 24-8c, e. Why do you think nearsightedness can be corrected with glasses containing biconcave lenses? How would you correct farsightedness?

In human beings, both eyes focus on only one set of objects—a characteristic called *single binocular vision*. The term *convergence* refers to a medial movement of the 2 eyeballs so that they are both directed toward the object being viewed. The nearer the object, the greater the degree of convergence necessary to maintain single binocular vision. Hold a pencil or pen about 2 feet from your nose and focus on its point. Now slowly bring the pencil toward your nose. At some moment you should suddenly see 2 pencil points, or a blurring of the point. Observe your partner's eyes when he does this test. Images are actually focused upside down on the retina. They also undergo mirror reversal. That is, light reflected from the right side of an object hits the left side of the retina and vice versa. Note in Figure 24-8b that reflected light from the top of the object crosses light from the bottom of the object and strikes the retina below the central fovea. Reflected light from the bottom of the object crosses light from the top of the object and strikes the retina above the central fovea.

The reason why we do not see a topsy-turvy world is that the brain learns early in life to coordinate visual images with the exact location of objects. The brain stores memories of reaching and touching objects and automatically turns visual images right-side-up and right-side-around.

From the rods and cones, impulses are transmitted through bipolar neurons to ganglion cells. The cell bodies of the ganglion cells lie in the retina and their axons leave the eye via the optic nerve (Figure 24-9). The axons pass through the *optic chiasma*, a crossing point of the optic nerves. Some

**211**

fibers cross to the opposite side. Others remain uncrossed. Upon passing through optic chiasma, the fibers, now part of the *optic tract*, enter the brain and terminate in the thalamus. Here the fibers synapse with the neurons whose axons pass to the visual centers located in the occipital lobes of the cerebral cortex.

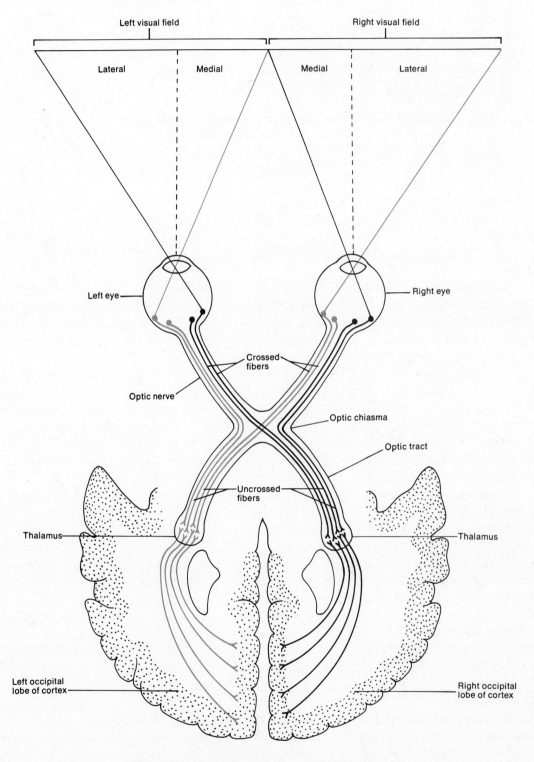

**Figure 24-9.** Afferent pathway for visual impulses.

**CONCLUSIONS**

1. Name the accessory organs of the eye.

2. Do the eyebrows have a specific function?

3. Which structures lubricate the eyeball?

4. What is a sty?

5. What is the lacrimal apparatus?

6. What are the 3 principal layers of the eyeball?

7. What are the dendrites of the photoreceptor neurons called?

8. How do they differ besides their shape?

9. When using the Snellen Chart why are both eyes tested separately?

10. Would you rest tired eyes by looking at objects far away or close by?

11. Why can't you see when light strikes the blind spot?

12. Did you notice that the object reappeared as you continued to move the diagram towards your eye? Explain.

13. What would you expect to happen if your eyes failed to accommodate?

14. As you become older does your near point accommodation increase or decrease? Explain.

15. What changes occur when you move from a dark room to a brightly lit one?

16. Does the pupillary reflex occur in both eyes when light is shone in only one of them?

17. Why does the pupil close in the presence of a bright light?

18. In the test where you moved the pencil toward your nose, why did the pencil become blurred or why did you see 2 points?

19. What muscles control the movement of the eyes during convergence?

20. Explain why visual images are right-side-up even though the image is upside-down on the retina.

21. What is the crossing point of the optic nerves called?

22. Where are the visual centers located in the brain?

## PART IV  AUDITORY SENSATIONS AND EQUILIBRIUM

A. In addition to containing receptors for sound waves, the *ear* also contains receptors for equilibrium. Anatomically the ear is subdivided into 3 principal regions: (1) the outer ear, (2) the middle ear, and (3) the inner ear.

Using your textbook as a reference, label Figure 24-10.

*Sound waves* result from the alternate compression and decompression of air. They originate from a vibrating object and travel through air in much the same way that waves travel over the surface of water.

The events involved in the physiology of hearing sound waves are listed below. While reading this succession of events, you will want to make constant reference to Figure 24-11.

1. Sound waves that reach the ear are directed by the pinna into the external auditory canal.
2. When the waves strike the tympanic membrane, the alternate compression and decompression of the air causes the membrane to vibrate.
3. The central area of the tympanic membrane is connected to the malleus, which also starts to vibrate. The vibration is then picked up by the incus, which transmits the vibration to the stapes.
4. As the stapes moves back and forth, it pushes the oval window in and out.
5. The movement of the oval window sets up waves in perilymph.
6. As the window bulges inward, it pushes the perilymph of the scala vestibuli up into the cochlea.
7. This pressure pushes the vestibular membrane inward and increases the pressure of the endolymph inside the cochlear duct.
8. The basilar membrane gives under the pressure and bulges out into the scala tympani.
9. The sudden pressure in the scala tympani pushes the perilymph toward the round window causing it to bulge back into the middle ear. Conversely, as the sound wave subsides, the stapes moves backward, and the procedure is reversed.
10. When the basilar membrane vibrates, the hair cells of the spiral organ are moved against the tectorial membrane. In some unknown manner, the movement of the hairs stimulates the dendrites of neurons at their base, and sound waves are converted into nerve impulses.

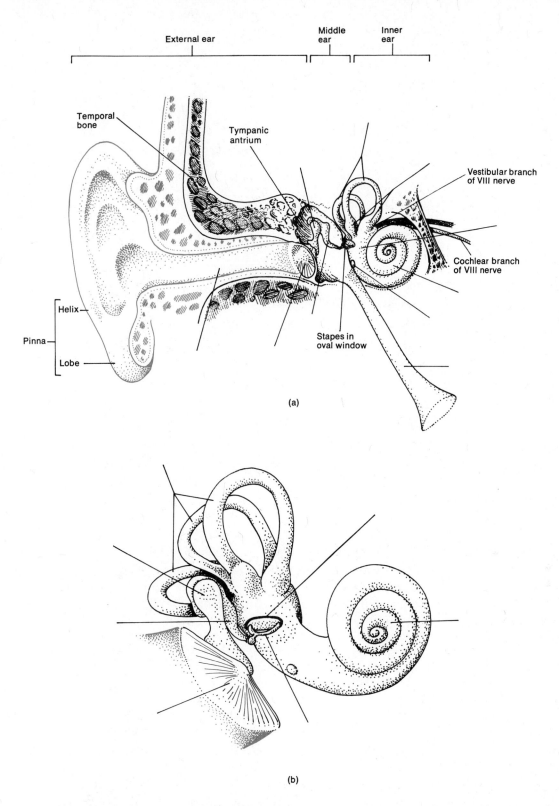

**Figure 24-10.** Structure of the auditory apparatus. (a) Divisions of the ear into external, middle, and inner portions. (b) Details of the middle ear.

11. The impulses are then passed on to the auditory branch of the vestibulocochlear nerve (VIII, see Figure 20-1) and the medulla. Here some impulses cross to the opposite side and finally travel to the auditory area of the temporal lobe of the cerebral cortex.

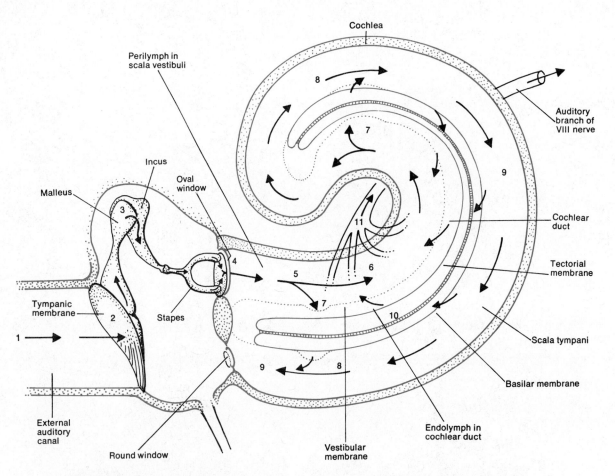

**Figure 24-11.** The physiology of hearing. Follow the text very carefully to understand the meaning of the indicated numbers.

To test for hearing impairment, the subject places a cotton plug in one ear and closes his eyes. The student partner then holds a watch next to the other ear and slowly moves it away. The subject indicates when the watch can no longer be heard. Measure and record this distance. This test is repeated 3 times and the average of the distances is calculated. The other ear is tested in a similar manner. Compare your results with those of your partner.

Left ear _____ _____ _____ Average _____
Right ear _____ _____ _____ Average _____

The inner ear is located within the temporal bone of the cranium. Therefore, any bone of the cranium can conduct sound to the cochlea, if the stimulus has a high enough intensity.

Strike a tuning fork with a rubber mallet and place the vibrating fork upon the following bones: (1) temporal, (2) parietal, (3) frontal, (4) occipital.

Keep the vibrating fork on these bones until you can barely hear it, and then put the fork next to your ear. Notice whether sound is conducted better in bone or in air.

B. The term *equilibrium* has 2 meanings. One kind of equilibrium, called *static equilibrium*, refers to the orientation of the body (mainly the head) relative to the ground. The second kind of equilibrium, called *dynamic equilibrium*, is the maintenance of the position of the body (mainly the head) in response to sudden movements or to a change in the rate or direction of movement. The receptor organs for equilibrium are the saccule, utricle, and semicircular ducts.

The utricle and saccule each contain within their walls sensory hair cells that project into the cavity of the membranous labyrinth (Figure 24-12). The hairs are coated with a gelatinous layer in which particles of calcium carbonate, called *otoliths*, are embedded. When the head tips downward, the otoliths slide with gravity in the direction of the ground. As the particles move, they exert a downward pull on the gelatinous mass which, in turn, exerts a downward pull on the hairs and makes them bend. The movement of the hairs stimulates the dendrites at the base of their hair cells. The impulse is then transmitted to the temporal lobe of the brain through the vestibular branch of the vestibulocochlear nerve (VIII). The utricle and saccule are considered to be sense organs of static equilibrium. They provide information regarding the orientation of the head in space and are essential for the maintenance of posture.

Let us now consider the role of the semicircular ducts in maintaining dynamic equilibrium. The 3 semicircular ducts are positioned at right angles to each other in 3 planes—frontal (the superior duct), sagittal (the posterior duct), and lateral (the lateral duct). This positioning permits correc-

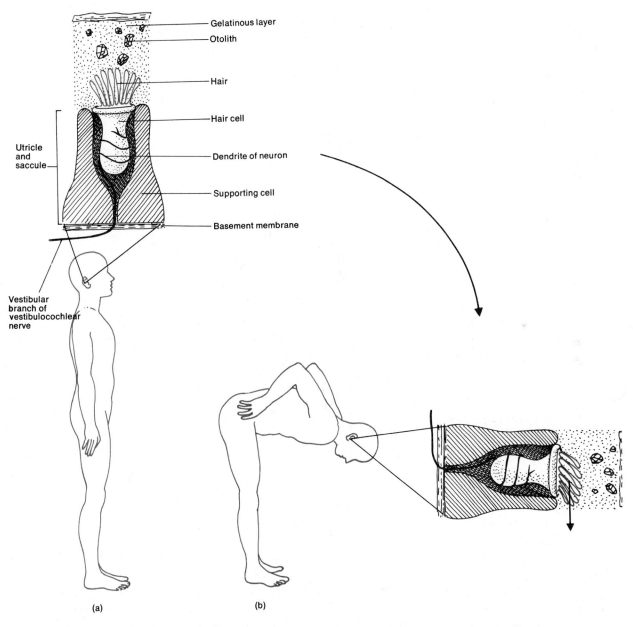

**Figure 24-12.** The utricle and saccule and static equilibrium. As the person bends, the otoliths drop downward in the direction of the ground, pulling on the gelatinous mass, which slides over the hairs. The bending of the hairs causes changes in the hair cells that stimulate the neurons.

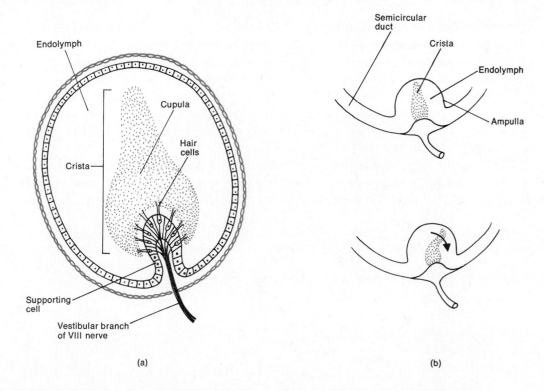

Figure 24–13 labels:
Endolymph
Cupula
Hair cells
Crista
Supporting cell
Vestibular branch of VIII nerve
(a)

Semicircular duct
Crista
Endolymph
Ampulla
(b)

**Figure 24–13.** The semicircular ducts and dynamic equilibrium. (a) Enlarged aspect of a crista. (b) Cupula at rest (above) and in movement (below). When the endolymph in the semicircular duct moves, the cupula is displaced. The impulse is picked up by the vestibular branch of the vestibulocochlear nerve and relayed to the brain.

tion of an imbalance in 3 planes. In the ampulla, the dilated portion of each duct, there is a small elevation called the *crista* (Figure 24-13a). Each crista is composed of a group of hair cells covered by a mass of gelatinous material called the *cupula*. When the head moves, the endolymph in the semicircular ducts flows over the hairs and bends them as water in a stream bends the plant life growing at its bottom. The movement of the hairs stimulates sensory neurons, and the impulses pass over the vestibular branch of the vestibulocochlear nerve (VIII; Figure 24-13b). The impulses then reach the temporal lobe of the cerebrum, before they are sent to the muscles that must contract in order to maintain body balance in the new position.

Equilibrium can be tested by using a few simple procedures. Balance is tested by having the subject stand perfectly still with his hands at his sides and his feet close together. Note any swaying movements. It might be helpful if the subject stands in front of a light, so that swaying movements can be observed by his shadow. Now have the subject repeat this test with his eyes closed. Note any swaying movements.

The next test serves to evaluate the semicircular canals. The subject sits on a stool, legs up on the stool rung, and the stool is revolved for a few seconds and suddenly stopped. The subject will experience the sensation that the stool is still rotating, which means that the semicircular canals are functioning properly.

A cold test serves to evaluate the integrity of the semicircular canals. When a cold swab is placed in one ear, it increases the density of the endolymph of the semicircular canal adjacent to that ear. This increase in the density of the endolymph stimulates the hair cells within the semicircular canals, causing a sensation of rotation called *nystagmus*. Examine the semicircular canals of each ear individually.

Place a cotton swab in an ice bath for several minutes and then carefully insert it in one of your ears, noting the results

**218**

**CONCLUSIONS**

1.  What are the two types of sensations received by the receptors of the ear?

2.  The ear is anatomically subdivided into what principal regions?

3.  Where do sound waves come from?

4.  Explain why an infection of the middle ear may be associated with a sore throat.

5.  Where in the brain is the center for perception of sound located?

6.  Where in the brain is the center for equilibrium?

7.  In which medium, bone or air, is sound better conducted?
8.  Which bone conducts sound best? Why?

9.  Define the 2 kinds of equilibrium.

10. Which are the receptor organs for equilibrium?

11. The particles of calcium carbonate which are embedded in the hair cells of the saccule and utricle are called

12. In the balance test, was a larger swaying movement produced with the eyes open or closed? Why?

## PART V   DISORDERS

For each of the following disorders describe the cause, symptoms, and treatment (where applicable).

cataracts

glaucoma


conjunctivitis


trachoma


Meniere's disease


impacted cerumen


## MEDICAL TERMINOLOGY

Define each of the following terms:
a. achromatopsia

b. ametropia

c. blepharitis

d. eustachitis

e. keratitis

f. keratoplasty

g. labyrinthitis

h. myringitis

i. otalgia

j. otitis

k. ptosis

l. retinoblastoma

m. strabismus

h. tinnitus

**STUDENT ACTIVITY**

List all of the receptors mentioned in this module in a column on the left; alongside of each one, in a middle column, give their locations; and, in a column on the right, mention their basic functions. For example:

| Receptor | Location | Function |
|---|---|---|
| Meissner's corpuscles | Papillae of skin | Detect touch |

# MODULE 25
# HISTOLOGY AND PHYSIOLOGY OF THE ENDOCRINE SYSTEM

**OBJECTIVES**

1. To discuss the function of the endocrine system in maintaining homeostasis

2. To define an endocrine gland and identify its relationship with a target organ

3. To define the anatomical and physiological relationship between the pituitary gland and hypothalamus

4. To list the 6 hormones of the adenohypophysis, their target organs, and their functions

5. To define the source of hormones stored by the neurohypophysis, their target organs, and their functions

6. To define a negative feedback mechanism

7. To discuss how thyroxin is synthesized, stored, and transported by thyroid follicles

8. To identify the physiological effects of thyroxin, thyrocalcitonin, and the parathyroid hormone

9. To distinguish the effects of adrenal cortical mineralocorticoids, glucocorticoids, and gonadocorticoids on physiological activities

10. To identify the function of the adrenal medullary secretions as supplements to sympathetic responses

11. To compare the roles of glucagon and insulin in the control of blood sugar levels

12. To identify the physiological effects of the hormones secreted by the pineal gland

13. To define the role of the thymus in antibody production

You have learned how the nervous system controls the body through electrical impulses that are delivered over neurons. Now we will look at the body's other great control system, the endocrine system. The endocrine organs affect bodily activities by releasing chemical messengers, called hormones, into the bloodstream. The nervous and endocrine systems coordinate their activities like an interlocking supersystem. Certain parts of the nervous system routinely stimulate or inhibit the release

222

of hormones. The hormones, in turn, are quite capable of stimulating or inhibiting the flow of particular nerve impulses.

## PART I  ENDOCRINE SYSTEM

The body contains 2 different kinds of glands: exocrine and endocrine. *Exocrine glands* secrete their products into ducts. The ducts then carry the secretions into body cavities or to the external surface of the body. Examples are sweat, sebaceous, mucous, and digestive glands. *Endocrine glands*, by contrast, secrete their products into the extracellular space around the secretory cells. The secretion passes through the membranes of cells lining blood vessels and into the blood. Since they have no ducts, endocrine glands are also called *ductless glands*.

The endocrine glands of the body are the pituitary, thyroid, parathyroids, adrenals, pancreas, ovaries, testes, pineal, and thymus. The endocrine glands are organs that together make up the *endocrine system*. Locate and label all of the organs of the endocrine system in Figure 25-1.

The secretions of endocrine glands are called *hormones*, the term hormone meaning to set in motion. A hormone may be a protein, an amine, or a steroid. The one thing that all hormones have in common, no matter what kind, is that they maintain homeostasis by changing the physiological activities of cells. A hormone may stimulate changes in the cells of an organ or in groups of organs. These

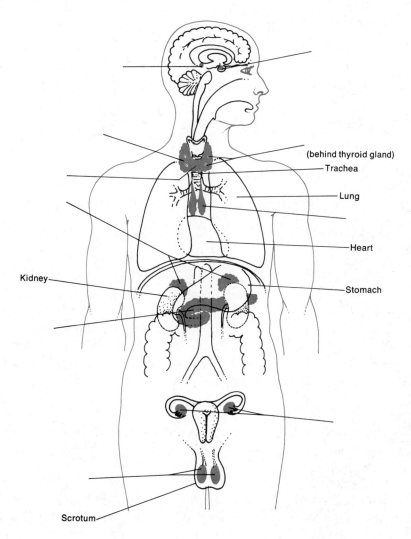

(behind thyroid gland)
Trachea
Lung
Heart
Kidney
Stomach
Scrotum

**Figure 25-1.** Location of the endocrine glands.

are called the *target organs* of a hormone. Or, the hormone may directly affect the activities of all the cells in the body.

## CONCLUSIONS

1. Which 2 systems of the body coordinate their activities like an interlocking supersystem?


2. What is the difference between endocrine and exocrine glands?




3. Name the endocrine glands of the body.



4. What are hormones?


5. What different types of hormones are there chemically?


6. What do they all have in common?



7. What are target organs?



## PART II  ENDOCRINE GLANDS

The hormones of the *pituitary gland*, also called the *hypophysis*, regulate so many body activities that the pituitary has been nicknamed the "master gland." It lies in the sella turcica of the sphenoid bone and is attached to the hypothalamus of the brain via a stalklike structure. This structure is called the *infundibular stalk* (See Figure 25-2).

The pituitary is divided structurally and functionally into an anterior lobe called the *adenohypophysis*, and a posterior lobe called the *neurohypophysis*, both of which are connected to the hypothalamus. The anterior lobe contains many glandular epithelium cells and forms the glandular part of the pituitary. The posterior lobe contains neurons, which form the neural part of the pituitary (See Figure 25-2).

The adenohypophysis releases hormones that regulate a whole range of bodily activities, from growth to reproduction. However, the release of these hormones is either stimulated or inhibited by chemical secretions that come from the hypothalamus of the brain. This is one hookup between the nervous system and the endocrine system.

The glandular cells of the anterior lobe secrete any one of 6 hormones. The glandular cells themselves are called *acidophils*, *basophils*, or *chromophobes*, depending on the way their cytoplasm reacts to laboratory stains. The acidophils, which stain pink, secrete growth hormone and prolactin. The basophils stain darkly and release the other 4 hormones. These are: thyroid stimulating hormone; adrenocorticotrophic hormone; follicle stimulating hormone; and luteinizing hormone. The chromo-

**224**

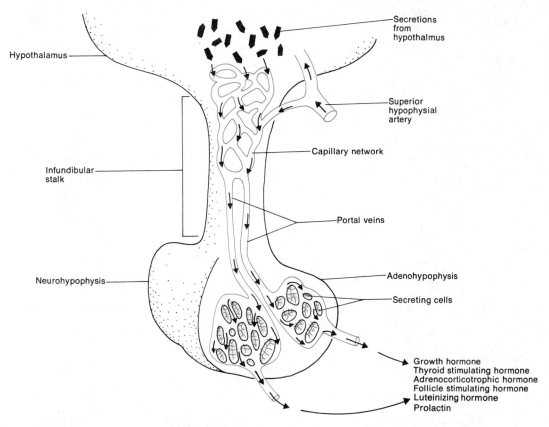

**Figure 25–2.** Blood supply of the adenohypophysis. Note that chemicals from the hypothalamus pass into the blood which is circulating through the adenohypophysis.

Labels in figure:
Hypothalamus
Infundibular stalk
Neurohypophysis
Secretions from hypothalmus
Superior hypophysial artery
Capillary network
Portal veins
Adenohypophysis
Secreting cells
Growth hormone
Thyroid stimulating hormone
Adrenocorticotrophic hormone
Follicle stimulating hormone
Luteinizing hormone
Prolactin

phobes may be involved in the secretion of the adrenocorticotrophic hormone. Except for growth hormone, all the secretions are called *trophic hormones*, which means that their target organs are other endocrine glands. Prolactin, follicle stimulating hormone, and luteinizing hormone are also called *gonadotrophic hormones* because they regulate the functions of the gonads, which are the endocrine glands that produce sex hormones.

## HORMONES OF THE ADENOHYPOPHYSIS

1. *Growth Hormone*, also called somatotropin and somatotrophic hormone (STH)—Its main function is to act upon the hard and soft tissues of the body to increase their rate of growth and to maintain their size once growth is attained.

   Growth hormone regulation is an illustration of a *negative feedback mechanism*. A *feedback system* is any circular situation in which information is fed back into the system, and may be either positive or negative. If the system reverses the direction of the initial condition it is *negative*, while if the initial condition continues ahead on stimulation and is not reversed it is *positive*. Most of the feedback systems of the body are negative feedback systems.

2. *Thyroid Stimulating Hormone* (TSH)—This hormone is also called thyrotropin, and its function is to stimulate the synthesis and secretion of the hormones produced by the thyroid gland.

3. *Adrenocorticotrophic Hormone* (ACTH)—This hormone has a dual function. Its trophic function is to control the production and secretion of the adrenal cortex hormones. It also acts directly on all body cells by increasing their catabolism of fats.

4. *Follicle Stimulating Hormones* (FSH)—In the female this hormone is transported to the ovaries where it stimulates the development of a follicle and an egg each month. FSH also stimulates cells in the follicles of the ovaries to secrete estrogens, or female sex hormones. In the male, FSH stimulates the testes to produce sperm and to secrete testosterone, a male sex hormone.

**225**

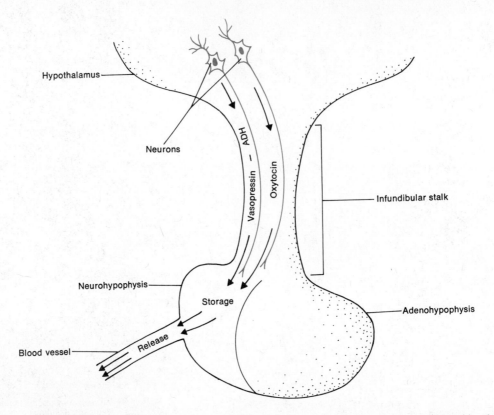

**Figure 25-3.** Innervation and function of the neurohypophysis. Hormones produced by the hypothalamus travel to the neurohypophysis where they are stored. At the appropriate time, impulses conducted over the neurons stimulate the neurohypophysis to release the stored hormones into the blood.

5. *Luteinizing Hormone (LH or ISCH)*—The luteinizing hormone is called luteotropin and LH in the female and interstitial cell stimulating hormone (ICSH) in the male. In the female it stimulates the ovary to release the developed egg, stimulates development of the corpus luteum, and prepares the uterus for implantation of a fertilized egg. It also stimulates the secretion of progesterone (another female sex hormone) from the corpus luteum and readies the mammary glands for milk secretion. In the male, ICSH stimulates the interstitial cells in the testes to develop and secrete testosterone.

6. *Prolactin*—Prolactin or lactogenic hormone, together with other hormones, initiates milk secretion by the mammary glands.

The posterior lobe, or neurohypophysis, is not an endrocrine gland in the strict sense, since it does not synthesize hormones. The posterior lobe consists of cells called *pituicytes*, which are similar in appearance to the connective tissue neuroglia of the nervous system. Running from the hypothalamus, down the infundibular stalk to the posterior lobe, are neurons that are vital to the functioning of the neurophypophysis (Figure 25-3).

The hypothalamus produces 2 hormones, oxytocin and vasopressin-ADH (anti-diuretic hormone). These hormones travel along the outside surface of the neurons to the posterior lobe of the pituitary, where they are stored and later released into the blood.

## HORMONES OF THE NEUROHYPOPHYSIS

1. *Oxytocin*—This hormone stimulates the contraction of the smooth muscle cells in the pregnant uterus and the contractile cells around the ducts of the mammary glands (Figure 25-4). It is released in large quantities just prior to giving birth.

2. *Vasopressin-ADH*—This hormone has 2 principal physiological activities. One of these is to

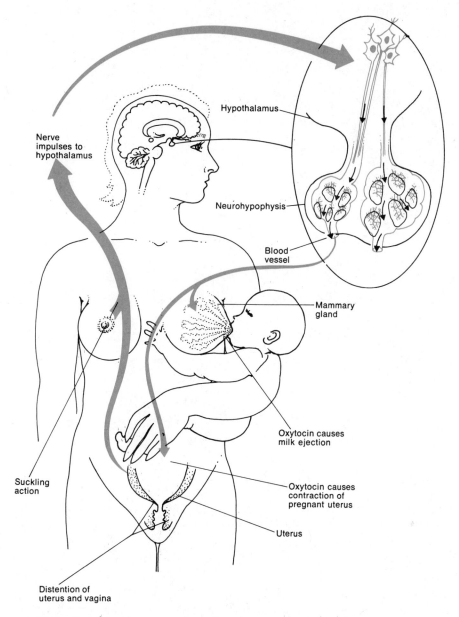

**Figure 25-4.** Effects and control of oxytocin of the neurohypophysis.

cause a rise in blood pressure by bringing about constriction of arterioles. The more important activity is its effect on urine volume. Vasopressin-ADH causes the kidneys to remove water from newly forming urine and return it to the bloodstream, preventing excessive urine production.

Examine a prepared slide of the pituitary gland, drawing a few cells of each lobe.

## CONCLUSIONS

1. Give 2 other names for the pituitary gland.


2. Where exactly is it located?

3. How is the pituitary gland divided?

4. List the cells of the anterior lobe and the hormones that they secrete.

5. Describe a feedback system, and both types.

6. Give the main function for each of the hormones of the anterior pituitary gland.

7. Why is the posterior lobe of the pituitary gland not considered an endocrine gland?

8. Describe its cells and its relationship to the hypothalamus.

9. Explain how the hormones of the posterior lobe are released into the bloodstream.

10. Name the posterior lobe hormones and their functions.

## THYROID GLAND

The double-lobed organ located just below the voicebox (larynx) is called the *thyroid gland*. The 2 lobes lie one on either side of the windpipe (trachea) and are connected by a mass of tissue called an *isthmus* (Figure 25–1). Histologically, the thyroid is composed of spherical-shaped sacs called *thyroid follicles* (Figure 25–5a). The walls of each follicle are formed by a layer of simple cuboidal epithe-

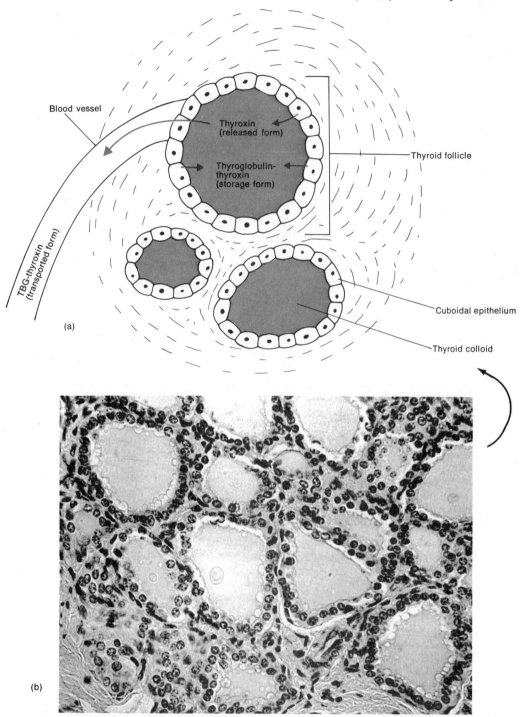

**Figure 25–5.** The thyroid gland. (a) Histology of thyroid tissue and the storage, release, and transportation of thyroxin. (b) Microscopic view of thyroid tissue. *(b. Courtesy of Edward J. Reith, Atlas of Descriptive Histology, Edward J. Reith and Michael H. Ross, Harper & Row, Publishers, Inc., New York, N.Y., 1970.)*

lium. This layer manufactures the 2 main hormones produced by the gland, *thyroxin* and *thyrocalcitonin*. The interior of each sac is filled with *thyroid colloid,* a stored form of the thyroid hormones.

One of the unique features of the thyroid gland is its ability to store its hormones and release them in a steady flow over a long period of time. For example, the principal hormone, thyroxin, is synthesized from iodine and an amino acid called tyrosine. If the body has no immediate need for the hormone, it combines with *thyroglobulin,* a protein secreted by the follicle cells, and is stored in the colloid. When demand for thyroxin occurs, the hormone splits apart from the thyroglobulin and is released into the blood. There, it combines with a plasma protein called *thyroid-binding globulin* or *TBG.* The thyroxin is released from TBG as it enters tissue cells.

The major function of thyroxin is to control the rate of metabolism. It also increases the rate at which carbohydrates are burned, and it stimulates cells to break down proteins for energy instead of using them for building processes. At the same time, thyroxin decreases the breakdown of fats. The overall effect is to increase catabolism. It produces energy and raises body temperature as heat energy is given off. This is called the *calorigenic effect.* Thyroxin is also an important factor in the regulation of tissue growth and tissue development. It acts as a diuretic, increases the reactivity of the nervous system, and increases the heart rate.

The second thyroid hormone produced by the thyroid gland is thyrocalcitonin. It is involved in the homeostasis of blood calcium level. It lowers the amount of calcium in the blood by inhibiting bone breakdown and by accelerating the absorption of calcium by the bones. If thyrocalcitonin is administered to a person with normal blood calcium levels, it causes *hypocalcemia,* or low blood calcium level. If thyrocalcitonin is given to an individual with *hypercalcemia* (high blood calcium level), the level returns to normal. It is suspected that the blood calcium level directly controls the secretion of thyrocalcitonin in a simple negative feedback system which does not involve the pituitary gland (Figure 25-6).

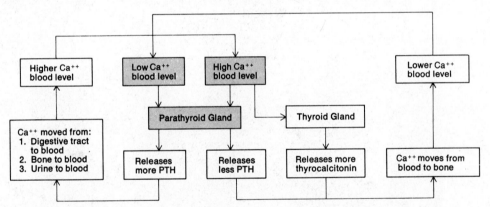

**Figure 25-6.** Regulation of the secretion of the parathyroid hormone and thyrocalcitonin.

Examine a prepared slide of the thyroid gland. Identify the thyroid follicles containing the thyroid colloid, the cuboidal epithelial cells, and draw a representative section of the gland.

**CONCLUSIONS**

1. Locate the thyroid gland and describe it histologically.

2. What is the unique feature of the thyroid gland?

3. From what is thyroxin synthesized?

4. What is thyroid-binding globulin (TBG)?

5. Give 5 functions of thyroxin.

6. What is the calorigenic effect?

7. What does thyrocalcitonin do for the body?

8. Define hypocalcemia and hypercalcemia.

## PARATHYROID GLANDS

Embedded on the posterior surfaces of the lateral lobes of the thyroid are small, round masses of tissue called the *parathyroid glands*. Typically, 2 parathyroids are attached to each thyroid lobe. The parathyroids are abundantly supplied with blood and are innervated by the autonomic nervous system. Histologically, the parathyroids contain 2 kinds of epithelial cells. The first kind, called *principal* or *chief cells*, are believed to be the major synthesizers of the parathyroid hormone. The other kind of cell is called an *oxyphil cell*, believed to synthesize a reserve capacity of hormone.

Parathyroid Hormone (PTH) controls the homeostasis of ions in the blood, especially the homeostasis of calcium and phosphate ions. PTH decreases blood phosphate level and increases blood calcium level. As far as blood calcium level is concerned, PTH and thyrocalcitonin are antagonists. PTH secretion is not controlled by the pituitary gland. When the calcium ion level of the blood falls, more PTH is released (Figure 25-6). Conversely, when the calcium ion level of the blood rises, less PTH (and more thyrocalcitonin) is secreted. Once again, this is an example of a negative feedback control system.

Examine a prepared slide of parathyroid tissue, and draw a representative area under high power.

## CONCLUSIONS

1. Locate the parathyroid gland.

2. What are the cells of the parathyroid gland called?

3. What is the function of PTH?

4. Is PTH controlled by the pituitary gland? Explain.

## THE ADRENAL GLANDS

The body has 2 *adrenal glands* and they are located one on top of each kidney. Each adrenal gland is structurally and functionally differentiated into 2 sections: the outer *adrenal cortex*, which makes up the bulk of the gland, and the inner *adrenal medulla*.

232

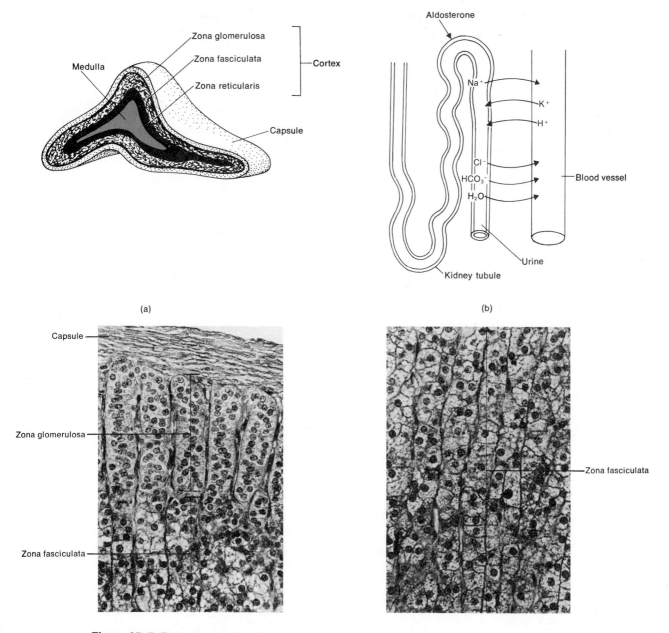

**Figure 25-7.** The adrenal gland. (a) Cross section of the gland showing the various zones. (b) Effects of aldosterone. *(Photographs courtesy of Edward J. Reith, Atlas of Descriptive Histology, Edward J. Reith and Michael H. Ross, Harper & Row, Publishers, Inc., New York, N.Y., 1970.)*

The adrenal cortex is subdivided into 3 zones. Each zone has a different cellular arrangement and secretes different hormones (Figure 25-7a). The outer zone is called the *zona glomerulosa*, and these cells secrete a group of hormones called *mineralocorticoids*.

These hormones help control electrolyte homeostasis, particularly the concentrations of sodium and potassium. Although the adrenal cortex secretes at least 3 different hormones classified as mineralocorticoids, one of these hormones is responsible for about 95 percent of the mineralocorticoid activity. The name of this hormone is *aldosterone*. Aldosterone acts on the tubule cells in the kidneys and causes them to increase their reabsorption of sodium (Figure 25-7b). It also decreases reabsorption of potassium. The sodium reabsorption leads to the elimination of $H^+$ ions, the retention of $Na^+$, $Cl^-$, and $HCO_3^-$, and water retention.

The middle zone is called the *zona fasciculata* and secretes *glucocorticoid hormones*. The glucocorticoids are a group of hormones that are largely concerned with normal metabolism and the ability of the body to resist stress. Three examples are *hydrocortisone (cortisol)*, *corticosterone*, and *cortisone*. Of the 3, hydrocortisone is the most abundant.

The glucocorticoids have the following effects on the body:

1. Glucocorticoids work with other hormones in promoting normal metabolism. They make sure enough energy is provided. They promote the breakdown of carbohydrates to glucose and encourage the movement of fats from storage depots to all the cells, where they are catabolized for energy.
2. Glucocorticoids work in many ways to provide resistance to stress. One of the more obvious ways is their hyperglycemic effect. A sudden increase in available glucose makes the body more alert and gives the body energy for combatting a range of stressors, such as fright, temperature extremes, high altitude, bleeding and infection.
3. Glucocorticoids decrease the blood vessel dilatation and edema associated with inflammations. They are thus anti-inflammatories.

The inner zone is called the *zona reticularis* and synthesizes sex hormones, chiefly male hormones called *androgens*. Both male and female *gonadocorticoids* or sex hormones are secreted, but the amount of sex hormones secreted is usually so small that it is insignificant, except in cases of hypersecretion.

The adrenal medulla consists of hormone-producing cells, called *chromaffin cells*, which surround large blood-containing sinuses (Figure 25-7a). Dispersed among the chromaffin cells are ganglions of the sympathetic division of the autonomic nervous system. The secretion of hormones from the medulla is directly controlled by the autonomic nervous system, and innervation by the preganglionic fibers allows the gland to respond very rapidly to a stimulus.

The 2 principal hormones synthesized by the adrenal medulla are epinephrine and norepinephrine. Epinephrine constitutes about 80 percent of the total secretion of the gland and is more potent in its action than norepinephrine. Both hormones are responsible for the "fight-or-flight" response, and help the body resist stress like the glucocorticoids, but, unlike them, the medullary hormones are not essential for life.

Epinephrine increases blood pressure by increasing the heart rate and by constricting the blood vessels. It accelerates the rate of respiration, dilates respiratory passageways, decreases the rate of digestion, increases the efficiency of muscular contractions, increases blood sugar level, and stimulates cellular metabolism.

Examine a prepared slide of the adrenal medulla and cortex. Make drawings of representative cells from each area.

**CONCLUSIONS**

1. Where are the adrenal glands located?

2. Name the 3 zones of the adrenal cortex and the groups of hormones that they each secrete.

3. Name the groups of hormones and list examples of each.

4. Correlate the adrenal medulla with the autonomic nervous system.

5. What hormones are produced by the adrenal medulla? What particular effect is their responsibility?

6. List some physiological effects of epinephrine.

**PANCREAS**

Because of its functions, the pancreas can be classified as both an endocrine and an exocrine gland. The endocrine portion of the pancreas consists of clusters of cells called *islets of Langerhans*. Two kinds of cells are found in these clusters: (1) *alpha cells*, which comprise about 25 percent of the islet cells and secrete the hormone glucagon, and (2) *beta cells*, which constitute about 75 percent of the islet cells and secrete the hormone insulin.

The endocrine secretions of the pancreas—glucagon and insulin—are concerned with regulation of the blood sugar level.

The product of the alpha cells is *glucagon*, a hormone whose principal physiological activity is to increase the blood glucose level. Glucagon does this by accelerating the conversion of liver glycogen into glucose. The liver then releases the glucose into the blood, and the blood sugar level rises. Secretion of glucagon is directly controlled by the level of blood sugar.

The beta cells of the islets produce a hormone called *insulin*. This hormone increases the buildup of proteins in cells. But its chief physiological action is opposite to that of glucagon. Insulin decreases blood sugar level. This is accomplished in 2 ways. First, insulin accelerates the transport of glucose from the blood into body cells, especially into the cells of the liver and muscles. Second, insulin accelerates the conversion of glucose into glycogen. The regulation of insulin secretion, like that of glucagon secretion, is directly determined by the level of sugar in the blood.

Examine a prepared slide of the pancreas, locate the islets of Langerhans, and draw a few of the cells.

## OVARIES AND TESTES

The female gonads, called the *ovaries*, are paired oval-shaped bodies located in the pelvic cavity (See Figure 25-1). The ovaries produce female sex hormones that are responsible for the development and maintenance of the female sexual characteristics. Along with the gonadotrophic hormones of the pituitary, the sex hormones also regulate the menstrual cycle, maintain pregnancy, and ready the mammary glands for lactation.

Examine a prepared slide of ovarian tissue. Use low power and draw a Graafian follicle.

The male has 2 oval-shaped glands, called *testes*, that lie in the scrotum. The testes produce the male sex hormones that stimulate the development and maintenance of the male sexual characteristics.

Examine a prepared slide of the testis. Locate the interstitial cells and seminiferous tubules. Examine them under high power or oil immersion. Draw a few representative areas.

## PINEAL GLAND

The cone-shaped gland located in the roof of the third ventricle is known as the *pineal gland*, or *epiphysis cerebri*. The pineal gland starts to degenerate at about age 7, and in the adult it is largely fibrous tissue. Although many anatomical facts concerning the pineal gland have been known for years, its physiology is still somewhat obscure. One hormone secreted by the pineal gland is *melatonin*, which appears to affect the secretion of hormones by the ovaries. One of the functions of the pineal gland might very well be regulation of the activities of the sexual endocrine glands, particularly the menstrual cycle.

Some evidence also exists that the pineal secretes a second hormone called *adrenoglomerulotropin*. This hormone may stimulate the adrenal cortex to secrete aldosterone. Still other functions attributed to the pineal gland are the secretion of a growth-inhibiting factor and the secretion of a hormone called *serotonin* that is involved in normal brain physiology.

## THYMUS GLAND

Usually a bilobed organ, the *thymus gland* is located in the upper mediastinum behind the sternum and between the lungs. The gland is conspicuous in the infant, and at about 2 years of age it attains its largest relative size. Around puberty, the thymic tissue, which consists primarily of a type of white blood cell, is replaced by fat. By the time the person reaches maturity, the gland has atrophied. The thymus is believed to secrete a hormone that enables the white blood cells to produce antibodies for the defense of the body.

Examine a prepared slide of thymus tissue, and draw a representative area as seen under high power.

## CONCLUSIONS

1. Differentiate between types of cells and their secretions in the pancreas.

2. What are their physiological functions?

3. How are both secretions regulated?

4. Compare the functions of the female sex hormones and the male sex hormones.

5. Where are these hormones produced?

6. Where is the pineal gland located?

7. Name its hormones and their possible functions.

8. Locate the thymus gland.

9. What is the function of the thymus gland?

**STUDENT ACTIVITY**

Following is a chart listing the endocrine glands discussed in this module. The glands are listed from 4 points of reference: top, bottom, left, and right. Draw arrows to indicate how 2 or more glands are related, and write the name of the relationship near the arrow. If a gland bears no relationship to any others do not draw an arrow. One example is shown in the chart.

Pituitary  Thyroid  Parathyroids  Adrenals  Pancreas  Ovaries  Testes  Pineal  Thymus

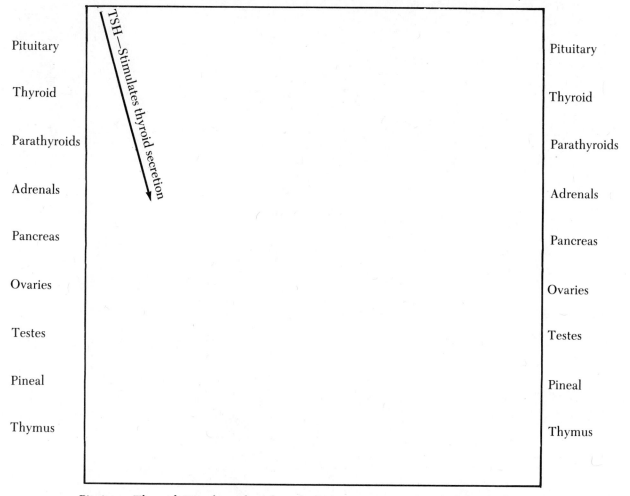

Pituitary

Thyroid

Parathyroids

Adrenals

Pancreas

Ovaries

Testes

Pineal

Thymus

Pituitary

Thyroid

Parathyroids

Adrenals

Pancreas

Ovaries

Testes

Pineal

Thymus

TSH—Stimulates thyroid secretion

Pituitary  Thyroid  Parathyroids  Adrenals  Pancreas  Ovaries  Testes  Pineal  Thymus

# MODULE 26
# THE ENDOCRINE SYSTEM: DISORDERS

**OBJECTIVES**

1. To describe pituitary dwarfism, Simmond's disease, giantism, and acromegaly and list the clinical symptoms of each

2. To name 4 abnormalities of thyroid secretion and list the clinical symptoms of each

3. To identify the principal effects of abnormal secretion of the parathyroid hormone on calcium metabolism

4. To compare the effects of hypo- and hypersecretions of adrenocortical hormones

5. To define medical terminology associated with the endocrine system

Disorders of the endocrine system, in general, are based upon the under- or overproduction of hormones. The term *hyposecretion* describes an *underproduction*, whereas the term *hypersecretion* means an *overproduction*. The pituitary gland produces many hormones. All these hormones, with the exception of the growth hormone, directly control the activities of other endocrine glands. It is hardly surprising, then, that hypo- or hypersecretion of a pituitary hormone produces widespread and complicated abnormalities.

## DISORDERS OF THE PITUITARY

Among the clinically interesting disorders related to the adenohypophysis are those involving the growth hormone. Growth hormone builds up cells, particularly those of bone tissue. If the growth hormone is hyposecreted during the growth years, bone growth is slow, and the epiphyseal plates close before normal height is reached. This is the condition called *pituitary dwarfism*. Other organs of the body also fail to grow, and the pituitary dwarf is childlike in many physical respects. Treating the condition requires administration of human growth hormone during childhood, before the epiphyseal plates close. If secretion of growth hormone is normal during childhood, but lower than normal during adult life, a rare condition called *pituitary cachexia* (Simmond's disease) occurs. The tissues of a person with Simmond's disease waste away, or atrophy. The victim becomes quite thin and shows signs of premature aging. The atrophy occurs because the person is not receiving enough growth hormone to stimulate the protein-building activities that are required for replacing cells and cell parts.

Hypersecretion of the growth hormone produces completely different disorders. For example, hyperactivity during childhood years results in giantism, which is an abnormal increase in the length of long bones. Hypersecretion during adulthood is called *acromegaly*. Acromegaly cannot produce further lengthening of the long bones because the epiphyseal plates are already closed. Instead, the bones of the hands, feet, cheeks, and jaws thicken. Other tissues also grow. For instance, the eyelids, lips, tongue, and nose enlarge, and the skin thickens and furrows, especially on the forehead and soles of the feet.

The principal abnormality associated with dysfunction of the neurohypophysis is *diabetes insipidus*. This disorder should not be confused with diabetes mellitus, which is a disorder of the pancreas and is characterized by sugar in the urine. Diabetes insipidus is the result of a hyposecretion of vaso-pressin-ADH, usually caused by damage to the neurohypophysis or to the hypothalamus. Symptoms of the disorder include excretion of large amounts of urine and subsequent thirst. Diabetes insipidus is treated by administering vasopressin-ADH.

## CONCLUSIONS

1. What part of the pituitary produces dramatic differences in the size of the individual?

    *Growth hormone — Adenohypophysis*

2. What hormone is involved?

    *Growth hormone*

3. What particular tissue is especially affected by a hypo- or hypersecretion of growth hormone? What specific part of this tissue? *Bone / Epiphyseal plates*

4. What is the main abnormality associated with dysfunction of the neurohypophysis? Name 2 outstanding symptoms of this disorder. *diabetes insipidus / excretion of large amts urine, subsequent thirst*

## DISORDERS OF THE THYROID

Hyposecretion of the thyroid hormone during the growth years results in a condition called *cretinism*. Two outstanding clinical symptoms of the cretin are dwarfism and mental retardation. The first is caused by failure of the skeleton to grow and mature. The second is caused by failure of the brain to develop fully. Cretins also exhibit retarded sexual development and a yellowish skin color. Flat pads of fat develop, giving the cretin a characteristic round face and thick nose; a large, thick, protruding tongue; and protruding abdomen.

Hypothyroidism during the adult years produces a disorder called *myxedema*. The name refers to the fact that thyroxin is a diuretic. Lack of thyroxin causes the body to retain water. One of the hallmarks of myxedema is an edema that causes the facial tissues to swell and look puffy. Another symptom caused by the retention of water is an increase in blood volume that frequently causes high blood pressure. Like the cretin, the person with myxedema also suffers slow heart rate, low body temperature, muscular weakness, general lethargy, and a tendency to gain weight easily. Myxedema occurs 8 times more frequently in females than in males. Its symptoms are abolished by the administration of thyroxin.

Hypersecretion of thyroid hormone gives rise to a condition called *exophthalmic goiter* (Graves' disease). This disease, like myxedema, is also more frequent in females, affecting 8 females to every one male. One of its primary symptoms is an enlarged thyroid, called a *goiter*, which may be 2 to 3 times its original size. The 2 other characteristic symptons are an edema behind the eye, which causes the eyes to "pop out" (*exophthalmos*), and an abnormally high metabolic rate. The high

**241**

metabolic rate produces a range of effects that are generally opposite to those of myxedema. The person has an increased pulse, the body temperature is high, and the skin is warm, moist, and flushed. The thyroxin also increases the responsiveness of the nervous system. The usual methods for treating hyperthyroidism are administering drugs that suppress thyroxin synthesis or surgically removing a part of the gland.

The term goiter simply means an enlargement of the thyroid gland. It is a symptom of many thyroid disorders. It may also occur if the gland does not receive enough iodine to produce sufficient thyroxin for the body's needs. This is called *simple goiter*. Simple goiter is most often caused by a lower-than-average amount of iodine in the diet. However, it may also develop if iodine intake is not increased during certain conditions that put a high demand on the body for thyroxin. Such conditions are frequent exposure to cold, and high fat and protein diets.

The iodine in thyroxin and thyrocalcitonin that is bound to TBG is called protein-bound iodine or PBI. Under normal circumstances, the amount of PBI in the blood is fairly constant—4 to 8 micrograms PBI/100 milliliters of blood. This amount can easily be measured. For this reason, PBI is a good index of thyroid hormone secretion and is often used as a tool to diagnose suspected thyroid malfunction.

## CONCLUSIONS

1. Differentiate between the 2 conditions arising from undersecretion of thyroxin.

   *Hypothyroidism - Cretin - Occurs during growing years (dwarfism / mental retardation)*
   *Hypothyroidism - myxedema - occurs in adults*

2. List the main symptoms of each condition.

   *Cretin - Dwarfism & mental retardation (stomach protruding, large thick protruding tongue, large nose & round face)*
   *myxedema - eyeballs protruding, high blood pressure, weight gain, slow heart rate, general lethargy, low body temp.*

3. What conditions are produced by oversecretion of thyroid hormone.

   *Exophthalmic goiter*
   *simple goiter*

4. Differentiate between these 2 conditions.

   *Exophthalmos - eyes pop out*
   *abnormally high metabolic rate. = high body temp, nervousness, warm moist skin, flushed, increased pulse rate.*

5. How is a suspected thyroid malfunction diagnosed?

   *By protein-bound iodine or PBI*

6. What is the normal PBI?

   *4-8 micrograms PBI/100 milli of blood*

## DISORDERS OF THE PARATHYROIDS

A deficiency of calcium caused by *hypoparathyroidism* causes neurons to depolarize without the usual stimulus. As a result, nervous impulses increase and result in muscle twitches, spasms, and convulsions. This condition is called *tetany*. The effects of hypocalcemic tetany are observed in the *Trousseau* and *Chvostek* signs.

The Trousseau sign is observed when the binding of a blood pressure cuff around the upper arm produces contraction of the fingers and inability to open the hand. The Chvostek sign is a contracture of the facial muscles elicited by tapping the facial nerves at the angle of the jaw. Hypoparathyroidism results from surgical removal of the parathyroids or from parathyroid damage caused by parathyroid disease, infection, hemorrhage, or mechanical injury.

Hyperparathyroidism causes demineralization of bone. This condition is called *osteitis fibrosa cystica* because the areas of destroyed bone tissue are replaced by cavities that fill with fibrous tissue. The bones thus become deformed and are highly susceptible to fracture. Hyperparathyroidism is usually caused by a tumor in the parathyroids.

## CONCLUSIONS

1. If there is a calcium deficiency in the body, what are its effects?

   *nervous impulses which result in muscle spasms & twitches & convulsions.*

2. What are 2 tests or signs used for the diagnosis of calcium deficiency?

   *Trousseau - Chvostek*

3. Describe each.

   *Trousseau is observed when a band is placed around the upper arm which produces contraction of the fingers & inability to open the hand.*

   *Chvostek - contraction of facial muscles by tapping nerves at the angle of the jaw.*

4. What is the main effect of hyperparathyroidism?

   *demineralization of the bone. The main effect is known as osteitis fibrosa cystica*

## DISORDERS OF THE ADRENALS

Hypersecretion of the *mineralocorticoid* aldosterone results in a decrease in the body's potassium concentration. If potassium depletion is great enough, neurons cannot depolarize and muscle paralysis results. Hypersecretion also brings about excessive retention of sodium and water. The water increases the volume of the blood and causes high blood pressure. It also increases the volume of the interstitial fluid, producing edema.

Disorders associated with *glucocorticoids* include Addison's disease and Cushing's syndrome. Hyposecretion of glucocorticoids results in the condition called *Addison's disease*. Clinical symptoms include hypoglycemia, which leads to muscular weakness, mental lethargy, and weight loss. In addition, increased potassium blood levels and decreased sodium blood levels lead to low blood pressure and dehydration. *Cushing's syndrome* is a hypersecretion of glucocorticoids, especially hydrocortisone and cortisone. The condition is characterized by the redistribution of fat. This results in spindly legs accompanied by a characteristic "moon face," "Buffalo Hump" on the back, and pendulous abdomen. The individual also bruises easily, and wound healing is poor.

The *adrenogenital syndrome* results from overproduction of sex hormones, particularly the male androgens, by the adrenal cortex. Hypersecretion in male infants and young male children results in an enlarged penis. In young boys, it also causes premature development of male sexual characteristics.

Hypersecretion in adult males is characterized by overgrowth of body hair, enlargement of the penis, and increased sexual drive. Hypersecretion in young girls results in premature sexual development. Hypersecretion in both girls and women usually produces a receding hairline, baldness, an increase in body hair, deepening of the voice, muscular arms and legs, small breasts, and an enlarged clitoris.

Tumors of the chromaffin cells, called *pheochromocytomas*, cause hypersecretion of the medullary hormones. The oversecretion causes high blood pressure, high levels of sugar in the blood and urine, an elevated basal metabolism rate, nervousness, and sweating. Since the medullary hormones create the same effects as does sympathetic nervous stimulation, hypersecretion puts the individual into a prolonged version of the "fight-or-flight" response.

## CONCLUSIONS

1. What pathology is involved if there is a severe depletion of potassium concentration in the body?

Hypersecreation of the mineralocoticoid aldostrone

2. What 2 diseases are involved in either a hypo- or hypersecretion of glucocorticoids?

Addisions, Cushings

3. Compare these 2 disorders.

Addisions - Hyposecreation of the glucocorticoids causing hypoglycemia (mental lethargy, muscle weakness & weight loss.)

Cushings - Hypersecreation of the glucocorticoids, hydrocortisone & cortisone. spindly legs "moon face" "Buffalo hump"

4. What is the adrenogenital syndrome?

Adrenogenital syndrome results from over production of sex hormones. Particularly in the male androgens.

5. What is a pheochromocytoma? Tumors of the chromaffin cells causing hypersecretion of the medullary hormones,

6. What are its symptoms?

Oversecretion causes high blood pressure, high levels of sugar in blood & urine, elevated basal rate, nervousness & sweating.

## DISORDERS OF THE PANCREAS

Hyposecretion of insulin results in a number of clinical symptoms referred to as *diabetes mellitus*. Typically an inherited disease, diabetes mellitus is caused by the destruction or malfunction of the beta cells. Among the symptoms are hyperglycemia and excretion of glucose in the urine as hyperglycemia increases. There is also an inability to reabsorb water, resulting in increased urine production, dehydration, loss of sodium, and thirst.

*Hyperinsulinism* is much rarer than hyposecretion and is generally the result of a malignant tumor in an islet. The principal symptom is a decreased blood glucose level, which stimulates the secretion of epinephrine, glucagon, and the growth hormone. As a consequence, anxiety, sweating, tremor, increased heart rate, and weakness occur. Moreover, brain cells do not have enough glucose to function efficiently. This leads to mental disorientation, convulsions, unconsciousness, shock, and eventual death as the vital centers in the medulla are affected.

## CONCLUSIONS

1. What is the disorder associated with hyposecretion of insulin?  Diabetes Mellitus.

2. List some outstanding symptoms of this disease.  hyperglycemia - inability to reabsorb H₂O. dehydration, loss of Na ✦ thirst.

3. Explain hyperinsulinism.
Caused usually by a tumor in an islet, decreased blood glucose level, anxiety, sweating, mental disorientation, convulsions, unconsciousness, shock ✦ eventual death.

## MEDICAL TERMINOLOGY

Define each of the following terms:

a. acromegaly - Hypersecretion during adulthood.

b. Addison's disease - Hypo-secretion of glucocorticoids

c. cretinism - hyposecretion of thyroid hormone during growth years

d. Cushing's disease - Hypersecretion of glucocorticoids.

e. diabetes insipidus - dysfunction of the neurohypophysis

f. diabetes mellitus - Hyposecretion of insulin - dysfunction of pancreas

g. exophthalmic goiter - over secretion of thyroid hormone

h. simple goiter - Enlargement of thyroid gland caused by not enough iodine

i. myxedema - hypothyroidism in adult yrs.

j. pituitary cachexia - Simmonds disease - secretion of growth hormone is normal during childhood but slows during adult yrs.

k. pituitary dwarfism - when growth hormone is hyposecreted during growth yrs.

l. pituitary giantism - acromegaly hyper secretion during adult yrs

## STUDENT ACTIVITY

Separate all of the endocrinological disorders in this module into 2 main groups: one group that exhibits mainly outstanding morphological (physical) abnormalities, as compared with the other that exhibits mainly outstanding physiological abnormalities. List the diseases, their main symptoms, and the gland involved.

# MODULE 27
# MICROSCOPIC AND DIAGNOSTIC FEATURES OF BLOOD

**OBJECTIVES**

1. To identify the formed elements and plasma constituents of blood

2. To compare the origins of the formed elements in blood and the reticuloendothelial cells

3. To describe the structure of erythrocytes and their function in the carriage of oxygen and carbon dioxide

4. To define erythropoiesis, identify the factors related to erythrocyte production and destruction, and to do a red blood cell count

5. To describe the importance of a reticulocyte count in the diagnosis of abnormal rates of erythrocyte production

6. To perform the following tests: red blood cell volume (hematocrit), sedimentation rate, and hemoglobin estimation

7. To list the structural features and types of leucocytes and to do a white blood cell count

8. To describe the importance of a differential count and to do a differential white cell count

9. To discuss the role of leucocytes in phagocytosis and antibody production

10. To discuss the structure of thrombocytes and explain their role in blood clotting

11. To list the components of plasma and identify their importance

The red body fluid that flows through all the vessels except the lymph vessels is called *blood*. Blood is an exceedingly complex liquid that performs a number of functions vital to the maintenance of homeostasis.

Microscopically, blood is composed of 2 portions: plasma, which is a liquid containing dissolved substances, and formed elements, which are cells and cell-like bodies suspended in the plasma.

In clinical practice, the most common classification of the *formed elements* of the blood is the following:

1. Erythrocytes, or red blood cells
2. Leucocytes, or white blood cells
   a. Granular leucocytes (granulocytes)
      (1) Neutrophils
      (2) Eosinophils
      (3) Basophils
   b. Agranular leucocytes (agranulocytes)
      (1) Lymphocytes
      (2) Monocytes
3. Thrombocytes, or platelets

## PART I  FORMED ELEMENTS

We shall now consider some of the more important facts about each of the formed elements. Where do these formed elements come from? The process by which blood cells are formed is called *hematopoiesis*. In the adult the production areas are known precisely. Bone marrow (myeloid tissue) is responsible for the production of red blood cells, granular leucocytes, and platelets. On the other hand, lymphatic tissue—which includes spleen, tonsils, and lymph nodes—is responsible for the production of agranular leucocytes.

During embryonic life certain epithelial cells, called *primitive reticular cells*, wander throughout the body and later become trapped in certain organs. Some of the primitive reticular cells move into red bone marrow and are transformed into *hemocytoblasts*, immature cells that are eventually capable of developing into mature blood cells (Figure 27-1). The hemocytoblasts in red bone marrow first develop into distinct cells that are later transformed into specific mature cells. For example, the hemocytoblasts develop into (1) *rubriblasts* that go on to form mature red blood cells, (2) *myeloblasts* that go on to form mature neutrophils, eosinophils, and basophils, and (3) *megakaryoblasts* that go on to form mature platelets. Other hemocytoblasts become entrapped in lymphatic tissue. The fate of these hemocytoblasts differs from that of the hemocytoblasts in the red bone marrow. Hemocytoblasts in lymphatic tissue develop into 2 distinct cells that are later transformed into other mature blood cells. These are (1) *lymphoblasts* that eventually form lymphocytes, and (2) *monoblasts* that eventually form monocytes.

Mature blood cells do not live in the bloodstream for very long. Their disintegrating bodies pose the danger of clogging small blood vessels, so certain cells clear away their bodies after they die. These cells are called *reticuloendothelial cells* and are also formed from primitive reticular cells. However, unlike the hemocytoblasts in bone marrow or lymphatic tissue that produce mature blood cells, reticuloendothelial cells have a different destiny. They enter the spleen, tonsils, lymph nodes, liver, and other organs and become highly specialized for phagocytosis. The reticuloendothelial cells that lie in the lymph nodes are particularly active in destroying microbes and their toxins. The reticuloendothelial cells in the liver and spleen concentrate more on ingesting dead blood cells.

### ERYTHROCYTES

Microscopically, red blood cells, or *erythrocytes*, appear as biconcave discs averaging about 7.7 microns in diameter (see Figure 27-1). Mature red blood cells are quite simple in structure. They lack a nucleus and can neither reproduce nor carry on extensive metabolic activities. The interior of the cell contains some cytoplasm, protein, lipid substances (including cholesterol), and a red pigment called hemoglobin. Hemoglobin, which constitutes about 33 percent of the cell volume, is responsible for the red color of blood.

The function of the erythrocytes is to combine with oxygen and carbon dioxide and transport them through the blood vessels. The shape of a red blood cell also increases its carrying capacity. A biconcave structure has a much greater surface area than, say, a sphere or cube.

As we mentioned, a red blood cell does not live long. For some reason, its cell membrane becomes fragile, and the cell is nonfunctional in about 120 days. The blood, however, contains inordinate numbers of these cells. A healthy male has 5.5 to 7 million red blood cells per cubic millimeter of blood. A healthy female has 4.5 to 6 million red blood cells per cubic millimeter. To maintain normal quantities of erythrocytes, the body must produce new mature cells at the astonishing rate of 2 million per second. In the adult, production takes place in the red bone marrow in the spongy bone of the cranium, ribs, sternum, bodies of vertebrae, and epiphyses of the humerus and femur. The process by which erythrocytes are formed is called *erythropoiesis*. Once the erythrocyte

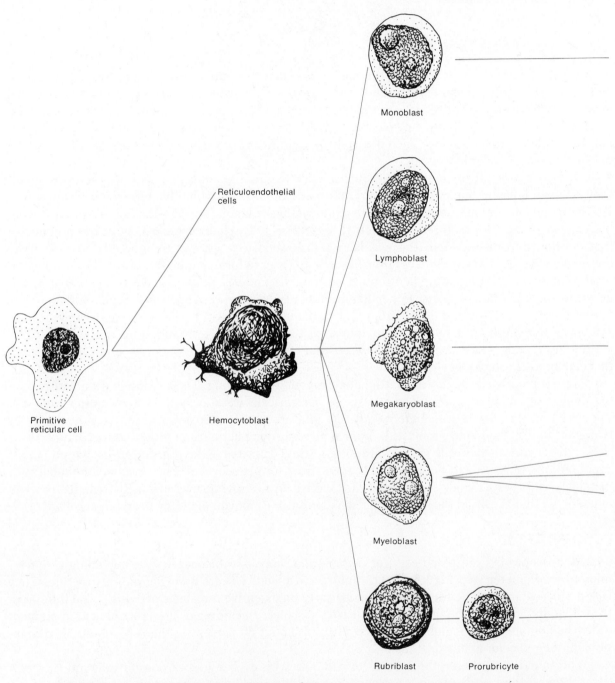

**Figure 27–1.** Origin of the formed elements in blood and reticuloendothelial cells. Consult text for details.

is formed, it circulates through the blood vessels until its life span comes to an end. Aged erythrocytes are destroyed by reticuloendothelial cells in the liver and spleen. The hemoglobin molecules are split apart; the iron is reused and the rest of the molecule is converted into other substances for reuse or elimination.

Normally erythropoiesis and red cell destruction proceed at the same pace. But if the body suddenly needs more erythrocytes, or if erythropoiesis is not keeping up with red blood cell destruction, a homeostatic mechanism steps up erythrocyte production. The stimulus for the homeostatic mechanism is oxygen deficiency within the cells of the kidney. This is not surprising, because the chief

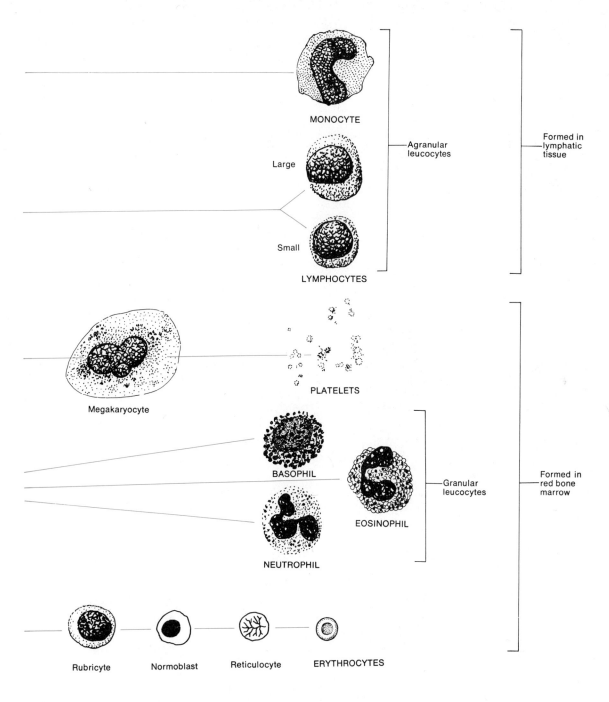

MONOCYTE

Large

Small

LYMPHOCYTES

Agranular leucocytes

Formed in lymphatic tissue

Megakaryocyte

PLATELETS

BASOPHIL

EOSINOPHIL

NEUTROPHIL

Granular leucocytes

Formed in red bone marrow

Rubricyte        Normoblast        Reticulocyte        ERYTHROCYTES

**Figure 27–1 continued.**

function of the erythrocytes is to deliver oxygen. As soon as the kidney cells become oxygen deficient, they release a hormone called *erythropoietin*. This hormone circulates through the blood to the red bone marrow, where it stimulates hemocytoblasts to develop into red blood cells.

A diagnostic test that informs the physician about the rate of erythropoiesis is the *reticulocyte count*. Some of the normoblasts and reticulocytes are normally released into the bloodstream before they become mature red blood cells (see Figure 27-1). If the number of reticulocytes in a sample of blood is less than 0.5 percent of the number of mature red blood cells in the sample, erythropoiesis is occurring too slowly. A low reticulocyte count might confirm a diagnosis of nutritional or pernicious anemia. Or, it might indicate a kidney disease that prevents the kidney cells from producing erythropoietin. If the reticulocyte number is more than 1.5 percent of the mature red blood cells, erythropoiesis is abnormally rapid. A raft of problems may be responsible for a high reticulocyte count. Among these are most types of anemia, oxygen deficiency, and uncontrolled red blood cell production caused by a cancer in the bone marrow. If the individual has been suffering from a nutritional or pernicious anemia, the high count may indicate that treatment has been effective, and the bone marrow is making up for lost time.

The procedure for finger puncture is as follows:
1. Obtain 70 percent alcohol, lancet, and cotton.
2. Thoroughly cleanse the third or fourth finger with alcohol.
3. Holding your finger firmly, take the lancet and use a quick, jabbing motion. (Do not begin the motion from a great distance away from the finger. The closer you are and the quicker the motion, the less discomfort that results.
4. Finally, place a piece of cotton between the finger and thumb, and press together firmly.

The procedure for using pipettes and tubing is as follows:
When using red and white cell pipettes for counting it is important that the blood be pipetted accurately. The aspirator tube is attached to the top of the pipette and the other end is held in the mouth in order to draw the blood into the pipette. When you have successfully drawn the blood up to the specified mark, placing the tongue over the end of the tube will prevent the blood from draining out of the pipette. None of the sample should leak out into the diluting fluid, because any count performed afterward would then be inaccurate because of contamination (see Figure 27-2).

The procedure for filling the counting chamber is as follows:
1. Obtain a counting chamber and coverglass.
2. Clean the chamber thoroughly.
3. Place the coverglass on the chamber.
4. Using pipette filled with proper dilution, place the tip of the pipette at the notch of the center-raised area of the chamber.
5. *Slowly* let the dilution out of the pipette and fill the center area (moat). It is imperative that the dilution does not overflow the moat. This would result in an inaccurate cell count (see Figure 27-3).

A. *Red Blood Cell Count.* The purpose of a red blood cell count is to determine the number of circulating red blood cells in the body. Red blood cells carry oxygen to all tissues, so it is imperative that a drastic change in the red cell count be dealt with immediately.

A decrease in red cells can result from a variety of conditions including impaired cell production, increased cell destruction, and acute blood loss. When the red cell count is increased a condition called *polycythemia* results. Severe anemias (decrease) can result in fatalities, due to an overall alteration in homeostasis.

The procedure for red cell diluting is as follows:
1. Fill the pipette exactly to the 0.5 mark with blood from a finger puncture.
2. Wipe excess blood from the tip of the pipette without disturbing the column of blood. Recheck the column.

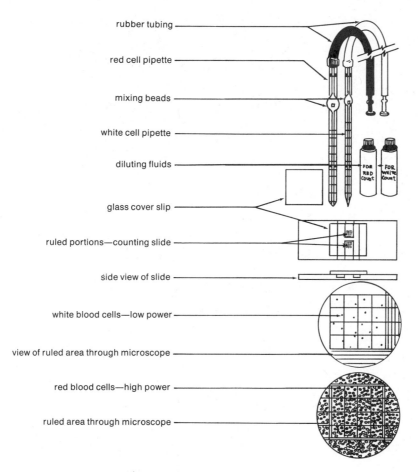

rubber tubing

red cell pipette

mixing beads

white cell pipette

diluting fluids

FOR RBD Count    FOR WHITE Count

glass cover slip

ruled portions—counting slide

side view of slide

white blood cells—low power

view of ruled area through microscope

red blood cells—high power

ruled area through microscope

**Figure 27–2.** Various parts of a hemocytometer and red and white pipettes used for blood counts.

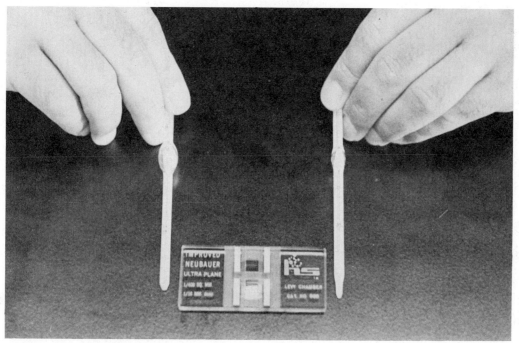

INPROVED NEUBAUER ULTRA PLANE

**Figure 27–3.** Hemocytometer with an RBC pipette on the left and a WBC pipette on the right. *(Photograph courtesy of Lenni Patti.)*

3. Fill the pipette to the 101 mark with diluting fluid.
4. Shake for 3 minutes, covering both ends of the pipette.
5. Discard a third of the fluid in the pipette.
6. Fill the counting chamber as in previously described manner.
7. Allow the cells to settle in the chamber (approximately one minute).
8. Using the 45x lens (highpower) count the cells in each of the 5 squares (E.F.G.H.I) as shown in Figure 27-4.
9. Multiply the results by 10,000 to obtain the number of red blood cells in one cubic millimeter of blood.

   B. *Red Blood Cell Volume (Hematocrit)*. The packed cell volume (PCV) measurement is a good method for the routine testing for anemia. The percentage of blood volume occupied by the red blood cells is called the *hematocrit* or *packed cell volume*. When a tube of blood is centrifuged the erythrocytes will pack into the bottom part of the tube with the plasma on top. Now it is a simple matter to determine the percentage.

   In men the normal range is between 40 and 54 percent, with an average of 47 percent. In women the normal range is between 37 and 47, with an average of 42 percent.

   The method described is a micro-method requiring only a drop of blood.

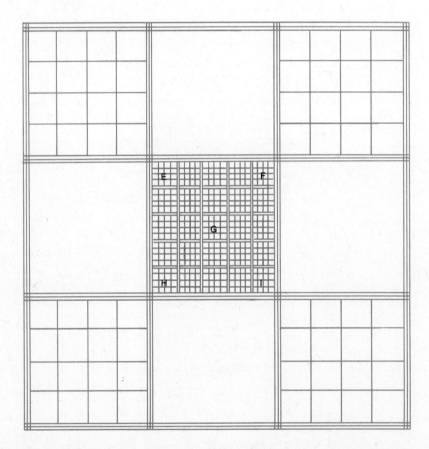

**Figure 27–4.** Hemocytometer (counting chamber). E.F.G.H.I.—area counted in red cell count. The large square containing the smaller squares E, F, G, H and I is seen in microscopic field under 10X magnification. In order to count the smaller squares (E, F, G, H, I), the 45X lens is used and therefore E, F, G, H or I will encompass the entire microscopic field.

The procedure for testing red blood cell volume is as follows:

1. A free flow of blood from the tip of the finger is produced (see Figure 27-5).
2. The marked end (red) of the capillary tube is placed into the drop of blood. Blood is drawn two thirds of the way into the tube. The tube fills rapidly if the open end is held downward from the source of the blood.
3. The blood end of the tube is then sealed with seal-ease.
4. The tube is placed into the centrifuge making sure that the sealed end is against the ring of rubber at the circumference. Properly balance the tubes in the centrifuge (see Figure 27-6).
5. Secure the inside cover and fasten down the outside cover.
6. Turn on the centrifuge and set the timer for 4 minutes.
7. Determine the hematocrit by placing the centrifuged tube in the tube reader. Follow instructions on the reader, and record your results on the laboratory report (see Figure 27-7).

C. *Sedimentation Rate.* If citrated blood is allowed to stand vertically in a tube for a period of time, the red blood cells will fall to the bottom of the tube, leaving clear plasma in the upper portion. The distance that the cells fall in one hour can be measured and is called the *sedimentation rate.* This rate is greater during menstruation, pregnancy, and most infections. If this rate is high, it may be an indication of tissue destruction in some part of the body; therefore, it is considered a valuable non-specific diagnostic tool. The normal rate for adults is 0 to 6 millimeters, for children 0 to 8 millimeters. The following method requires only one drop of blood and is called the Landau micro-method.

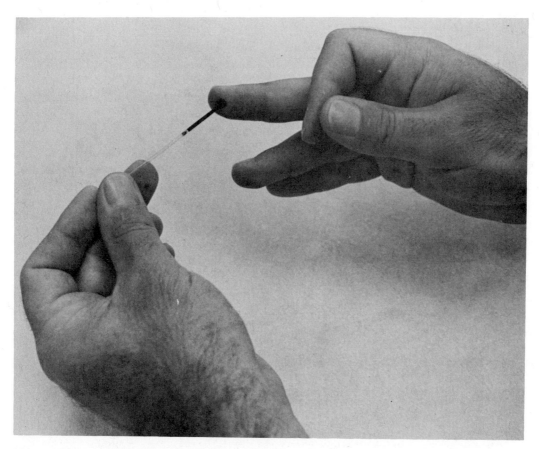

**Figure 27-5.** A heparinized capillary tube is filled with blood from a freely flowing finger puncture. The tube is used for Hematocrit determination. *(Photograph courtesy of Lenni Patti.)*

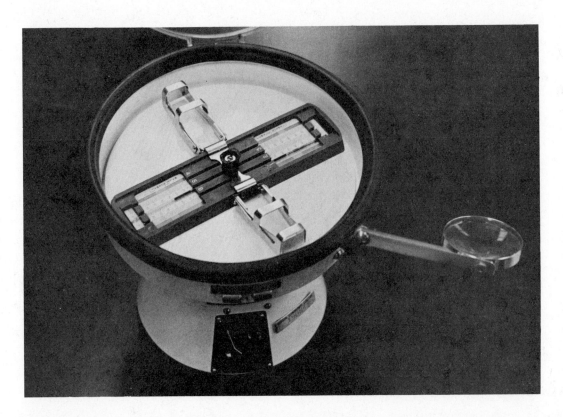

**Figure 27–6.** Centrifuge used for spinning blood to determine Hematocrit. *(Photograph courtesy of Lenni Patti.)*

**Figure 27–7.** A Micro-Hematocrit Tube Reader. *(Photograph courtesy of Lenni Patti.)*

The procedure for determining sedimentation rate is as follows:

1. Using the mechanical suction device, draw the sodium citrate up to the first line that completely encircles the pipette.
2. Draw free-flowing blood up into the pipette until it reaches the second line. Care must be taken to avoid air bubbles. (If air bubbles are drawn into the blood, carefully expel the mixture onto a clean microscope slide and draw it up again.)
3. The 2 fluids are drawn up into the bulb and mixed by expelling them into the lumen of the tube. This is done *6 times*. If any air bubbles appear, use the procedure described in step 2 above.
4. Adjust the top level of the blood as close to zero as possible. It is very difficult to get it to stop exactly at zero.
5. Remove the suction device by placing the lower end of the pipette on the index finger of the left hand before removing the device off the other end. The blood is pulled out of the pipette if the lower end is not completely sealed.
6. Place the lower end of the pipette on the base of the pipette rack and the opposite end at the top of the rack. The tube must be *exactly perpendicular*. Record the time at which it is put in the rack.
7. One hour later, measure the distance from the top of the plasma to the top of the red blood cells with an accurate millimeter scale, and record your results on the laboratory report (see Figure 27-8).
8. Rinse the pipette with cleaning solutions.

D. *Hemoglobin Estimation.* Since the amount of hemoglobin in the blood is an accurate indication of the oxygen-carrying potential of the blood, various methods have been developed for its measurement. The Tallqvist method is one of the oldest and one of the most inaccurate. In this method a piece of blotting paper that has been saturated with a drop of blood is compared with a color chart to determine the percentage of hemoglobin.

A better method is to compare an acid-tested sample of blood with a color standard called a *hemoglobinometer*. The procedure outlined here is the Sahli-Adams method.

The procedure for measuring hemoglobin is as follows:

1. Fill the graduated tube to the 2 gm. mark with one percent hydrochloric acid.
2. Produce a free flow of blood and draw it into the pipette to the 20 cubic millimeter mark. Wipe off the tip of the pipette and blow the contents of the pipette into the acid solution in the test tube.

**Figure 27–8.** One type of Sedimentation Tube Rack. *(Photograph courtesy of Lenni Patti.)*

Draw the solution up into the pipette 2 to 3 times, expelling all fluid into the tube. Use the pipette cleaning solution to get it completely clean.

3. Place the tube in the hemoglobinometer and let stand for 10 minutes.
4. Compare the color of your test sample with the color standard. If the mixture is darker, add distilled water, drop by drop, mixing the solution after each addition. Keep adding water until both colors are identical. While mixing, care should be taken that no solution is lost.
5. When the 2 colors are identical, read the grams of hemoglobin per 100 ml. of blood on the side of the tube. The normal range for both men and women is between 13.5 and 17.5 grams. Record your results on the laboratory report.

## CONCLUSIONS

1. What is the name of the process by which all blood cells are formed?

2. Name the different parts of the body responsible for blood cell production.

3. Hemocytoblasts are immature cells capable of developing into mature blood cells. What are the 3 intermediate forms that these hemocytoblasts develop into in bone marrow? What type of cells do these forms eventually develop into?

4. How does the body get rid of old, disintegrating mature blood cells?

5. What is another name for red blood cells?

6. What is their shape?

7. Where does blood get its red color?

8. What is the basic function of the red blood cell?

9. What is the life span of a red blood cell?

10. What is the average number of red blood cells per cubic mm. of blood in a healthy male and female?

11. What is the name of the process by which red blood cells are formed by the body?

12. Normally, red blood cell production and red blood cell destruction proceed at the same pace. If the body, however, suddenly needs more red blood cells, what is the stimulus that keeps the body in homeostasis?

13. What happens to the kidneys when they become oxygen deficient?

14. How are the red blood cells affected?

15. What is the name of the diagnostic test that physicians use to check on the rate of erythropoiesis?

16. Explain this test in terms of high and low count, and the possible consequences.

17. What is the significance of the sedimentation rate test?

18. What are normal sedimentation rate values?

19. Of what value is measuring the amount of hemoglobin in the blood?

## LEUCOCYTES

Unlike red blood cells, *leucocytes*, or *white blood cells*, have nuclei and do not contain hemoglobin. In addition, they are far less numerous, averaging from 5000 to 9000 cells per cubic millimeter. Red blood cells, therefore, outnumber white blood cells about 700 to one. Leucocytes fall into 2 major groups (see Figure 27-1). The first group contains the *granular leucocytes*. These develop from hemocytoblasts in the bone marrow. They have granules in the cytoplasm and possess lobed nuclei. Three kinds of granular leucocytes exist. These are the *neutrophils*, the *eosinophils*, and the *basophils*. The second principal group of leucocytes, called the *agranular leucocytes*, originate from hemocytoblasts in lymphatic tissue. When they are placed under a light microscope, no cytoplasmic granules can be seen. Their nuclei are more or less spherical. The 2 kinds of agranular leucocytes are *lymphocytes* and *monocytes*.

The general function of the leucocytes is to combat inflammation and infection. Some leucocytes are actively *phagocytotic*. This means that they can ingest bacteria and dispose of dead matter. Most leucocytes also possess, to some degree, the ability to crawl through the minute spaces between the cells that form the walls of blood vessels and through connective and epithelial tissue. This movement, the same kind that is exhibited by amoebas, is called *diapedesis*.

The diagnosis of an injury or infection within the body may involve a *differential count*. In this procedure, the number of each kind of white cell in 100 white blood cells is counted. A normal differential count might appear as follows:

| | |
|---|---|
| Neutrophils | 60-70% |
| Eosinophils | 2-4% |
| Basophils | 0.5-1% |
| Lymphocytes | 20-25% |
| Monocytes | 3-8% |
| Total | 100% |

In interpreting the results of a differential count, particular attention is paid to the neutrophils. The neutrophils are the most active in response to tissue destruction. More often than not, a high neutrophil count indicates that the damage is caused by invading bacteria. An increase in the number of monocytes generally indicates a chronic (of long duration) infection such as tuberculosis. It is hypothesized that monocytes take longer to reach the site of infection than neutrophils, but once they arrive they do so in larger numbers and destroy more microbes than do neutrophils. High eosinophil counts typically indicate allergic conditions since eosinophils are believed to combat allergens, the causative agents of allergies.

Some leucocytes fight infection in another extremely important way. These are the lymphocytes, which are involved in the production of antibodies. *Antibodies* are special proteins that inactivate antigens. An *antigen* is any type of protein that the body is not capable of synthesizing. Antigens, in other words, are "foreign" proteins that are introduced into the body in any number of ways. For example, many of the proteins that make up the cell structures and enzymes of bacteria are antigens.

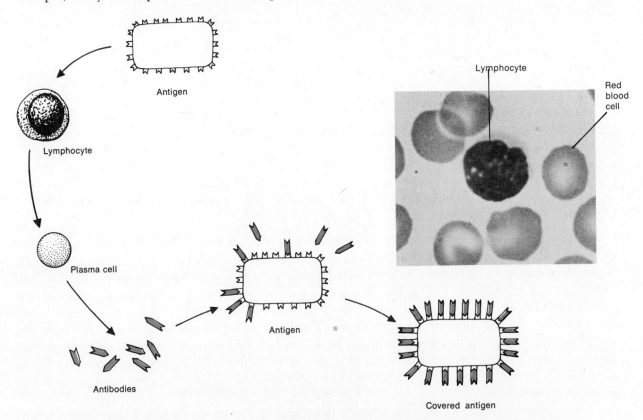

**Figure 27-9.** Antigen-antibody response. An antigen entering the body stimulates a lymphocyte to develop into an antibody-producing plasma cell. The antibodies attach to the antigen, cover it, and render it harmless. Photomicrograph of lymphocyte x 1800. *(Photomicrograph courtesy of Edward J. Reith, from Atlas of Descriptive Histology, Edward J. Reith and Michael H. Ross, Harper & Row, Publishers, Inc., New York, N.Y., 1970.)*

The toxins released by bacteria are also antigens. When antigens enter the body, they react chemically with substances in the lymphocytes and stimulate some of the lymphocytes to become *plasma cells* (Figure 27-9). The plasma cells then produce antibodies, which are globulin-type proteins that attach to antigens, much as enzymes attach to substrates.

Even when the body is healthy, the leucocytes actively ingest bacteria and debris. However, a leucocyte can phagocytose only a certain number of substances before the substances interfere with the normal metabolic activities of the leucocyte and bring on its death. Consequently, the life span of a leucocyte is very short. During times of health, some white blood cells will live a couple of days. During a period of infection they may live for only a few hours.

A. *White Blood Cell Count.* This procedure is done to determine the number of circulating white blood cells in the body. Since white blood cells are a vital part of the body's immune defense system, it is important that any abnormalities in the white cell count should be carefully noted.

An increase in number (leucocytosis) may result from such conditions as bacterial infection, viral infection, metabolic disorders, chemical and drug poisoning, and acute hemorrhage. A decrease (leucopenia) in the white cell count may result from typhoid infection, measles, infectious hepatitis, tuberculosis, and cirrhosis of the liver.

The procedure for white blood cell volume is as follows:

1. Fill the pipette exactly to the 0.5 mark with blood from a finger puncture.
2. Wipe excess blood from tip of pipette without disturbing the column of blood.
3. Fill the pipette up to the 11 mark with diluting fluid.
4. Shake for 2 minutes.
5. Discard 2 to 3 drops to clear the stem of the pipette of diluting fluid.
6. Fill the counting chamber as in the previously described method.
7. Using 10x lens, count the cells in each of the 4 corner squares (A, B, C, D, in Figure 27-10) of the chamber.
8. Multiply the results by 50 to obtain the amount of circulating white blood cells per cubic millimeter of blood.

B. *White Cell Count Differential.* The purpose of white cell count differential is to determine the percentages of each of the normally circulating white cells including neutrophils, eosinophils, basophils, lymphocytes, and monocytes. Significant elevations of different types of white blood cells usually indicate specific pathological conditions. For example, lymphocytes predominate in specific leukemias and in infectious mononucleosis. An increase in eosinophils may be seen in allergic reactions. An elevated percentage of monocytes may result from parasitic infections and leukemia. An increase in basophils is rare and denotes a specific type of leukemia.

The procedure for white blood cell count is as follows:
1. Place one drop of blood on a glass slide.
2. Use a second slide as a spreader.

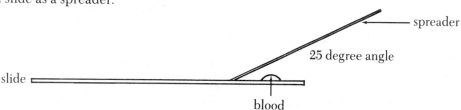

3. Draw the spreader toward the drop of blood. The blood should fan out to the edges of the spreader slide.
4. Keeping the spreader at a 25-degree angle, press the edge of the spreader firmly against the slide and push rapidly over the entire length of the slide. The drop of blood will thin out toward the end of the slide.

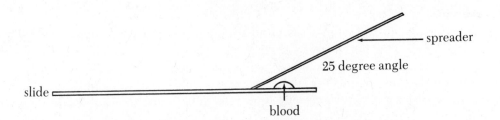

spreader

25 degree angle

slide

blood

Draw the spreader in this direction first. ———————>
To make the smear, push the spreader in this direction. <———————
5. Let the smear dry.
6. To stain the slide follow this procedure:
   Place the slide on staining rack.
   Cover the entire slide with Wright's stain.
   Let it stand for one minute.
   Add 25 drops of buffer solution, mix it completely with Wright's stain.
   Let it stand for 8 minutes.
   Wash it off completely with distilled water.
7. Let the slide dry before counting.
8. To proceed with counting the cells, use the area of the slide where the blood is thinnest. The cells are to be counted under an oil-immersion lens. A total of 100 white cells are counted.

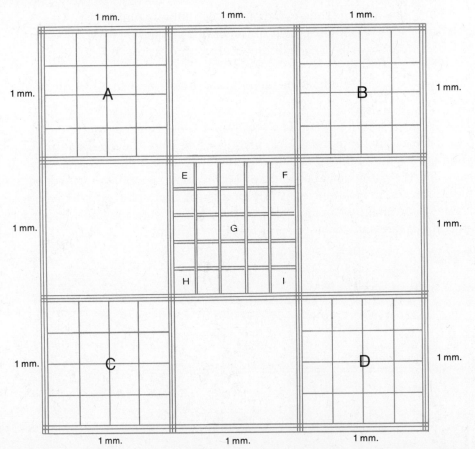

**Figure 27-10.** Hemocytometer (counting chamber) for white blood cell count (as seen through scanning lens). A, B, C, D—Areas counted in white cell count (when viewed through 10X lens, 1 square mm. is seen in microscopic field). Areas A, B, C or D equal 1 square mm., therefore, a total of 4 square mm. is counted in white cell count.

## CONCLUSIONS

1. How do white blood cells differ from red blood cells?

2. What is the normal ratio of red blood cells to white blood cells?

3. What is the average number of white blood cells per cubic millimeter of blood?

4. Give the 2 groups of white blood cells and the types of cells each group contains.

5. What is the general function of leucocytes?

6. What is diapedesis?

7. What is a differential count?

8. Of what value is it?

9. What are the percentages of each type of leucocyte in a normal differential count?

10. Which leucocyte is most active in response to tissue destruction?

11. Which type if increased indicates a chronic infection such as tuberculosis?

12. Which type is associated with allergic conditions?

13. Explain antibodies and antigens.

14. Explain the relationship between antibodies and lymphocytes.

15. What is the life span of a leucocyte?

## THROMBOCYTES

If a hemocytoblast in bone marrow does not become an erythrocyte or granular leucocyte, it may develop into still another kind of cell, called a megakaryoblast (see Figure 27–1), that is, transformed into a *megakaryocyte*. Megakaryocytes are large cells whose cytoplasm breaks up into fragments. Each fragment becomes enclosed by a piece of the cell membrane and is called a *thrombocyte* or *platelet*. Platelets are disc-shaped cell fragments without a nucleus. They average from 2 to 4 mm. in diameter. Between 250,000 and 400,000 platelets appear in each cubic millimeter of blood.

The function of the platelets is to prevent fluid loss by initiating a chain of reactions that results in blood clotting. Like the leucocytes, platelets have a short life span, probably only about one week. This short life span is due to the fact that platelets are "used up" in clotting, and that their contents are too simple to carry on much metabolic activity.

## PART II   PLASMA

When the formed elements are removed from blood, a straw-colored liquid called *plasma* is left. About 7 to 9 percent of the solutes are proteins. Some of these proteins are also found elsewhere in the body, but when they occur dissolved in blood, they are called *plasma proteins*. Albumins, which constitute the majority of plasma proteins, are responsible for the viscosity or thickness of blood. Along with the electrolytes, albumins also regulate blood volume by preventing all the water in the blood from diffusing into the interstitial fluid. Recall that water moves by osmosis from an area of low solute (high water) concentration to an area of high solute (low water) concentration. Globulins, which are antibody proteins released by plasma cells, form a small component of the plasma proteins. Gamma globulin is especially well known because it is able to form an antigen-antibody complex with the proteins of the hepatitis and measles viruses and tetanus bacterium. Fibrinogen, a third plasma protein, takes part in the blood-clotting mechanism, along with the platelets.

### CONCLUSIONS

1. What is another name for thrombocyte?

2. How many thrombocytes are found in each cubic millimeter of blood?

3. What is their function?

4. What is the life span of a thrombocyte. Explain your answer.

5. When the formed elements of the blood are removed, what remains?

6. What constitutes most of the plasma proteins?

7. What are the functions of plasma proteins?

8. Which globulin in the plasma is well known and why?

9. What is the significance of the plasma protein, fibrinogen?

**STUDENT ACTIVITY**

Using colored crayons or pencil crayons, draw and label all of the different blood cells. Be very accurate in drawing and coloring the granules and the nuclear shapes, since both of these are involved in the proper recognition and identification of the various types of cells.

# MODULE 28
# BLOOD CLOTTING, GROUPING AND DISORDERS

## OBJECTIVES

1. To describe the stages involved in blood clotting

2. To identify the factors that promote and inhibit blood clotting

3. To contrast a thrombus and an embolus

4. To define clotting and bleeding times, and to perform these tests

5. To define the ABO and Rh blood-grouping classifications

6. To define the antigen-antibody reaction as the basis for ABO blood grouping

7. To define the antigen-antibody reaction of the Rh blood-grouping system

8. To define erythroblastosis fetalis as a harmful antigen-antibody reaction

9. To contrast the causes of hemorrhagic, hemolytic, aplastic, and sickle-cell anemia

10. To compare the clinical symptoms of polycythemia and leukemia

11. To identify the clinical symptoms of infectious mononucleosis

12. To define medical terminology associated with blood

This module considers blood clotting, blood grouping, and their disorders. Under normal circumstances, blood maintains its liquid state as long as it remains in the vessels. If, however, it is drawn from the body, it first becomes very thick and then forms a soft jelly. The gel eventually separates from the liquid component. The straw-colored liquid component, called *serum*, is simply plasma minus its clotting proteins. The gel is called a *clot* and consists of a network of insoluble fibers in which the cellular components of blood are trapped.

The surfaces of erythrocytes contain genetically determined antigens called *agglutinogens*. These proteins are responsible for the 2 major blood-group classifications: the ABO group and the Rh system.

## PART I  BLOOD CLOTTING

The clotting process may be initiated when blood comes in contact with a rough surface. Drops of blood, like drops of any fluid, will stick to a rough surface. In the case of blood, the roughened area may be the cut end of a vessel or a deposit of cholesterol-type lipids. It is assumed that when the platelets stick to the surface, their membranes rupture, and chemicals contained within the platelets are released. In the presence of $Ca^{++}$ ions, the platelet chemicals are transformed through a series of reactions to a substance called *thromboplastin*. During the second step in the formation of the clot, thromboplastin and $Ca^{++}$ serve as catalysts to convert a plasma protein called *prothrombin* into *thrombin*, an enzyme. In the final step, thrombin catalyzes the conversion of another plasma protein, called *fibrinogen*, into *fibrin*. Fibrin is the insoluble network of fine threads in which the cellular elements of blood are trapped. This is the clot. In summary, the 3 phases of clot formation are

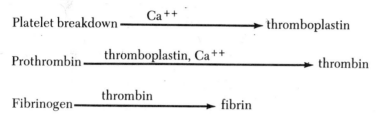

Clot formation is a vital mechanism that controls excessive loss of blood from the body. Unwanted clotting may be brought on by the formation of cholesterol-containing plaques on the walls of the blood vessels. These are the plaques of arteriosclerosis, or "hardening of the arteries." They supply a rough surface that is perfect for the adhesion of platelets and are often the sites of clotting. Clotting may also be encouraged by applying a thrombin or fibrin spray, a rough surface such as gauze, or heat.

In general, any chemical substance that prevents clotting is an *anticoagulant*. Examples of anticoagulants are heparin, dicumarol, and the citrates and oxalates. Heparin is a quick-acting anticoagulant that is extracted from donated human blood. The pharmaceutical preparation dicumarol may be given to patients who are thrombosis prone. Dicumarol is isolated from sweet clover and acts as an antagonist to vitamin K. Vitamin K is not involved in the actual clot formation, but it is required for the synthesis of prothrombin, which occurs in the liver. Dicumarol is slower acting than heparin, and it is used primarily as a preventative. The citrates and oxalates are used by laboratories and blood banks to prevent blood samples from clotting. These substances react with calcium to form insoluble compounds. In this way, the blood calcium is tied up and is no longer free to catalyze the conversion of prothrombin to thrombin.

Clotting within an unbroken blood vessel is referred to as *thrombosis*. The clot itself is called a *thrombus*. A thrombus may dissolve, or it may remain intact and interfere with circulation. In the latter case, it may cause damage to tissues by cutting off the oxygen supply. Equally serious is the possibility that the thrombus will become dislodged and be carried with the blood to a smaller vessel. In a smaller vessel, the clot will get stuck and may block the circulation to a vital organ. A blood clot, bubble of air, or piece of debris that is transported by the blood stream is called an *embolus*. When an embolus becomes lodged in a vessel and cuts off circulation, the condition is called an *embolism*.

### CLOTTING TESTS

The time required for blood to coagulate, usually from 5 to 15 minutes, is known as *clotting time*. This time is used as an index of a person's blood-clotting properties. One method of determining clotting time involves taking a sample of blood from a vein and placing one milliliter of the blood into each of 3 Pyrex tubes. The tubes are then submerged in a water bath at 37° centigrade (98.6° Fahrenheit). Every 30 seconds they are examined for the formation of a clot. The clotting process is initiated when the platelets break up upon coming into contact with the glass tubing. When the clot adheres to the walls of the tube, the end point is reached, and the time is recorded. Blood taken from individuals with hemophilia clots very slowly or not at all.

## CLOTTING TIME (SLIDE METHOD)

1. Place a few drops of blood on a glass slide (start stopwatch when bleeding begins)
2. Pass an applicator stick or needle through the blood until a fibrin strand appears on the needle.
3. Stop the watch when the strand appears.

## BLEEDING TIME

The time required for the cessation of bleeding from a small skin puncture is called bleeding time. This procedure is usually accomplished by puncturing the ear lobe. As the droplets of blood escape, they are dried by gently touching the wound with filter paper. When the paper is no longer stained, the bleeding has stopped. Normally, bleeding time varies from one to 4 minutes. Unlike coagulation time, which involves only the breakdown of platelets, and fibrin formation, bleeding time also involves constriction of injured blood vessels, as well as all 3 steps of clot formation.

## COAGULATION TIME

Blood normally clots in between 2 to 6 minutes. During this period of time, *fibrin*, the essential substance of a blood clot, is produced. The simplest way to determine one's clotting time is to fill a capillary tube with blood and break it at intervals to see how long it takes for fibrin to form.

1. Puncture the finger to expose a free flow of blood and *record the time immediately*. Place one end of the capillary tube into the drop of blood, holding the open end below the drop of blood so that the force of gravity aids the capillary action.
2. At approximately one-minute intervals, break off small portions of the tubing by scratching the glass capillary with a file first. *Note: Separate the broken ends slowly and gently* when looking for coagulation. Coagulation has occurred when threads of fibrin span the gap between the broken ends. Record the time as the interval from the blood's first appearance on the finger to the formation of fibrin.

## PROTHROMBIN TIME (TILT TUBE METHOD)

The prothrombin time is a test of the tissue, or extrinsic clotting system. It measures the clotting time of plasma in an optimal concentration of tissue extract.

The normal prothrombin time is 11 to 14 seconds. Prothrombin, which is synthesized in the liver, is one of the essential clotting factors in blood. Deficiencies in prothrombin levels (prolonged time) can usually be attributed to either liver disease or drug therapy. In cases of cardiac distress, anticoagulant drugs are given in order to retard excess clotting within the vascular system. The level of these drugs in the body is usually monitored daily by performing a prothrombin time.

1. Warm some simplastin (a commercially prepared tissue thromboplastin) to 37° centigrade by placing 0.2 ml. in a test tube and placing it in a rack in water bath.
2. Incubate the plasma for 3 minutes at 37° centigrade.
3. Pipette 0.1 ml. of plasma and blow into the simplastin. Start the watch immediately.
4. Tilting the tube gently, watch for the fibrin clot. (At repeated intervals place tube back in the water bath while tilting.)
5. Stop the watch when the fibrin clot appears.

## CONCLUSIONS

1. In the blood-clotting process, which plasma proteins are affected?

2. What are they converted into?

266

3. What are the 3 steps in clot formation?

4. How can clotting be encouraged? List 4 different ways.

5. Give 3 examples of anticoagulants.

6. Distinguish between a thrombus and an embolus.

7. Distinguish between clotting time and bleeding time.

8. What does a prolonged prothrombin time usually indicate?

## PART II  BLOOD GROUPING OR TYPING
### ABO GROUP

The ABO blood grouping is based upon 2 agglutinogens, which are symbolized as *A* and *B* (Figure 28-1). Individuals whose erythrocytes manufacture only agglutinogen A are said to have blood type A. Those who manufacture only agglutinogen B are type B. Individuals who manufacture both A and B are typed AB. Others, who manufacture neither, are called type O.

| AGGLUTINOGEN OR ERYTHROCYTE | |
|---|---|
| MEMBRANE | BLOOD TYPE |
| A | A |
| B | B |
| AB | AB |
| Neither A nor B | O |

The percentages of individuals possessing these 4 blood types are not equally distributed. The incidence of the various types in the Caucasian population in the United States is as follows: type A (41 percent), type B (10 percent), type AB (4 percent), and type O (45 percent).

The blood plasma of many people contains genetically determined antibodies referred to as *agglutinins*. These are antibody a (anti-A), which attacks agglutinogen A, and antibody b (anti-B) which attacks B. The antibodies formed by each of the 4 blood types are shown in Figure 28-1.

Note that an individual does not have antibodies that attack the antigens of his own erythrocytes.

**267**

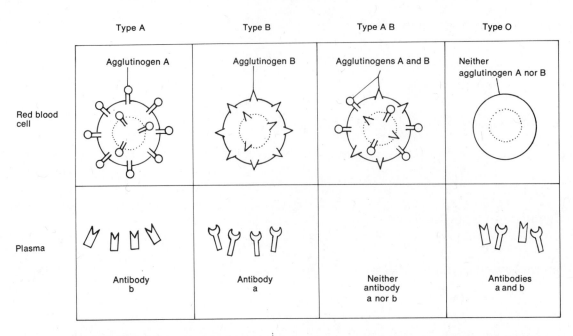

| Type A | Type B | Type A B | Type O |
|---|---|---|---|
| Agglutinogen A | Agglutinogen B | Agglutinogens A and B | Neither agglutinogen A nor B |

Red blood cell

Plasma

| Antibody b | Antibody a | Neither antibody a nor b | Antibodies a and b |

**Figure 28-1.** Agglutinogens (antigens) and agglutinins (antibodies) involved in the ABO blood grouping system.

For instance, a person with blood type A does not have antibody a. But every person has an antibody against any agglutinogen that he himself does not synthesize. For example, suppose type A blood is accidentally given to a person who does not have A agglutinogens. The individual's body recognizes that the A protein is foreign and therefore treats it as an antigen. Antibody a's rush to the foreign erythrocytes, attack them, and cause them to *agglutinate* (clump). Hence the names agglutinogen and agglutinins. This is another example of an antigen-antibody response. Any invading agent that can produce an immune response is called an antigen. Antigens usually stimulate the formation of *antibodies*.

When blood is given to a patient care must be taken to ensure that the individual's antibodies will not attack the donated erythrocytes and cause clumping. The destruction of the donated cells will not only undo the work of the transfusion, but the clumps can block vessels and cause serious problems that may lead to death.

Referring again to the agglutinogen-agglutinin-chart in Figure 28-1, note that a person can receive blood from others in his blood group. Type AB blood obviously does not have a or b antibodies, and neither A nor B antigens are foreign to this AB person. This means that AB individuals can receive all 4 types of blood without any danger of clumping. Thus, AB blood is called the *universal recipient*. Type O blood is called the *universal donor* because it does not have agglutinogens to act as antigens in another person's body.

## THE Rh SYSTEM

When blood is transfused, the technician must make sure that the donor and recipient blood types are safely matched—not only for ABO group type, but also for Rh type.

The *Rh system* is so named because it was first worked out in the blood of the Rhesus monkey. Like the ABO grouping, the Rh system is based upon agglutinogens that lie on the surfaces of erythrocytes. Individuals whose erythrocytes contain the Rh agglutinogens are designated as $Rh^+$. Those who lack Rh agglutinogens are designated as $Rh^-$. It is estimated that 90 percent of all citizens of the United States are $Rh^+$, whereas 10 percent are $Rh^-$.

Under normal circumstances, human plasma does not contain anti-Rh antibodies. However, if an $Rh^-$ person receives $Rh^+$ blood, his body starts to make anti-Rh antibodies that will remain in his blood. If a second transfusion of $Rh^+$ blood is given later, the previously formed anti-Rh antibodies will react against the donated blood, and a severe reaction may occur. One of the most common problems with Rh incompatibility arises from pregnancy. During delivery, some of the fetus' blood is apt to leak from the afterbirth into the mother's bloodstream. If the fetus is Rh$^+$ and the woman Rh$^-$, the mother will make anti-Rh antibodies. If the woman becomes pregnant again, her anti-Rh antibodies will make their way into the bloodstream of the baby. If the baby is Rh$^-$, no problem will occur since he does not have the Rh antigen. If he is Rh$^+$, an antigen-antibody response called *hemolysis* may occur in his blood. Hemolysis means a breakage of erythrocytes resulting in the liberation of hemoglobin. The hemolysis brought on by fetal-maternal incompatibility is called *erythroblastosis fetalis.* When a baby is born with erythroblastosis, all his blood is slowly drained and replaced with antibody-free blood. Moreover, modern technology has made it possible to transfuse blood into the unborn child if erythroblastosis is suspected. More important, though, is the fact that erythroblastosis can be prevented with RhoGAM Rh$_0$ (D) Immune Globulin (Human), a commercially available drug that is administered to Rh$^-$ mothers right after each delivery. RhoGAM prevents the mother's blood from forming antibodies against the fetal Rh antigens that are released during delivery. Thus, the fetus of the next pregnancy is protected. In the case of an Rh$^+$ mother and an Rh$^-$ child, there are no complications since the fetus cannot make antibodies.

## ABO AND Rh BLOOD GROUPING

The purpose of this exercise is to accurately type blood samples according to the ABO and Rh blood group systems.

The procedure for ABO grouping is as follows:
1. Divide a glass slide in half with a marker and label the left side "A" and the right side "B."
2. On the left side, place one drop of anti-A serum, and on the right side place one drop of anti-B serum.
3. Next to the drops of anti-sera place one drop of blood obtained by finger puncture, being careful not to mix samples on the left and right sides.
4. With an applicator stick or a toothpick, mix the blood on the left side with the anti-A serum, and proceed to mix the blood on the right with the anti-B, using a separate toothpick or the opposite side of the original one.
5. Gently tilt the slide back and forth and observe for one minute.
6. Record results—a " + " for positive clumping (agglutination) and " − " for negative clumping.

The procedure for Rh grouping is as follows:
1. Place one drop of anti-D serum on a slide.
2. Add one drop of blood and mix with a toothpick.
3. Gently tilt the slide and observe for one minute.
4. Record results using " + " for positive clumping and " − " for negative clumping.

The following table interprets the slide agglutination tests according to positive and negative results:

Table 28–1. BLOOD-GROUPING INTERPRETATION.

| REACTION WITH | | | BLOOD GROUP AND TYPE |
|---|---|---|---|
| ANTI-A | ANTI-B | ANTI-D | |
| + | − | + | A Positive |
| + | − | − | A Negative |
| − | + | + | B Positive |
| − | + | − | B Negative |
| + | + | + | AB Positive |
| + | + | − | AB Negative |
| − | − | + | O Positive |
| − | − | − | O Negative |

Table 28–2. INCIDENCE OF HUMAN BLOOD GROUPS IN THE U.S.

| BLOOD GROUPS (PERCENTAGES) | | | | |
|---|---|---|---|---|
| | O | A | B | AB |
| Caucasians | 45 | 41 | 10 | 4 |
| Blacks | 48 | 27 | 21 | 4 |
| Japanese | 31 | 38 | 22 | 9 |
| Chinese | 36 | 28 | 23 | 13 |
| American Indians | 23 | 76 | 0 | 1 |
| Hawaiians | 37 | 61 | 2 | 0.5 |

**CONCLUSIONS**

1. How many blood groupings or types are there? What are they?

2. Where is the agglutinogen located?

3. What are the percentages of the various blood types of the Caucasian population in the United States?

4. Where are the agglutinins located?

5. What could cause agglutination or clumping in a person's body?

6. What is the danger of agglutination?

7. Explain universal recipient and universal donor.

8. Where did the name Rh come from?

9. What is the ratio of Rh positive to Rh negative people in the United States?

10. What is erythroblastosis fetalis?

11. What is hemolysis?




12. How was erythroblastosis fetalis treated years ago?




13. How is it treated today?




Record the blood-grouping results of the class, and see how the percentages compare with the normal values in Table 28-2.

## PART III    BLOOD DISORDERS

The various blood disorders that can arise affect different portions of the blood. With anemia, for example, the patient may have an abnormally low number of red blood cells, whereas with mononucleosis white blood cells are affected. There are numerous blood disorders, and all may have wide-ranging effects.

### ANEMIA

The term *anemia* means that the number of functional red blood cells or their hemoglobin content is below normal. Consequently, the erythrocytes are unable to transport enough oxygen from the lungs to the cells. Anemia has many causes. The most common are a lack of iron, lack of certain amino acids, or the lack of vitamin $B_{12}$.

An excessive loss of erythrocytes through bleeding is called *hemorrhagic anemia*. Common causes are large wounds, stomach ulcers, and excessive menstrual bleeding. The term *hemolytic anemia* comes from the word *hemolysis*, the rupturing of erythrocyte cell membranes. The cell is destroyed, and its hemoglobin pours out into the plasma. Agents that may cause hemolytic anemia are parasites, toxins, and antibodies from incompatible blood, such as those which develop with an $Rh^-$ mother and $Rh^+$ fetus in erythroblastosis fetalis.

Destruction or inhibition of the red bone marrow results in *aplastic anemia*. Typically, the marrow is replaced by fatty tissue, fibrous tissue, or tumor cells. Toxins and certain medications are causes. Many of the medications inhibit the enzymes involved in hemopoiesis.

The erythrocytes of a person with *sickle cell anemia* manufacture an abnormal kind of hemoglobin. When the erythrocyte gives up its oxygen to the interstitial fluid, it tends to lose its integrity in places of low oxygen tension and forms long, stiff rodlike structures that bend the erythrocyte into a sickle shape (Figure 28-2), which gives the disorder its name. The sickle cells rupture easily. And even though erythropoiesis is stimulated by the loss of the cells, it cannot keep pace with the hemolysis. The individual consequently suffers from a hemolytic anemia that reduces the amount of oxygen which can be supplied to his tissues. Prolonged oxygen reduction may eventually cause extensive tissue damage. Furthermore, because of the shape of the sickle cells they tend to get stuck in blood vessels—a situation that can cut off blood supply to an organ altogether.

**271**

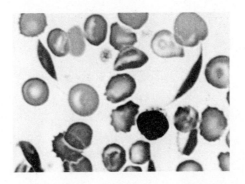

**Figure 28-2.** Microscopic appearance of erythrocytes in sickle cell anemia. (*Courtesy of Carolina Biological Supply Company.*)

Sickle cell anemia is inherited. The gene that is responsible for the disorder also seems to give the individual immunity against malaria. This theory is corroborated by the fact that sickle cell genes are found primarily among populations, or descendants of populations, which live in the malaria belt around the world. This includes parts of Mediterranean Europe and subtropical Africa and Asia. A person with only one of the sickling genes is said to have sickle cell trait. He has a high resistance to malaria—a factor that may have tremendous survival value for him—but he does not develop the anemia. Only people who inherit a sickling gene from both parents experience sickle cell anemia.

## POLYCYTHEMIA

The term *polycythemia* refers to a condition characterized by an abnormal increase in the number of red blood cells. Increases of 2 to 3 million cells per cubic millimeter are considered to be polycythemic. The disorder is harmful because the thickness of the blood (viscosity) is greatly increased due to the extra red blood cells, and viscosity contributes to a tendency to thrombosis and hemorrhage. It also causes a rise in blood pressure. The tendency to thrombosis results from too many red blood cells piling up as they try to enter smaller vessels. The tendency to hemorrhage is caused by hyperemia (unusually large amount of blood in an organ part) in all organs.

There are 2 basic types of polycythemia; *primary* and *secondary*. The primary type is characterized by an overactivity of the red bone marrow and by an enlarged liver and spleen. Its cause is unknown. The secondary type is the result of a lack of oxygen in the arteries of people suffering from chronic cardiac or pulmonary disease. Other causes include kidney tumors or cysts, which would increase secretion of the hormone aldosterone resulting in increasing blood volume; liver cancer, which could interfere with the liver's normal function of dispersing of millions of old red blood cells daily; and very high altitudes.

## LEUKEMIA

This disorder is also called "cancer of the blood." *Leukemia* is an uncontrolled, greatly accelerated production of white cells. Many of the cells fail to reach maturity. As with most cancers, the symptoms and the cause of death do not result so much from the cancer cells themselves as from the interference of the cancer cells with normal body processes. The accumulation of cells leads to abnormalities in organ functions. For example, the anemia and bleeding problems commonly seen in leukemia result from the "crowding out" of normal bone marrow cells. This interferes with the normal production of red blood cells and platelets. The most common cause of death from leukemia is internal hemorrhaging, especially cerebral hemorrhage that destroys the vital centers in the brain. The second most frequent cause of death is uncontrolled infection. This happens because there is a lack of mature or normal white blood cells available to fight infection.

Therapy may temporarily stop the pathologic process. The abnormal accumulation of leucocytes may be reduced or even eliminated by using x-ray and antileukemic drugs. Partial or complete remissions may be induced, with some lasting as long as 15 years.

## INFECTIOUS MONONUCLEOSIS

This is a contagious disease with an unknown cause that occurs mainly in children and young adults. The trademark of the disease is an elevated white count with an abnormally high percentage of lymphocytes and mononucleocytes—hence, the name mononucleosis. As mentioned earlier in the chapter, an increase in the number of monocytes usually indicates a chronic infection. The various signs and symptoms include slight fever, sore throat, brilliant red throat and soft palate, stiff neck, cough, and malaise. The spleen may enlarge; and secondary complications involving the liver, heart, kidneys, and nervous system may develop.

There is no cure for mononucleosis, and treatment consists of watching for and treating any complications. Usually the disease runs its course in a few weeks, and the individual suffers no permanent ill effects.

## CONCLUSIONS

1. Define anemia.

2. Compare hemorrhagic anemia with hemolytic anemia.

3. Give some possible causes for each of these 2 disorders.

4. What is aplastic anemia and how is it caused?

5. Describe sickle cell anemia.

6. Compare polycythemia with leukemia.

7. What are some causes of polycythemia?

8. Explain infectious mononucleosis, with its signs and symptoms.

## MEDICAL TERMINOLOGY

Define each of the following terms:

a. corpuscles

b. embolus

c. hemolysis

d. hemorrhage

e. plasmolysis

f. septicemia

g. thrombus

h. venesection

## STUDENT ACTIVITY

Divide the disorders of the blood into 3 main categories: those that affect the red blood cells; those that affect the white blood cells; and those that involve abnormal blood clotting. List the disease, its cause, its main symptoms, and its treatment, for each of the main groups.

# MODULE 29
# STRUCTURE OF THE HEART
# AND BLOOD VESSELS

**OBJECTIVES**

1. To describe the location of the heart in the mediastinum

2. To distinguish between the structure and location of fibrous and serous pericardium

3. To contrast the structure of the epicardium, myocardium, and endocardium

4. To identify the blood vessels, chambers, and valves of the heart

5. To contrast the structure and function of veins, capillaries, and arteries

In this module we will see how the circulatory system can be likened to an intricate transportation network within the body. The blood vascular system consists of the blood, heart, and blood vessels. The overall function of the circulatory system is to keep things moving from one part of the body to another. This movement depends on the circulation of blood within a maze of vessels. The architecture of these vessels is analogous to an interstate highway system. Just as a highway system is designed to route vehicles from one destination to another, the vessels of the circulatory system are designed to transport blood to and from various parts of the body.

## PART I    THE HEART

Let us first examine the center of your body's highway system, the heart. This is the pump that maintains circulation in vessels. The heart is a hollow, muscular organ that pumps blood through the vessels. It is situated between the lungs in the mediastinum, and about two-thirds of its mass lies to the left of the midline of the body (Figure 29-1a). The heart is shaped like a blunt cone about the size of a closed fist. A roentgenogram of a normal heart is pictured in Figure 29-1b. Compare it with the adjacent diagram. Its pointed end, called the apex, projects downward, forward, and to the left, and lies above the central depression of the diaphragm. Its broad end, or base, projects upward, backward, and to the right, and lies just below the second rib. The major parts of the heart to be considered here are the pericardium, the walls and chambers, and the valves.

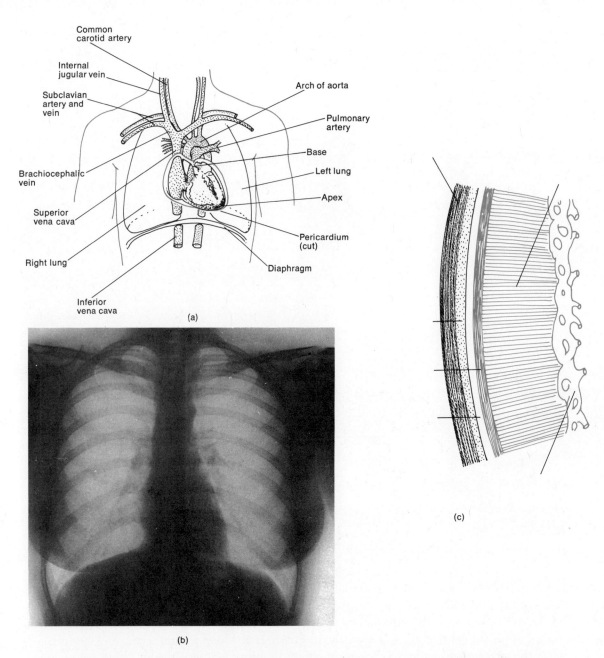

**Figure 29–1.** The heart. (a) Position of the heart in the thoracic cavity. (b) Roentgenogram of a normal heart. (c) Structure of the pericardium and heart wall. *(X-ray courtesy of William L. Leonard, Bergen Community College.)*

## PART II   THE PERICARDIUM

The heart is enclosed in a loose-fitting serous membrane called the *pericardium* (Figure 29–1a, c). The pericardium consists of 2 principal layers: an external fibrous layer that binds the heart in place, and an internal serous layer that secretes a lubricating fluid. The *fibrous layer* is composed of tough fibrous tissue. Its upper surface attaches to the large blood vessels that emerge from the heart. Its lower end attaches the heart to the diaphragm, and its anterior surface binds the heart to the sternum. The internal *serous layer* contains 2 subdivisions: *the parietal layer,* and the *visceral layer,* which adheres to the outside of the heart muscle. The visceral layer is also the outermost layer of the heart. Between the parietal and visceral layers of the serous pericardium is a small space, called the

**276**

*pericardial cavity*. This cavity contains *pericardial fluid*. As the heart beats, its surface continually moves against the outer layers of the pericardium. The fluid prevents friction between the membranes. An inflammation of the pericardium is called *pericarditis*.

## PART III   WALLS AND CHAMBERS

The wall of the heart (Figure 29-1c) is divided into 3 portions: (1) the epicardium, or external layer, (2) the myocardium, or middle layer, and (3) the endocardium, or inner layer. The *epicardium* is the same as the visceral layer of the pericardium, which has just been described. The *myocardium*, which is cardiac muscle tissue, comprises the bulk of the heart. As you will recall, cardiac muscle fibers are involuntary, striated, and branched, and the tissue is arranged in interlacing bundles of fibers. The myocardium is responsible for the actual contraction of the heart. Inflammation of the myocardium is referred to as *myocarditis*. The *endocardium* is a thin layer of endothelium pierced by tiny blood vessels and some bundles of smooth muscle. It lines the inside of the myocardium and covers the valves of the heart and the tendons that hold the valves open, and keeps them from opening into the atria. It is continuous with the endothelium of the large blood vessels of the heart. Inflammation of the endocardium is called *endocarditis*.

Completely label Figure 29-1c.

The interior of the heart is divided into 4 spaces or chambers, which receive the circulating blood (Figure 29-2). The 2 upper chambers are called the right and left *atria*. They are separated by a partition called the *interatrial septum*. The 2 lower chambers, the right and left *ventricles*, are separated by an *interventricular septum*. The right atrium drains blood from all parts of the body except the lungs. It receives the blood through 3 veins. One of these veins is the *superior vena cava*, which brings blood from the upper portion of the body. Another of the veins is the *inferior vena cava*,

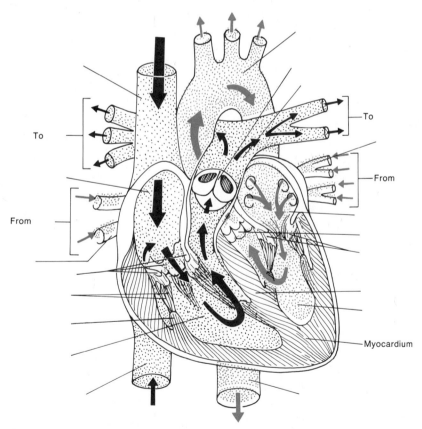

**Figure 29-2.** Internal anatomy of the heart. Shown here is a frontal section of the heart indicating the principal internal structures.

which brings blood from the lower portions of the body. The third vein is the *coronary sinus*, which drains blood from the vessels supplying the walls of the heart. The right atrium then squeezes the blood into the right ventricle, which pumps it into the *pulmonary artery*. The pulmonary artery then carries the blood to the lungs. In the lungs, the blood releases its carbon dioxide and takes on oxygen. It returns to the heart via 4 *pulmonary veins* that empty into the left atrium. The blood is then squeezed into the left ventricle and exits from the heart through the *aorta*. This large artery transports the blood to all parts except the lungs.

Completely label the missing parts in Figure 29-2.

## PART IV  VALVES

As each chamber of the heart constricts, it pushes a portion of blood into a ventricle or out of the heart through an artery. But as the walls of the chambers relax, some structure must prevent the blood from flowing back into the chamber. That structure is a *valve*.

*Atrioventricular valves* lie between the atria and their ventricles (Figure 29-3). The atrioventricular valve between the right atrium and right ventricle is called the *tricuspid valve* because it consists of 3 flaps, or cusps. These flaps are fibrous tissues that grow out of the walls of the heart and are covered with endocardium. The pointed ends of the cusps project into the ventricle. Cords called *chordae tendineae* connect the pointed ends to small *papillary muscles* (muscular columns) that are located on the inner surface of the ventricles. The chordae tendineae and their muscles keep the flaps pointing in the direction of the blood flow. As the atrium relaxes and the ventricle squeezes the blood out of the heart, any blood that is driven back toward the atrium is pushed between the flaps and the walls of the ventricle. This drives the cusps upward until their edges meet and close the opening. The atrioventricular valve between the left atrium and left ventricle is called the *bicuspid* or *mitral valve*. It has 2 cusps that work in the same way as the cusps of the tricuspid valve.

Each of the arteries that leave the heart has a valve that prevents blood from flowing back into

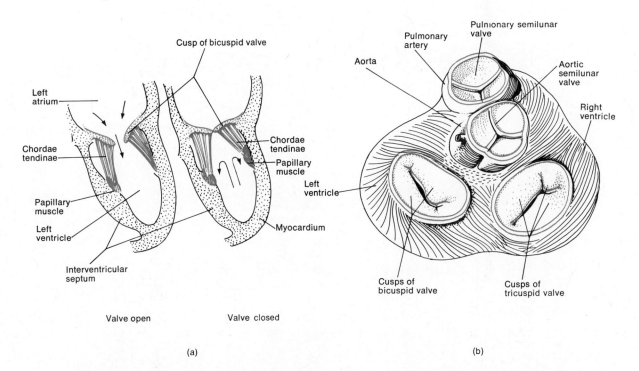

**Figure 29-3.** Valves of the heart. (a) Structure and function of the bicuspid valve. (b) Valves of the heart viewed from above. The atria have been removed to expose the tricuspid and bicuspid valves.

the heart. These are the *semilunar valves—semilunar* meaning half-moon or crescent-shaped. This refers to the shape of the cusps of the valves. The *pulmonary semilunar valve* lies in the opening where the pulmonary artery leaves the right ventricle. The *aortic semilunar valve* is situated at the opening between the left ventricle and the aorta (Figure 29-2). Both valves consist of 3 semilunar cusps. Each cusp is attached by its convex or inwardly curved margin to the wall of its artery. The free borders of the cusps curve outward and project into the opening inside the blood vessel (Figure 29-3b). Like the atrioventricular valves, the semilunar valves permit the blood to flow in only one direction. In this case, the flow is from the ventricles into the arteries.

## PART V   BLOOD VESSELS

The blood vessels form a network of tubes that carry blood away from the heart, transport it to the tissues of the body, and then return it to the heart. If you examine the general plan of these vessels in Figure 29-4, you will note that blood vessels are called either arteries, capillaries, or veins. *Arteries* are the vessels that carry blood from the heart to the tissues. Two large arteries leave the heart and divide into medium-sized vessels that head toward the various regions of the body. The medium-sized

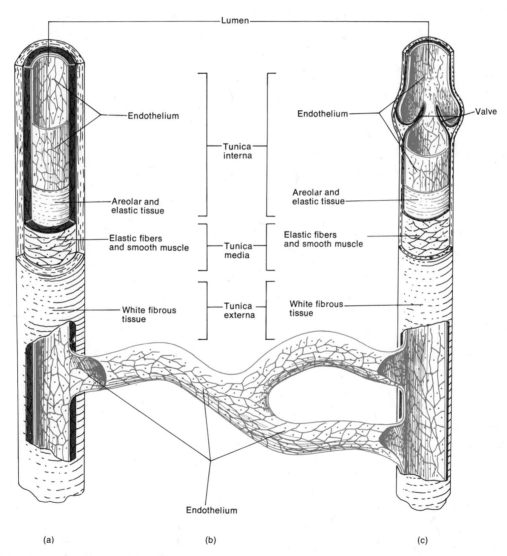

**Figure 29-4.** Structure of blood vessels. (a) Artery. (b) Capillary. (c) Vein. The relative size of the capillary is enlarged for emphasis.

arteries, in turn, divide into small vessels called *arterioles*. As the arterioles enter a tissue, they branch into countless numbers of microscopic vessels called *capillaries*. Through the walls of the capillaries, substances are exchanged between the blood and body tissues. Before leaving the tissue, groups of capillaries reunite to form small veins called *venules*. These, in turn, merge to form progressively larger tubes—the veins themselves. *Veins*, in other words, are blood vessels that convey blood from the tissues back to the heart.

## PART VI   ARTERIES

Arteries and veins are fairly similar in construction (Figure 29-4a). They both have walls constructed of 3 coats and a hollow inner core, called a *lumen*, through which the blood flows. Arteries, however, are considerably thicker and stronger than veins because the pressure in an artery is always greater than in a vein. The inner coat of an arterial wall is called the *tunica interna*. It is composed of a lining of endothelium (squamous epithelium) that is in contact with the blood. It also has an overlying layer of areolar connective tissue and an outer layer of elastic tissue. The middle coat, or *tunica media*, is usually the thickest layer. It consists of elastic fibers and smooth muscle. The outer coat, the *tunica externa*, is composed principally of white fibrous tissue. The fibrous tissue is tough and firm and prevents the artery from collapsing if it is cut.

As a result of the structure of the middle coat, especially, arteries have 2 very important properties: elasticity and contractility. When the ventricles of the heart contract and eject blood into the large arteries, the arteries expand to contain the extra blood volume. Then, as the ventricles relax, the elastic recoil of the arteries forces the blood onward. The contractility of an artery is a function of its smooth muscle. The smooth muscle is arranged in rings around the lumen, resembling somewhat the shape of a donut. As the muscle contracts, it squeezes the wall more tightly around the lumen and consequently narrows the area through which the blood flows. Such a decrease in the size of the lumen is called *vasoconstriction*. The nerves responsible for the action are termed *vasoconstrictor nerves*. Conversely, if all the muscle fibers relax, the size of the arterial lumen increases. This is called *vasodilation*, and the nerves mediating the response are called *vasodilator nerves*.

## PART VII   CAPILLARIES

Capillaries are microscopic vessels measuring about 0.01 millimeters in diameter. They connect arterioles with venules (Figure 29-4b). The function of the capillaries is to permit the exchange of nutrients and gases between the blood and interstitial fluid. The structure of the capillaries is admirably suited for this purpose. First, the capillary walls are composed of only a single layer of cells (endothelium). This means that a substance in the blood must diffuse through the plasma membranes of just one cell in order to reach the interstitial fluid. It should be noted again that this vital exchange of materials occurs only through capillary walls. The thick walls of arteries and veins present too great a barrier for diffusion to occur. Capillaries are also well suited to their function in that they branch to form an extensive *capillary network* throughout the tissue. The network increases the surface area through which diffusion can take place and thereby allows a rapid exchange of large quantities of materials.

## PART VIII   VEINS

Veins are composed of essentially the same 3 coats as arteries, but they have considerably less elastic tissue and smooth muscle (Figure 29-4c). However, veins do contain more white fibrous tissue. They are also elastic enough to adapt to variations in the volume and pressure of blood passing through them. If you cut a vein, you will notice that the blood leaves the vessel in an even flow rather than in the rapid spurts characteristic of arteries. This is because by the time the blood leaves the capillaries and moves into the veins, it has lost a great deal of its pressure. Most of the structural differences between arteries and veins reflect this pressure difference. For example, veins do not need to have walls that are as strong as the walls of their corresponding arteries. The low pressure in veins, however, has its disadvantages. For instance, when you stand, the pressure pushing blood up the veins in your legs is barely enough to balance the force of gravity pushing it back down. For this reason, many veins, especially those in the limbs, contain valves that prevent any backflow.

## CONCLUSIONS

1. Where exactly is the heart located?

2. The heart is enclosed in a loose-fitting serous membrane called the _____

3. What are the membrane's 2 principal layers?

4. What structures of the heart prevent friction between the membranes?

5. What are the 3 layers that compose the wall of the heart?

6. What are the 4 spaces or chambers of the heart?

7. How are the 2 lower chambers separated?

8. The right atrium receives blood through 3 veins. What are these called, and from what part of the body does each of them bring blood to the right atrium?

9. How does the blood get from the right atrium to the lungs?

10. How does the blood return to the heart from the lungs?

11. Into what chamber does the blood from the lungs empty?

12. What happens to the blood in the left atrium?

13. Name the valves on the right side of the heart, then the left side.

14. Explain the chordae tendinae and the papillary muscles.

15. What is their function?

16. What is the basic function of all the valves in the heart?

17. Name all the basic types of blood vessels in the body.

18. What are some main differences between arteries and veins?

19. What are some functions of capillaries?

20. What is a capillary network? What is its advantage?

**STUDENT ACTIVITY**

Draw a large model of the heart and carefully label the 4 chambers, the 4 valves in their proper places, and the major blood vessels as they enter and exit from the heart.

# MODULE 30
# CIRCULATORY ROUTES

**OBJECTIVES**

1. To identify the principal arteries and veins of systemic circulation

2. To describe the route of blood in coronary circulation

3. To describe the importance and route of blood involved in hepatic portal circulation

4. To identify the major blood vessels of pulmonary circulation

5. To contrast fetal and adult circulation

Now that you have studied the structure of the heart and blood vessels, you will investigate the course of blood through various regions of the body. **A circulatory route** is the flow of blood from the heart to a region of the body and back again to the heart. The circulatory routes that you will study in this module are: systemic, coronary, hepatic portal, pulmonary, and fetal.

## PART I   SYSTEMIC CIRCULATION

The flow of oxygenated blood from the left ventricle to all parts of the body (except the alveoli of the lungs) and the return of deoxygenated blood back to the right atrium is called **systemic circulation**. Its purpose is to carry oxygen and nutrients to body tissues and to remove carbon dioxide and wastes from them. All systemic arteries branch from the aorta; most veins of systemic circulation flow into either the superior or inferior vena cava.

### ARTERIES

On the next page is a listing of the divisions of the aorta and the major arterial branches from each division. Indicate next to each arterial branch the region of the body supplied. After you have completed this, refer to Figure 30-1 to determine the location of the arteries.

The *ascending aorta* gives rise to the coronary arteries. These will be studied later in the module. The *aortic arch* gives rise to 3 principal branches: the brachiocephalic (innominate) artery, the left common carotid artery, and the left subclavian artery (see Figure 30-1).

| DIVISION OF AORTA | ARTERIAL BRANCH | REGION SUPPLIED |
|---|---|---|
| Ascending aorta | Coronary | |
| Aortic arch | Brachiocephalic (innominate) → Right common carotid / Right subclavian<br>Left common carotid<br>Left subclavian | |
| Thoracic aorta | Intercostals<br>Superior phrenics<br>Bronchials<br>Esophageals<br>Inferior phrenics | |
| Abdominal aorta | Celiac → Hepatic / Left gastric / Splenic<br>Superior mesenteric<br>Suprarenals<br>Renals<br>Testiculars<br>Ovarians<br>Inferior mesenteric<br>Common iliacs → External iliacs / Internal iliacs | |

In Figure 30-2, label the following arteries: brachiocephalic, right subclavian, axillary, brachial, ulnar, radial, superficial volar arch, deep volar arch, and digital. Next to the name of each artery, indicate the region of the body supplied.

Before passing into the axilla, the right subclavian gives off a branch to the brain called the *right vertebral artery*. This artery unites with the left vertebral artery to form the basilar artery. Label these arteries in Figure 30-3 (also see Figure 30-4).

Branches of the left and right internal carotid arteries, together with the basilar artery, form an arterial circle at the base of the brain called the circle of Willis. From these vessels arise the arteries that supply the brain. Refer to Figure 30-4 and label the arteries of the circle of Willis.

The brachiocephalic gives off a branch called the right common carotid artery which ascends and divides at the upper level of the larynx into the right external carotid and right internal carotid arteries. Label these arteries in Figure 30-3. Next to each, indicate the parts of the body supplied.

The left common carotid, the second branch off the aortic arch, branches into the left external carotid and left internal carotid arteries. Refer to Figure 30-1 for their location.

The left subclavian, the third branch off the aortic arch, distributes blood to the left vertebral artery and the arteries of the left arm. Refer to Figure 30-1 for their location.

The *thoracic aorta* sends off numerous small arteries to the viscera (visceral branches) and skeletal muscles (parietal branches) of the chest (see Figure 30-5).

The *abdominal aorta* and its branches are shown in Figure 30-5. Next to the name of each part that you label, indicate the part of the body supplied.

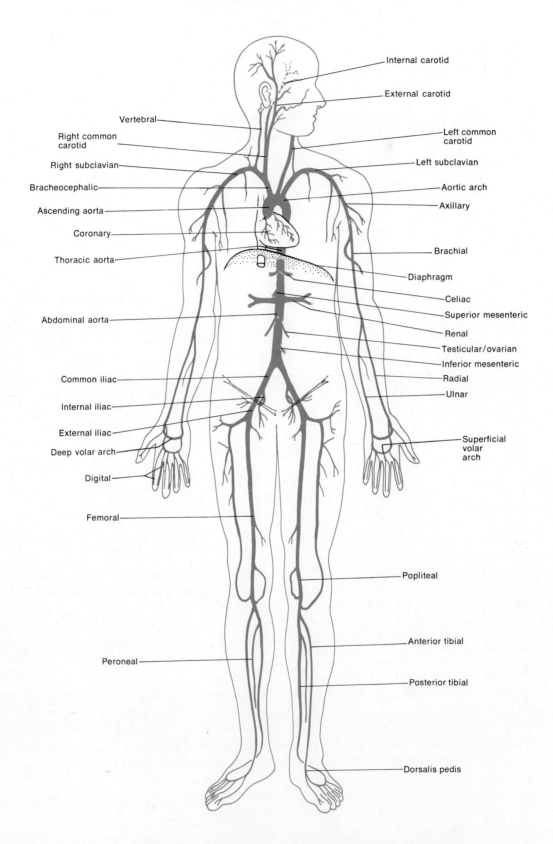

**Figure 30-1.** Anterior view of the aorta and its principal branches.

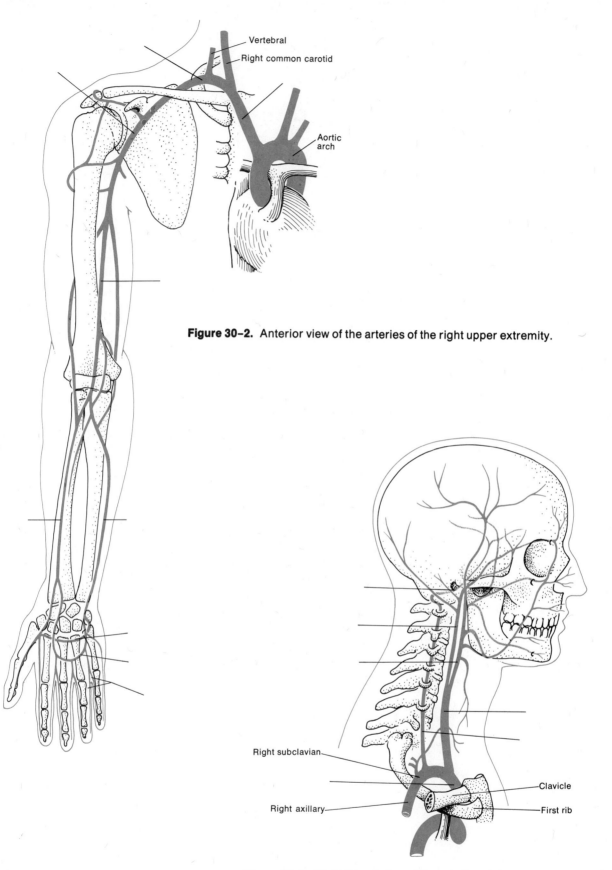

Vertebral

Right common carotid

Aortic arch

**Figure 30–2.** Anterior view of the arteries of the right upper extremity.

Right subclavian

Right axillary

Clavicle

First rib

**Figure 30–3.** Right lateral view of the arteries of the neck and head.

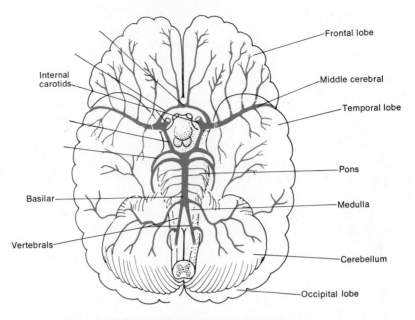

Internal carotids

Frontal lobe

Middle cerebral

Temporal lobe

Pons

Basilar

Medulla

Vertebrals

Cerebellum

Occipital lobe

**Figure 30–4.** Arteries of the base of the brain.

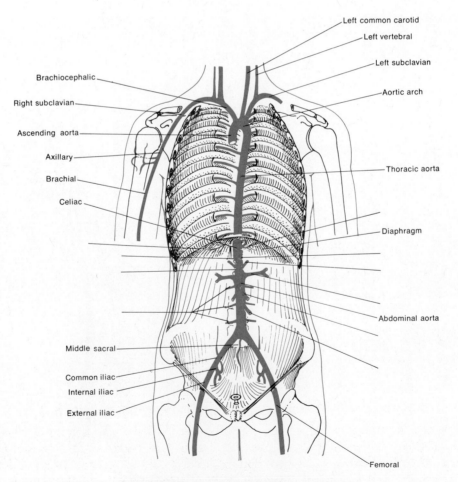

Left common carotid

Left vertebral

Left subclavian

Brachiocephalic

Aortic arch

Right subclavian

Ascending aorta

Axillary

Thoracic aorta

Brachial

Celiac

Diaphragm

Abdominal aorta

Middle sacral

Common iliac

Internal iliac

External iliac

Femoral

**Figure 30–5.** Thoracic and abdominal aorta.

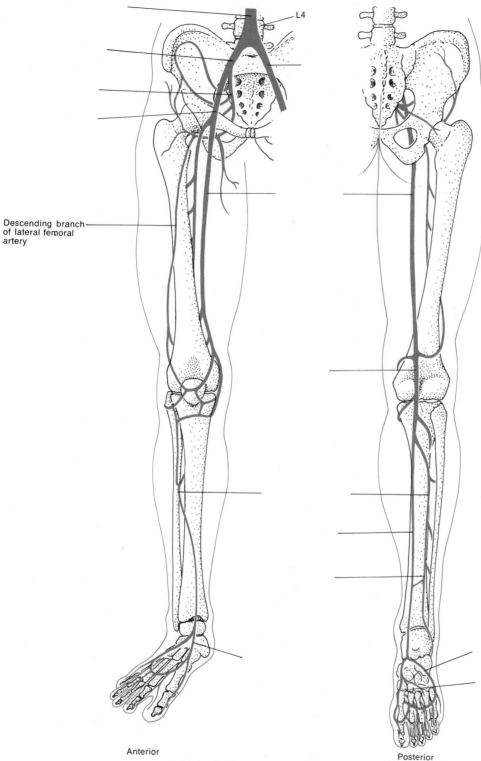

Descending branch
of lateral femoral
artery

L4

Anterior

Posterior

**Figure 30–6.** Arteries of the pelvis and lower extremities.

At about the fourth lumbar vertebra, the abdominal aorta divides into the right and left common iliac arteries. Each then divides into 2 branches, the internal iliac and external iliac arteries. Refer to Figure 30-6 and label the following arteries: abdominal aorta, common iliac, internal iliac, external iliac, femoral, popliteal, posterior tibial, peroneal, anterior tibial, dorsalis pedis, medial plantar, and lateral plantar. Next to each labeled artery, indicate the region of the body supplied.

## VEINS

Veins of systemic circulation are of 3 principal kinds: (1) deep veins, located deep within the body, that usually accompany arteries and have the same names as their corresponding arteries; (2) superficial veins, located just below the skin and visible through the skin; and (3) venous sinuses, thin-walled veins between the layers of the dura mater, one of the membranes that covers the brain.

The principal veins of the head and neck are the internal jugulars and external jugulars. The internal jugulars drain blood from the transverse (lateral) sinuses, superior sagittal sinus, inferior sagittal sinus, straight sinus, and sigmoid sinus. The internal jugulars join with the subclavian veins to form the brachiocephalic (innominate) veins. From here the blood flows into the superior vena cava. Label these veins in Figure 30-7 and indicate next to each label the regions of the body drained by the veins.

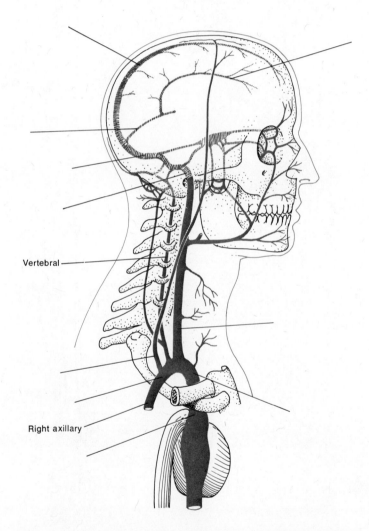

Vertebral

Right axillary

**Figure 30-7.** Right lateral view of the veins of the neck and head.

The deep veins of the upper extremities include the radial, ulnar, brachial, axillary, and subclavian. Label these veins in Figure 30-8. The superficial veins include the dorsal arch, cephalic, accessory cephalic, basilic, and median cubital. The axillary vein is a continuation of the basilic; at the first rib it becomes the subclavian. The subclavians unite with the internal jugulars to form the brachiocephalic veins. Label these veins in Figure 30-8 and indicate next to each label the regions of the body drained by the veins.

The principal veins of the thorax include the brachiocephalics and azygos. The 3 azygos veins are the azygos hemiazygos, and accessory hemiazygos. Label these veins and their subdivisions in Figure 30-9. Next to each vein indicate the region of the body that is drained.

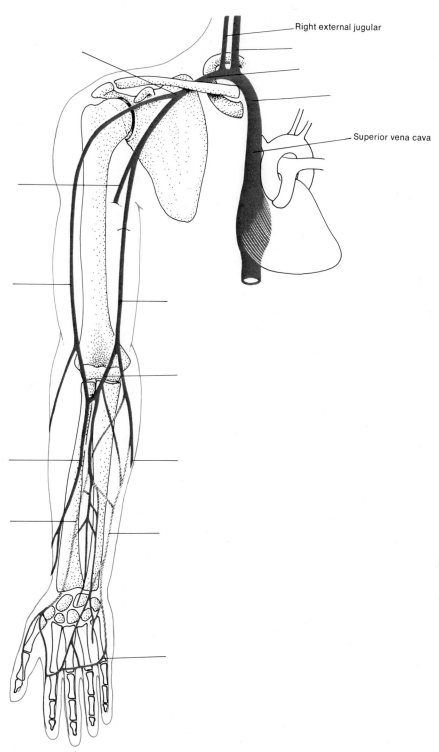

Right external jugular

Superior vena cava

**Figure 30-8.** Anterior view of the right upper extremity.

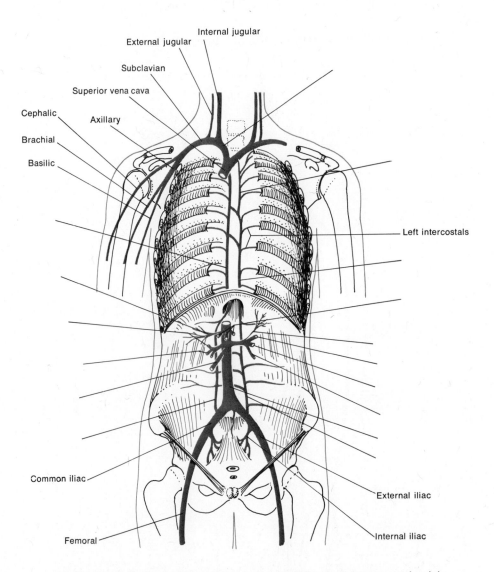

Internal jugular

External jugular

Subclavian

Superior vena cava

Cephalic

Axillary

Brachial

Basilic

Left intercostals

Common iliac

External iliac

Internal iliac

Femoral

**Figure 30-9.** Anterior view of the veins of the thorax, abdomen, and pelvis.

The veins of the abdomen and pelvis drain into the inferior vena cava. It is the largest vein in the body and is formed by the union of 2 common iliac veins that drain the legs and abdomen. Numerous small veins enter the inferior vena cava and, for the most part, they carry return flow from capillaries from branches of the abdominal aorta. Refer to Figure 30-9 and label the following veins: inferior vena cava, renal, testicular, ovarian, suprarenal, inferior phrenic, hepatic, and lumbar. Next to each vein indicate the region of the body that is drained.

Blood from the lower extremity is returned by deep and superficial veins. The principal superficial veins are the great saphenous and small saphenous. Both begin in the dorsal venous arch. The major deep veins are the posterior tibial, formed by the union of the medial and lateral plantar veins; peroneal; anterior tibial, a continuation of the dorsalis pedis; popliteal; and femoral. Label these veins in Figure 30-10 and next to each vein indicate the region of the body that is drained.

## CONCLUSIONS

1. What is the purpose of systemic circulation?

**292**

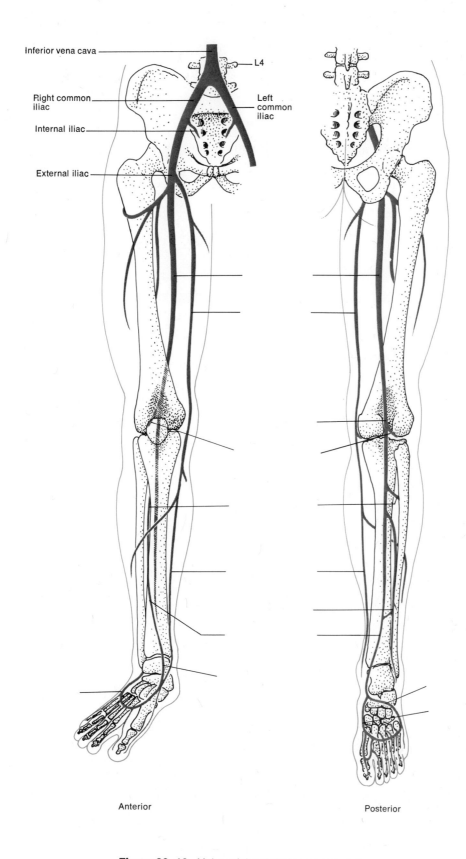

Inferior vena cava

Right common iliac

Internal iliac

External iliac

L4

Left common iliac

Anterior

Posterior

**Figure 30-10.** Veins of the right lower extremity.

2. What is anastomosis?


  Why is it important?

3. What is the function of the circle of Willis?

4. How are the subclavian veins related to the lymphatic system?

5. For each vessel listed, indicate the region supplied (if an artery) or the region drained (if a vein):

  a. coronary artery

  b. internal iliac veins

  c. lumbar veins

  d. renal artery

  e. superior phrenic artery

  f. external jugular vein

  g. left subclavian artery

  h. axillary vein

  i. brachiocephalic veins

  j. transverse sinuses

  k. hepatic artery

  l. inferior mesenteric artery

  m. suprarenal artery

  n. inferior phrenic artery

  o. great saphenous vein

  p. popliteal vein

  q. azygos vein

  r. internal iliac artery

  s. intercostal arteries

  t. cephalic vein


## PART II   CORONARY CIRCULATION

The wall of the heart, like any other tissue of the body, has its own blood supply. The flow of blood through the numerous vessels that pierce the myocardium is called the **coronary circulation**. It is a specialized part of systemic circulation.

The arteries supplying the heart with oxygenated blood include the left coronary, anterior descending, circumflex, right coronary, posterior descending, and marginal. Label these arteries in Figure 30-11a and next to each artery indicate the part of the heart supplied.

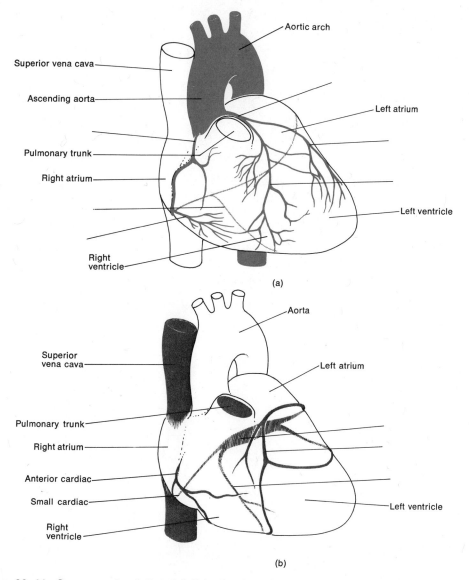

**Figure 30–11.** Coronary circulation. (a) Anterior view of arterial distribution. (b) Anterior view of venous drainage.

Deoxygenated blood is collected by a large vein, the coronary sinus, which empties into the right atrium. The principal veins that drain into the coronary sinus are the great cardiac and middle cardiac. Label these veins in Figure 30–11b and indicate the parts of the heart drained by each.

## CONCLUSIONS

1. Define ischemia.

2. What is angina pectoris?

3. How may it be caused?

4. How is angina pectoris treated?

5. Define myocardial infarction.



6. How may it be caused?



## PART III HEPATIC PORTAL CIRCULATION

Blood entering the liver is derived from 2 sources: (1) the hepatic artery delivers oxygenated blood from the systemic circulation, and (2) the portal vein delivers deoxygenated blood from the digestive organs. This flow of blood from the digestive organs through the liver before returning to the heart is referred to as **hepatic portal circulation**.

Label the vessels involved in hepatic portal circulation in Figure 30-12 and draw arrows to indicate the direction of flow.

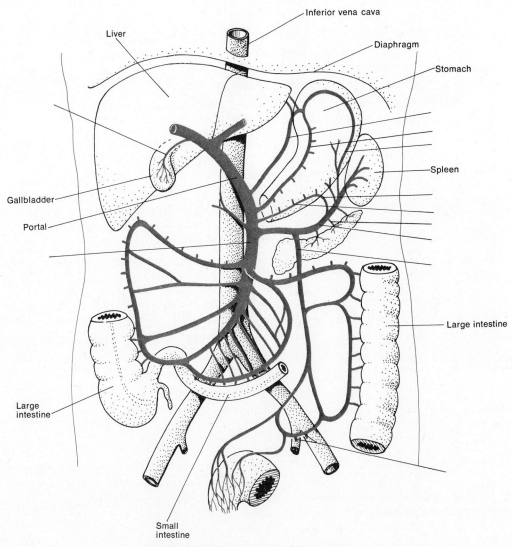

**Figure 30-12.** Hepatic portal circulation.

**CONCLUSIONS**

1.  What are the purposes of hepatic portal circulation?

2.  How does the blood from hepatic portal circulation reenter systemic circulation?

## PART IV    PULMONARY CIRCULATION

The flow of deoxygenated blood from the right ventricle to the alveoli of the lungs and the return of oxygenated blood from the alveoli of the lungs to the left atrium is called the **pulmonary circulation**. Label the blood vessels involved in pulmonary circulation in Figure 30–13.

**CONCLUSIONS**

1.  What is the purpose of pulmonary circulation?

2.  How is the pulmonary artery unlike any other artery of the body?

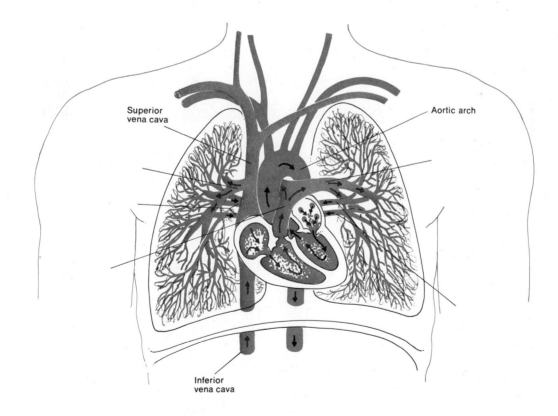

**Figure 30–13.** Pulmonary circulation.

**297**

3. How are the pulmonary veins unlike any other veins of the body?

4. How are pulmonary and systemic circulation related?

## PART V  FETAL CIRCULATION

In the fetus, the lungs, kidneys, and digestive organs are nonfunctional. Oxygen and nutrients are derived from maternal blood via diffusion through the placenta. Carbon dioxide and wastes are removed into maternal blood via diffusion through the placenta. This exchange of materials between fetal and maternal blood is called **fetal circulation**.

The scheme for fetal circulation is shown in Figure 30-14. Label the umbilical arteries, umbilical vein, ductus venosus, foramen ovale, and ductus arteriosus.

### CONCLUSIONS

1. What is the function of the umbilical arteries?

2. How is fetal blood oxygenated?

3. How is the fetal liver bypassed?

4. What is the function of the foramen ovale?

5. What is the ductus arteriosus?

6. What happens to the following structures after birth?

   a. umbilical arteries

   b. umbilical vein

   c. placenta

   d. ductus venosus

   e. foramen ovale

   f. ductus arteriosus

   g. umbilical cord

### STUDENT ACTIVITY

Trace a drop of blood from the aorta, through the common iliac artery, back to the heart, through the lungs, and end at the left ventricle. Be sure to name all the major arteries and veins through which the blood passes.

**298**

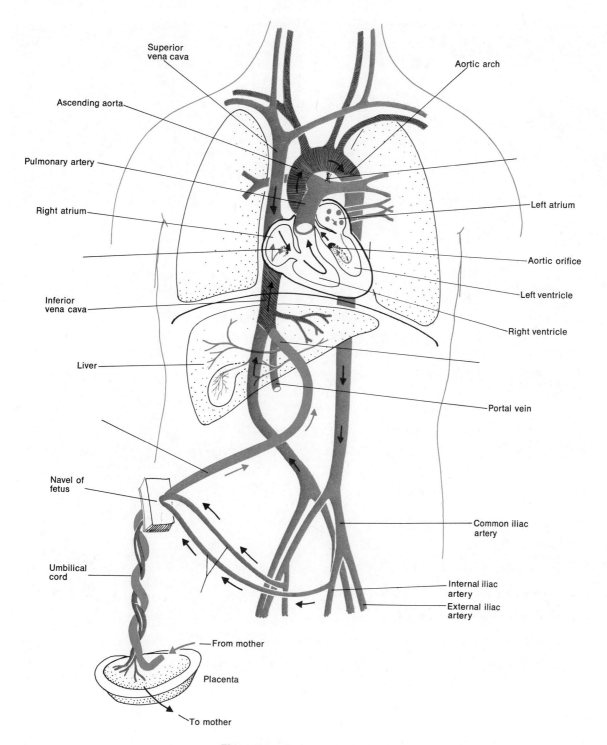

**Figure 30–14.** Fetal circulation.

# MODULE 31
# CARDIOVASCULAR PHYSIOLOGY

**OBJECTIVES**

1. To define systole and diastole as the 2 principal events of the cardiac cycle
2. To describe the events of the cardiac cycle as a function of time
3. To describe the sounds of the heart and their clinical significance
4. To describe the initiation and conduction of nerve impulse through the nodal system of the heart
5. To label and explain the deflection waves of a normal electrocardiogram
6. To define cardiac output and identify those factors that determine it
7. To define pulse and identify those arteries where pulse may be felt
8. To compare the several kinds of abnormal pulse rates
9. To define blood pressure
10. To describe one clinical method for recording systolic and diastolic pressure
11. To contrast the clinical significance of systolic, diastolic, and pulse pressures

After discussing some of the pertinent structural features of the circulatory system, this module looks at the system in action. We shall attempt to answer questions such as: What is a heartbeat? How is a heartbeat recorded? What factors regulate heartbeat? Why does blood flow through vessels? What are the meanings of pulse and blood pressure?

## PART I   THE CARDIAC CYCLE

In a normal heartbeat, the 2 atria contract simultaneously while the 2 ventricles relax. Then, when the 2 ventricles contract, the 2 atria relax. The term *systole* refers to the phase of contraction, and the term *diastole* refers to the phase of relaxation. A *cardiac cycle*, or complete heartbeat, consists of the systole and diastole of both atria, plus the systole and diastole of both ventricles followed by a short pause.

For purposes of discussion, we shall take the *atrial systole* as the starting point in the cardiac cycle (Figure 31-1a). During this period, the atria contract and force blood into the ventricles. Deoxygenated blood from the right atrium passes into the right ventricles through the open tricuspid valve, and oxygenated blood passes from the left atrium into the left ventricle through the open mitral valve. While the atria are contracting, the ventricles are in diastole. During *ventricular diastole* the ventricles are filling with blood, and the semilunar valves in the aorta and pulmonary artery are closed.

When atrial systole and ventricular diastole are completed, the events are reversed. That is the atria go into diastole, and the ventricles go into systole. During *atrial diastole*, deoxygenated blood from the various parts of the body enters the right atrium through the superior vena cava, inferior vena cava, and coronary sinus. Simultaneously, oxygenated blood from the lungs enters the left

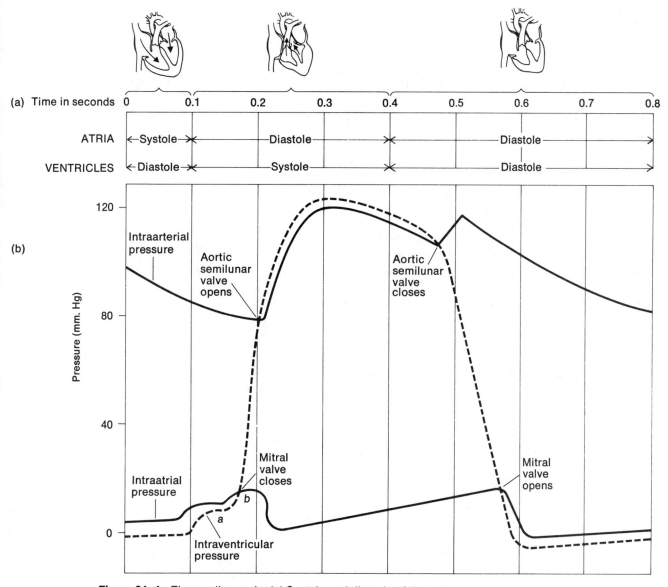

**Figure 31-1.** The cardiac cycle. (a) Systole and diastole of the atria and ventricles. (b) Intraatrial, intraventricular, and intraarterial pressure changes during the cardiac cycle. Note the relationship of the events of the cardiac cycle to time.

atrium through the pulmonary veins. During the first part of atrial diastole, the atrioventricular valves are closed, since the ventricles are in systole. In *ventricular systole,* the ventricles contract and force blood into their respective vessels. The right ventricle pumps deoxygenated blood to the lungs through the semilunar valve of the pulmonary artery. The left ventricle pumps oxygenated blood through the open semilunar valve of the aorta. At the end of the ventricular systole, the semilunar valves close, and the ventricles relax.

Notice that 2 phenomena control the movement of blood through the heart. These are the opening and closing of the valves and the contraction and relaxation of the myocardium. Both these activities occur without any direct stimulation from the nervous system. The valves are controlled by pressure changes that occur within each heart chamber. The contraction of the cardiac muscle is stimulated by nervelike tissue that lies in the walls of the heart. Impulses from the nervous system influence only the rate of the heartbeat.

## PART II   TIMING OF THE CYCLE

We can now relate the events of the cardiac cycle to time. If we assume that the average heart beats 72 times per minute, then each beat with its short pause requires about 0.8 second (Figure 31-1a). During the first 0.1 second, the atria are contracting, and the ventricles are relaxing. The atrioventricular valves are open, and the semilunar valves are closed. For the next 0.3 second, the atria are relaxing and the ventricles contracting. During the first part of this period, all valves are closed, and during the second part the semilunars are open. The last 0.4 second of the cycle is the relaxation, or quiescent, period. All chambers are in diastole. And for the first part of the quiescent period, all valves are closed. During the latter part of the relaxation period, the atrioventricular valves open, and blood starts draining into the ventricles. When the heart beats at a faster rate than normal, the quiescent period is shortened accordingly.

## PART III   SOUNDS OF THE HEART

If you place the bell of a stethoscope on the surface of the skin about an inch below and a little to the median side of the left nipple, you will hear 2 distinct sounds. The first sound, which can be described as a *lubb* (oo) sound, is a comparatively long, booming sound. The lubb is the sound of the atrioventricular valves closing soon after ventricular systole begins. The second sound, which is heard as a short, sharp sound, can be described as a *dup* (u) sound. Dup is the sound of the semilunar valves closing toward the end of ventricular systole. A pause about 2 times longer comes between the second sound and the first sound of the next cycle. Thus, the cardiac cycle can be heard as a lubb, dup, pause; lubb, dup, pause; lubb, dup, pause. This is the sound of the heartbeat. But note that it comes from the closure of the valves and not from the contraction of the heart muscle.

Heart sounds provide valuable information about the valves of the heart. If the sounds are peculiar, they are referred to as *murmurs.* Murmurs are frequently the noise made by a little blood bubbling back up into an atrium because of the failure of one of the atrioventricular valves to close properly. However, murmurs do not always indicate a valve problem, and many have no clinical significance.

Using the stethoscopes provided, identify the heart sounds of your laboratory partner.

### CONCLUSIONS
1. Define systole and diastole.

2. What is a cardiac cycle?

3. What are the 2 phenomena that control the movement of blood through the heart?

4. Both actions occur without any direct stimulation from the nervous system. How is this accomplished?

5. Does the nervous system function in the heartbeat at all?

6. How much time is involved in an average heartbeat?

7. What are the 2 heart sounds as detected by a stethoscope?

8. Of what value are heart sounds?

9. What are abnormal sounds called?

## PART IV   THE CONDUCTION SYSTEM

Skeletal and smooth muscle must receive impulses from the nervous system to initiate their contraction. Cardiac muscle is different. The heart is hooked up to the autonomic nervous system, but the autonomic neurons only increase or decrease the time it takes to complete a cardiac cycle. The walls of the chambers can go on contracting and relaxing without any direct stimulus from the nervous system. This is because the heart has a type of built-in private nervous system called the *conduction system*. The conduction system is composed of specialized tissues that generate the electrical impulses which stimulate the cardiac muscle fibers to contract. These tissues are the sino-atrial node, the atrioventricular node, the atrioventricular bundle, and the Purkinje fibers. The cells of the conduction system develop during embryological life from certain cardiac muscle cells. These cells lose their ability to contract and become specialists in impulse transmission.

A *node* is a compact mass of conducting cells. The *sinoatrial node,* known as the *SA node* or *pacemaker,* is located in the right atrium beneath the opening of the superior vena cava (Figure 31-2a). The SA node initiates each cardiac cycle and thereby sets the basic pace for the heart rate. This is why it is commonly called the pacemaker. However, the rate set by the SA node may be altered by nervous impulses from the autonomic nervous system or by certain hormones such as thyroid hormone or epinephrine. Once an electrical impulse is initiated by the SA node, the impulse

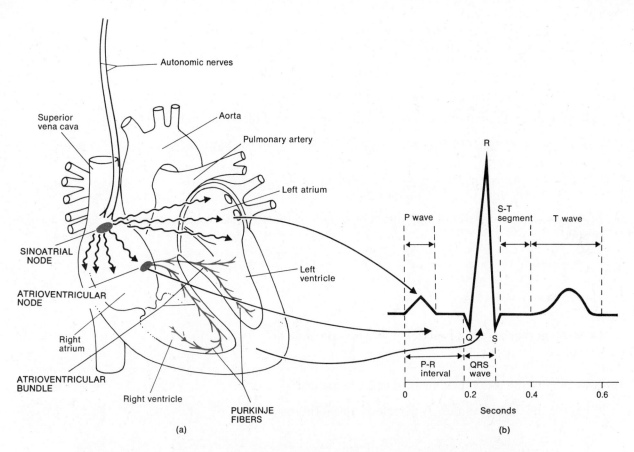

**Figure 31-2.** Conduction system of the heart. (a) Location of the nodes and bundles of the conduction system. (b) Recordings of a normal electrocardiogram. The significance of the recordings is as follows: P wave: passage of impulse from sinoatrial node through atria. P–R interval: time required for impulse to pass through atria, atrioventricular node, atrioventricular bundle, and Purkinje fibers. QRS wave: passage of the impulse through ventricles. S–T segment: time between end of spread of impulse and relaxation of ventricles. T wave: ventricular relaxation.

spreads out over both atria and causes them to contract. From here, the impulse passes to the *atrioventricular (AV) node*, located toward the bottom of the interatrial septum. From the AV node, a tract of conducting fibers called a *bundle* runs to the top of the interventricular septum and then down both sides of the septum. This is called the *atrioventricular bundle*, or *bundle of His*. The bundle of His distributes the charge over the medial surfaces of the ventricles. Actual contraction of the ventricles is stimulated by the *Purkinje fibers*. The Purkinje fibers are individual conducting cells that emerge from the bundle of His and pass into the cells of the myocardium.

## PART V   THE ELECTROCARDIOGRAM

Impulse transmission through the conduction system generates electrical currents that may be detected on the surface of the body. A recording of the electrical changes that accompany the cardiac cycle is called an *electrocardiogram (ECG)*. The instrument used to record the changes is an *electrocardiograph*.

Each portion of the cardiac cycle produces a different electrical impulse. These are transmitted from the electrodes to a recording needle that graphs the impulses as a series of up-and-down waves called *deflection waves*. In a typical record (Figure 31-2b), 3 clearly recognizable waves accompany each cardiac cycle. The first wave, called the *P wave*, is a small upward wave. It indicates the spread

**304**

of an impulse from the SA node over the surface of the 2 atria. A fraction of a second after the P wave begins, the atria contract. Following this, there is a complex called the *QRS wave*. It begins as a downward deflection, continues as a large, upright, triangular wave, and ends as a downward wave at its base. This deflection represents the spread of the electrical impulse through the ventricles. The third recognizable deflection is a dome-shaped wave called the *T wave*. This wave indicates ventricular repolarization (relaxation).

In reading an electrocardiogram, it is exceedingly important to note time relationships between various waves. For example, refer to Figure 31-2b, and note the P-R interval. This interval, measured from the beginning of the P wave to the beginning of the Q wave, represents the conduction time from the beginning of atrial excitation to the beginning of ventricular excitation. The P-R interval is the time required for an impulse to travel through the atria and atrioventricular node, to the atrioventricular bundle, and Purkinje fibers. The lengthening of this interval indicates partial blockage of conduction at the atrioventricular bundle. Other intervals and their significance are also indicated in the figure. The ECG is invaluable in diagnosing abnormal cardiac rhythms and conduction patterns, detecting the presence of fetal life, and determining multiple pregnancies.

## PART VI  CARDIAC OUTPUT

The heart is capable of living a rather independent life. But its output is nevertheless regulated by events occurring in the rest of the body. Cells must receive a certain amount of oxygenated blood each minute in order to maintain health and life. When they are very active, as during exercise, they need even more blood. When you are asleep, cellular need is reduced, and the heart cuts back on its output.

The amount of blood ejected per minute from the left ventricle into the aorta is called the *cardiac output*, or *minute volume*. Cardiac output is determined by 2 factors: (1) the amount of blood that is pumped by the left ventricle during each beat, and (2) the number of heartbeats per minute. The amount of blood ejected by a ventricle during each systole is called the *stroke volume*. In a resting adult, stroke volume averages 70 milliliters, and heart rate is about 72 beats per minute. The average cardiac output, then, in a resting adult is:

Cardiac output = stroke volume X ventricular systole/minute

$$= 70 \text{ ml. X } 72/\text{min.}$$

$$= 5,040 \text{ ml.}/\text{min.}$$

In general, any factor that increases the heart rate or increases its stroke volume tends to increase cardiac output. Factors that decrease the heart rate or its stroke volume tend to decrease cardiac output. If stroke volume falls dangerously low, the body can compensate to some extent by increasing the heartbeat and vice versa.

Using an electrocardiograph, record the electrical changes of the cardiac cycle for at least 2 male and 2 female students. Compare their deflection waves with the normal P-QRS-T waves in Figure 31-2. The electrocardiogram can be taken before and after moderate exercise, to detect any possible deflection wave differences.

### CONCLUSIONS

1. What is the conduction system of the heart?

2. Name these special tissues.

3. Define a node.

4. Which of the specialized tissues is called the "pacemaker" of the heart?
5. Why is it called the pacemaker of the heart?

6. Describe the conduction system of the heart in exactly the proper sequence.

7. How can we record the electrical impulses accompanying the cardiac cycle?

8. What are the various up-and-down impulses called?

9. What are the 3 clearly recognizable waves that accompany each cardiac cycle?

10. Of all the waves, which time interval is the most important?

11. Why is this interval so important?

12. What is the diagnostic value of the ECG?

13. Define cardiac output.

14. What 2 factors determine cardiac output?

15. Define stroke volume.

16. What is a normal stroke volume and heart rate in a normal resting adult?

## PART VII  PULSE

The alternate expansion and elastic recoil of an artery with each systole and diastole of the left ventricle is called the *pulse*. Pulse is strongest in the arteries closest to the heart. It becomes weaker as it passes over the arterial system, and it disappears altogether in the capillaries. The pulse may be felt in any artery that lies near the surface of the body and over a bone or other firm tissue. The radial artery at the wrist is most commonly used for this purpose. Other arteries that may be used for determining pulse are (1) the temporal artery, which is above and toward the outside of the eye, (2) the facial artery, which is at the lower jawbone on a line with the corners of the mouth, (3) the common carotid artery, which is on the one side of the neck, (4) the brachial artery along the inner side of the biceps of the arm, (5) the femoral artery near the pelvic bone, (6) the popliteal artery behind the knee, (7) the posterior tibial artery behind the medial malleolus, and (8) the dorsalis pedis artery over the instep of the foot.

The pulse rate is the same as the heart rate and averages between 70 and 80 per minute in the resting state. The term *tachycardia* is applied to a very rapid heart rate or pulse rate. *Tachy* means fast. The term *bradycardia, brady* meaning slow, indicates a very slow heart rate or pulse rate. In addition to recording the rate of the pulse, other factors should be noted. For example, the intervals between beats should be equal in length. If a pulse is missed at regular or irregular intervals, the pulse is said to be irregular. Also, each pulse beat should be of equal strength. Irregularities in strength may indicate a lack of muscle tone in the heart or arteries.

1. Take your pulse by placing the index and third finger of your right hand over the wrist of your left hand, just inside the large wrist bone at the base of the thumb. *Do not use your thumb*. If you use your thumb, you may find that you are getting your own pulse. Count the pulse beats for one minute and record your rate on the blackboard.
2. After the entire class has recorded their counts, rearrange them with the lowest at the bottom and the highest on top.
3. Determine the median pulse rate of the class by counting down halfway from the top. Record this result on the laboratory report.

## PART VIII  BLOOD PRESSURE

Although the term *blood pressure* may be defined as the pressure exerted by the blood on the walls of any blood vessel, in clinical settings it refers only to the pressure in the large arteries. Blood pressure is usually taken in the left brachial artery, and it is measured by a *sphygmomanometer*. The term *sphygmo* means pulse. A commonly used kind of sphygmomanometer (Figure 31–3a) consists of a rubber cuff attached by a rubber tube to a compressible hand pump or bulb. Another tube attaches to the cuff and to a column of mercury that is marked off in millimeters. This column measures the pressure. The cuff is wrapped around the arm over the brachial artery and inflated by squeezing the bulb. This causes a pressure on the outside of the artery (Figure 31–3b). The bulb is squeezed until the pressure in the cuff exceeds the pressure in the artery. At this point, the walls of the brachial artery are compressed tightly against each other, and no blood can flow through.

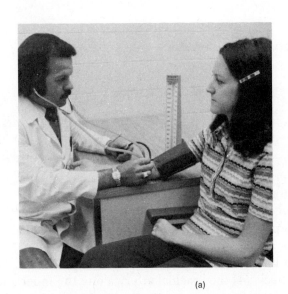

(a)

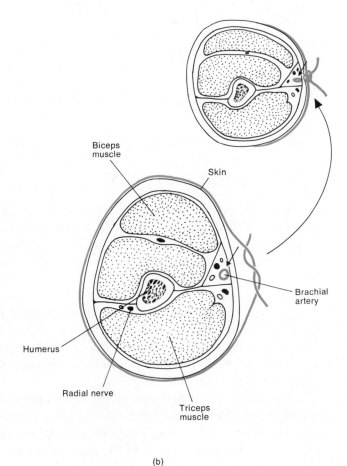

Biceps muscle

Skin

Humerus

Radial nerve

Triceps muscle

Brachial artery

(b)

**Figure 31–3.** Measurement of blood pressure. (a) Use of sphygmomanometer. (b) Pressure changes associated with the brachial artery. *(Courtesy of Lenny Patti.)*

Compression of the artery may be evidenced in 2 ways. First, if a stethoscope is placed below the cuff over the artery, no pulse can be heard. Secondly, no pulse can be felt by placing the fingers over the radial artery of the wrist.

Next, the cuff is deflated gradually until the pressure in the cuff equals the maximal pressure in the brachial artery. At this point, the artery opens, a spurt of blood passes through, and the pulse may be heard through the stethoscope. As the cuff pressure is further reduced, the sound suddenly becomes more faint or muffled. Finally, the sound becomes muffled and disappears altogether. When the first sound is heard, a reading on the mercury column is made. This sound corresponds to *systolic blood pressure*. This pressure is the force with which blood is pushing against arterial walls during ventricular contraction. The pressure recorded on the mercury column when the sounds suddenly become faint and muffled is called *diastolic blood pressure*. It measures the force of blood in arteries during ventricular relaxation. Whereas systolic pressure typically indicates the force of the left ventricular contraction, the diastolic pressure typically provides information about the resistance of blood vessels.

The average blood pressure of a young adult male is about 120 millimeters Hg systolic and about 80 millimeters Hg diastolic. For convenience and brevity, these pressures are indicated as 120/80. In young adult females, the pressures are 8 to 10 millimeters Hg less. The difference between the systolic and diastolic pressure is called *pulse pressure*. This pressure, which averages 40 millimeters Hg, provides information about the condition of the arteries. The higher the systolic pressure and the lower the diastolic pressure, the greater is the pulse pressure. The normal ratio of systolic pressure to diastolic pressure to pulse pressure is about 3:2:1.

Working in pairs, each student should take and record the blood pressure of his partner. The room should be as quiet as possible, since the sounds heard through the stethoscope are not very loud. The student may have to repeat the procedure more than once before being successful.

**CONCLUSIONS**

1. Define pulse.

2. Where is the pulse strongest?

3. Where is the pulse most commonly felt?

4. Why is this particular artery used so frequently?

5. Name and locate 6 other arteries that may be used for determining pulse.

6. What is a normal pulse-rate range?

7. Define tachycardia and bradycardia.

8. Define blood pressure.

9. Which blood vessel is usually used in taking blood pressure?

10. What is the name of the instrument that is used to determine blood pressure?

11. What are the sounds called in determining blood pressure?

12. What is an average blood pressure value for an adult?

13. Define pulse pressure.

**STUDENT ACTIVITY**

All the students in the class should have their pulse, blood pressure, and pulse pressure recorded on the blackboard or on a laboratory report sheet. The mean averages of these 3 values should then be calculated to see how close to the normal ranges the class comes.

# MODULE 32
# CARDIOVASCULAR DISORDERS

**OBJECTIVES**

1. To list the causes and symptoms of aneurysms, arteriosclerosis, and hypertension

2. To describe the diagnosis of arteriosclerosis by the use of angiography

3. To define inadequate blood supply, faulty architecture, and malfunctions of conduction as primary causes of heart trouble

4. To describe patent ductus arteriosus, septal defects and valvular stenosis as congenital heart defects

5. To list the 4 abnormalities of the heart present in tetralogy of Fallot

6. To define atrioventricular block, atrial flutter, atrial fibrillation, and ventricular fibrillation as abnormalities of the conduction system of the heart

7. To define circulatory shock and the circulatory shock cycle

8. To describe the use of hypothermia, the heart-lung bypass, artificial parts, and cardiac catheterization in cardiovascular surgery

9. To define medical terminology associated with the cardiovascular system

It should be obvious to you that the cardiovascular system is exceedingly complex and is distributed to nearly every part of the body. Therefore, it is not surprising that disorders of the cardiovascular system also affect many other parts of the body. This module will examine some of these disorders.

Diseases of the heart and blood vessels are the biggest single killers in the developed world. These diseases account for approximately 53 percent of all deaths. A recent comparison indicates that cardiovascular disease kills more people than cancer, accidents, pneumonia, influenza, and diabetes combined. Some of the cardiovascular problems involve aneurysms, arteriosclerosis, hypertension, and various heart disorders.

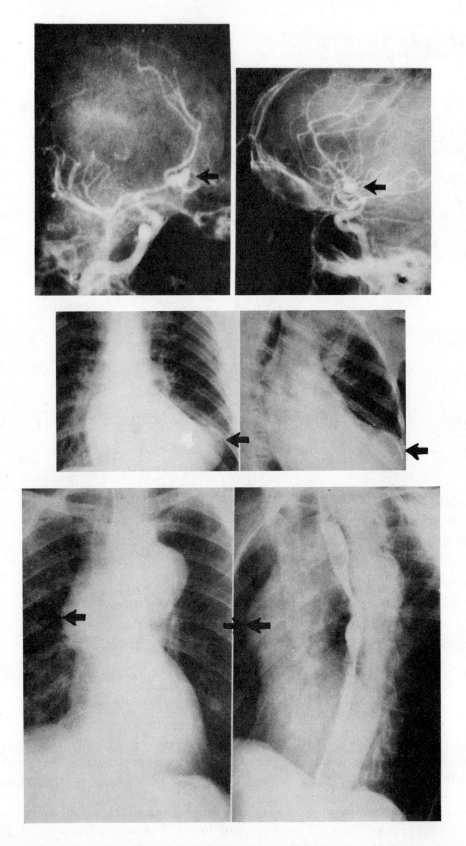

**Figure 32-1.** Aneurysms. (Top) Aneurysm of the anterior carebral artery. (Middle) Ventricular aneurysm. (Bottom) Aortic aneurysm. *(From Lester W. Paul and John H. Juhl, The Essentials of Roentgen Interpretation, 3d ed., Harper & Row, Publishers, Inc., New York, N.Y., 1972.)*

## PART I   ANEURYSM

A blood-filled sac formed by an outpouching in an arterial or venous wall is called an *aneurysm*. Aneurysms may occur in any major blood vessel of the body and include the following types:

1. Berry, which is a small aneurysm of a cerebral artery. If it ruptures, it may cause a hemorrhage below the dura mater (Figure 32-1 top). Hemorrhaging is one cause of stroke.
2. Ventricular, which is a focal dilatation of a ventricle of the heart (Figure 32-1 middle).
3. Aortic, which is a focal dilatation of the aorta (Figure 32-1 bottom).

## PART II   ARTERIOSCLEROSIS

A hardening of the arteries is described by the term *arteriosclerosis*. *Arterio* means artery, whereas *scler* means hard. One type of arteriosclerosis is responsible for the most important and prevalent of all clinical complications. In this type, the inner layer of the artery becomes thickened with soft fatty deposits, called *plaques*. The plaque looks like a pearly gray or yellow mound of tissue on the inside of the blood vessel wall. It usually consists of a core of lipid (mainly cholesterol) covered by a cap of fibrous (scar) tissue. As the plaques increase in size, they not only calcify, but they may also impede or cut off blood flow in affected arteries. This causes damage to the tissues supplied by these arteries. An additional danger is that the lipid core of the plaques may be washed into the bloodstream. There, it could become an embolus and obstruct small arteries and capillaries quite a distance away from the original site of formation. A third possibility is that the plaque will provide a roughened surface for clot formation.

Arteriosclerosis is generally a slow, progressive disease. It may start in childhood, and its development may produce absolutely no symptoms for 20 to 40 years or longer. Even if it reaches the advanced stages, the individual may feel no symptoms, and the condition may be discovered only at postmortem examination. Diagnosis during life is made possible by injecting radiopaque substances into the blood and then taking x-rays of the arteries. This technique is called *angiography* or *arteriography*. The film is called an arteriogram (Figure 32-2a). Figure 32-2b indicates the most common sites of arteriosclerotic plaques.

Animal experiments have given us considerable scientific information about the plaques. They begin as yellowish fatty streaks of lipids that appear under the tunica intima. It is possible to produce the streaks in many animals by feeding them a diet that is high in fat and cholesterol. This raises the blood lipid levels, a condition called *hyperlipidemia*. *Hyper* means over or above whereas *lipo* means fat. *Hyperlipidemia* is an important factor in increasing the risk of arteriosclerosis. Patients with high blood levels of cholesterol should be identified and treated with appropriate diet and drug therapy.

## PART III   HYPERTENSION

*Hypertension*, or high blood pressure, is the most common of the diseases affecting the heart and blood vessels. Statistics from a recent National Health Survey indicate that hypertension afflicts at least 17 million American adults and perhaps as many as 22 million.

*Primary hypertension*, or essential hypertension, is a persistently elevated blood pressure that cannot be attributed to any particular organic cause. Specifically, the diastolic pressure continually exceeds 95 millimeters Hg. Approximately 85 percent of all hypertension cases fit this definition. The other 15 percent represent what is called *secondary hypertension*. Secondary hypertension is caused by disorders such as arteriosclerosis, kidney disease, and adrenal hypersecretion. Arteriosclerosis increases blood pressure by reducing the elasticity of the arterial walls and by narrowing the space through which the blood can flow.

High blood pressure is of considerable concern because of the harm it can do to certain body organs such as the heart, brain, and kidneys if it remains uncontrolled for long periods. The heart is most commonly affected by high blood pressure. When pressure is high, the heart uses more energy

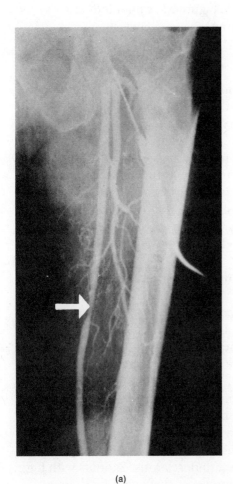

(a)

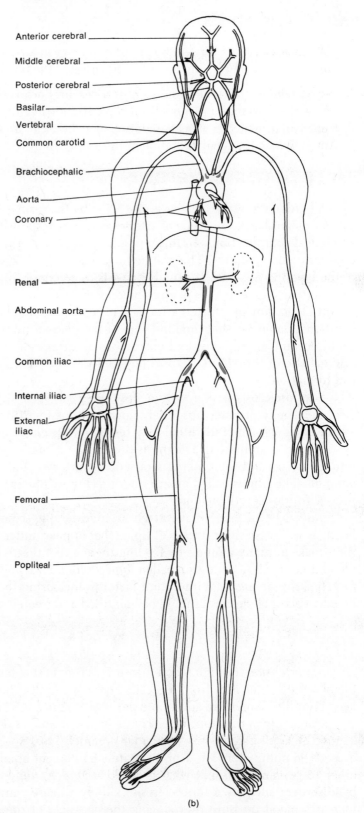

Anterior cerebral

Middle cerebral

Posterior cerebral

Basilar

Vertebral

Common carotid

Brachiocephalic

Aorta

Coronary

Renal

Abdominal aorta

Common iliac

Internal iliac

External
iliac

Femoral

Popliteal

(b)

**Figure 32–2.** Arteriosclerosis. (a) Femoral arteriogram showing an arteriosclerotic plaque (arrow) in the middle third of the thigh. (b) Common sites of arteriosclerotic plaques as seen in a very schematic diagram of the arterial system. The plaques are shown in color. *(Arteriogram courtesy of Lester W. Paul and John H. Juhl, The Essentials of Roentgen Interpretation, 3d ed., Harper & Row, Publishers, Inc., New York, N.Y., 1972.)*

in pumping. Because of the increased effort, the heart muscle thickens, and the heart becomes enlarged. The heart also needs more oxygen. If it cannot meet the demands put on it, angina pectoris or even myocardial infarction may occur. (These terms are defined later.) Continued high blood pressure may produce a cerebral vascular accident, or "stroke." In this case, severe strain has been imposed on the cerebral arteries that supply the brain. These arteries are usually less protected by the surrounding tissues than are the major arteries in other parts of the body. As a result, one or more of these weakened cerebral arteries may finally rupture, and a brain hemorrhage follows.

At present, the causes of primary hypertension are unknown. Medical science cannot cure it. However, almost all cases of hypertension, whether mild or very severe, can be controlled by a variety of effective drugs that reduce elevated blood pressure.

## CONCLUSIONS

1. Define an aneurysm.

2. What are the 3 types of aneurysms?

3. Define arteriosclerosis

4. What is a plaque in arteriosclerosis?

5. What are the 3 possible dangerous results of plaque formation?

6. What is the procedure for diagnosing arteriosclerosis during life?

7. How is this accomplished?

8. Name 8 major common arterial sites of arteriosclerotic plaques.

9. Define hyperlipidemia.

10. What is the relationship between arteriosclerosis and hyperlipidemia?

11. Hypertension is more commonly called

13.  Name some of the known causes of secondary hypertension.


14.  How is hypertension treated since it cannot be cured?


*Heart Trouble.* Generally, the immediate cause of heart trouble is one of the following: (1) failure of the heart's blood supply, (2) faulty heart architecture, or (3) failure of the conductivity. Of these 3 reasons, the first 2 are far more common than the third.

## PART IV    FAILURE OF BLOOD SUPPLY

Angina pectoris and myocardial infarction result from insufficient oxygen supply to the myocardium. Coronary artery disease takes about one in 12 of all Americans who die between the ages of 25 and 34. It claims almost one in 4 of all those who die between 35 and 44. It has been reported that 50 to 65 percent of all sudden deaths are due to coronary heart disease.

The majority of "heart problems" result from some foul-up in the coronary circulation. If a reduced oxygen supply weakens the cells, but does not actually kill them, the condition is called *ischemia. Angina pectoris* is ischemia of the myocardium. The name comes from the area in which the pain is felt. Remember that pain impulses originating from most visceral muscles are referred to an area on the surface of the body. Angina pectoris occurs when coronary circulation is somewhat reduced for some reason. Stress, which produces constriction of vessel walls, is a common cause. Equally common is strenuous exercise after a heavy meal.

Angina pectoris weakens the heart muscle, but it does not produce a full-scale heart attack. The simple remedy of taking nitroglycerin, a drug that dilates vessels and thereby increases the area of blood flow, brings coronary circulation back to normal and stops the pain of angina. Because repeated attacks of angina can weaken the heart and lead to serious heart trouble, angina patients are told to avoid activities and stress that bring on the attacks.

A much more serious problem is *myocardial infarction,* commonly called a "coronary" or "heart attack." *Infarction* means death of an area of tissue because of a drastically reduced or completely interrupted blood supply. Myocardial infarction results from a thrombus or embolus in one of the coronary arteries. The tissue on the far side of the obstruction dies, and the heart muscle loses at least some of its strength. The aftereffects depend partly on the size and location of the infarcted, or dead, area.

At least half of the deaths from myocardial infarction occur before the patient reaches the hospital. These early deaths could result from an irregular heart rhythm, which is called an *arrhythmia.* Sometimes this progresses to the stage called *cardiac arrest* or *ventricular fibrillation,* in which the heart stops functioning. An arrhythmia is an abnormal, irregular rhythm change of the heart, caused by disturbances in the conduction system. This abnormal rhythm of the heartbeat could result in cardiac arrest because in this condition the heart is not capable of supplying the oxygen demands of the body. Serious arrhythmias can be controlled, and the normal heart rhythm can be reestablished, if they are detected and treated early enough. Coronary care units have reduced hospital mortality rates from acute myocardial infarctions by about 30 to 20 percent by preventing or controlling serious arrhythmias.

## PART V    FAULTY ARCHITECTURE

Less than one percent of all new babies have a *congenital,* or *inborn, heart defect.* Even so, the total number in this country each year is estimated to be 30,000 to 40,000. Some of these infants may be able to live quite healthy and long lives without any need for repairing their hearts. But sometimes an inborn heart defect is so severe than an infant lives only a few hours.

**316**

Another common group of congenital problems are the septal defects. A *septal defect* is an opening in the septum that separates the interior of the heart into a left and right side. *Atrial septal defect* is a hole caused by the failure of the fetal foramen ovale to close off the 2 atria from one another. Because pressure in the right atrium is low, atrial septal defect generally allows a good deal of blood to flow from the left atrium to the right. This results in an overload of the pulmonary circulation, producing fatigability, increased respiratory infections, and growth failure, if it occurs early in life, because the systemic circulation may be deprived of a considerable portion of the blood destined for the organs and tissues of the body. *Ventricular septal defect* is caused by an abnormal development of the interventricular septum. Pressure is normally somewhat lower in the right ventricle than in the left, so the blood initially pours from the left ventricle to the right. Deoxygenated blood subsequently gets mixed with the oxygenated blood that is pumped into the systemic circulation. Consequently, the victim suffers cyanosis, a blue or dark purple discoloration of the skin. Cyanosis results from insufficient oxygen in the blood. It occurs whenever deoxygenated blood reaches the cells because of heart defect, lung defect, or suffocation. Septal openings can now be sewn shut or covered with synthetic patches.

A third defect is *valvular stenosis*. It is a narrowing, or stenosis, of one of the valves regulating blood flow inside the heart. Narrowing may occur in the valve itself, most commonly in the mitral valve, from rheumatic heart disease, or in the aortic valve from sclerosis or rheumatic fever. Or it may occur in an area near a valve. The seriousness of all types of stenoses stems from the fact that they all place a severe work load on the heart by making it work harder to push the blood through the abnormally narrow valve openings. As a result of mitral stenosis, blood pressure is increased, and angina pectoris and heart failure may accompany the progress of this disorder. The majority of stenosed valves are totally replaced with artificial valves developed in recent years.

The last congenital defect that we shall discuss is tetralogy of Fallot. Tetralogy of Fallot is a combination of 4 defects causing a "blue baby." These are: a ventricular septal opening, an aorta that emerges from both ventricles instead of solely from the left ventricle, a stenosed pulmonary semilunar valve, and an enlarged right ventricle (Figure 32-3). Because of the ventricular septal defect, both oxygenated and deoxygenated blood are mixed in the ventricles. However, the tissues of the body are much more starved for oxygen than are those of a child with simple ventricular septal defect.

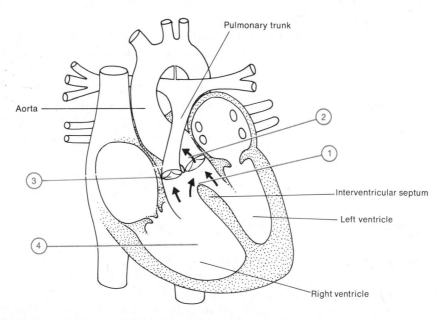

**Figure 32–3.** Tetralogy of Fallot. The 4 abnormalities associated with this condition are indicated by numbers. (1) Opening in the ventricular septum. (2) Origin of the aorta in both ventricles. (3) Stenosed pulmonary semilunar valve. (4) Enlarged right ventricle.

Because the aorta also emerges from the right ventricle and the pulmonary artery is stenosed, very little blood ever gets to the lungs and pulmonary circulation is bypassed almost completely. Today it is possible to completely cure cases of tetralogy of Fallot when the patient is of proper age and condition. Open-heart operations are performed in which the narrowed pulmonary valve is cut open and the septal defect is sealed with a Dacron patch.

Another disorder that arises from some fault in the structure of the heart is rheumatic fever. The symptoms of rheumatic fever mimic many other diseases, but the most common and most serious effects are on the heart. *Rheumatic fever* is basically an acute inflammatory complication of a streptococcal infection. The infection can affect one or more of 5 major sites: the joints (arthritis), the brain (chorea), the heart (carditis), the subcutaneous tissues, and the skin.

Rheumatic fever occurs most frequently during school age, with the majority of the attacks occurring between the ages of 4 and 18 years. Factors such as overcrowding and malnutrition have been implicated. And statistics indicate that this disease is the most common cardiac abnormality of school children.

The most common serious effect of rheumatic fever is an inflammation of the heart that leaves permanent structural defects in the heart valves and chordae tendineae. This inflammation may cause edema, thickening, fusion, or other destruction of the valves. This leads to stenosis or to failure of the valves to close properly. The pericardial sac can also be adversely affected. In addition, there is a chance that an embolism could develop if a piece of scar tissue became dislodged in the circulating blood.

The long-term prognosis depends directly on the cardiac severity of the initial attack. Except for carditis, all symptoms of rheumatic fever subside without residual effects.

## PART VI    FAULTY CONDUCTION

As noted earlier, the term arrhythmia refers to any variation in the rate, rhythm, or synchrony of the heart. It arises when electrical impulses through the heart are blocked at critical points in the conduction system. One such arrhythmia is called a *heart block*. Perhaps the most common blockage is in the atrioventricular node, which conducts impulses from the atria to the ventricles. This disturbance is called *atrioventricular (AV) block*. And it usually indicates a myocardial infarction, arteriosclerosis, rheumatic heart disease, diphtheria, or syphilis.

In a first-degree AV block, which can be detected only by the use of an electrocardiograph, the transmission of impulses from the atria to the ventricles is delayed. In a second-degree AV block, every second impulse fails to reach the ventricles so that the ventricular rate is about one-half that of the atrial rate. In a third-degree or complete AV block, impulses reach the ventricle at irregular intervals, and some never reach it at all. The result is that atrial and ventricle rates are out of synchronization (Figure 32-4a). With complete AV block, many patients may have vertigo, unconsciousness, or convulsions. Among the causes of AV block are excessive stimulation by the vagus nerves that depresses conductivity of the functional fibers, destruction of the AV bundle as a result of coronary infarct, arteriosclerosis, myocarditis, or depression caused by various drugs. Other heart blocks include intraatrial (IA) block, interventricular (IV) block, and bundle branch (BBB) block.

## PART VII    RHYTHMS INDICATING HEART TROUBLE

Two rhythms that indicate heart trouble are atrial flutter and fibrillation. In *atrial flutter* the atrial rhythm averages between 240 and 360 beats per minute. The condition is essentially very rapid atrial contractions accompanied by a second-degree AV block. It is typically indicative of severe damage to heart muscle. Atrial flutter usually becomes fibrillation after a few days or weeks. *Atrial fibrillation* is an asynchronous contraction of the atrial muscles that causes the atria to contract irregularly and still faster. An electrocardiogram of atrial fibrillation is shown in Figure 32-4b. Atrial flutter and fibrillation occur in myocardial infarction, acute and chronic rheumatic heart disease, and

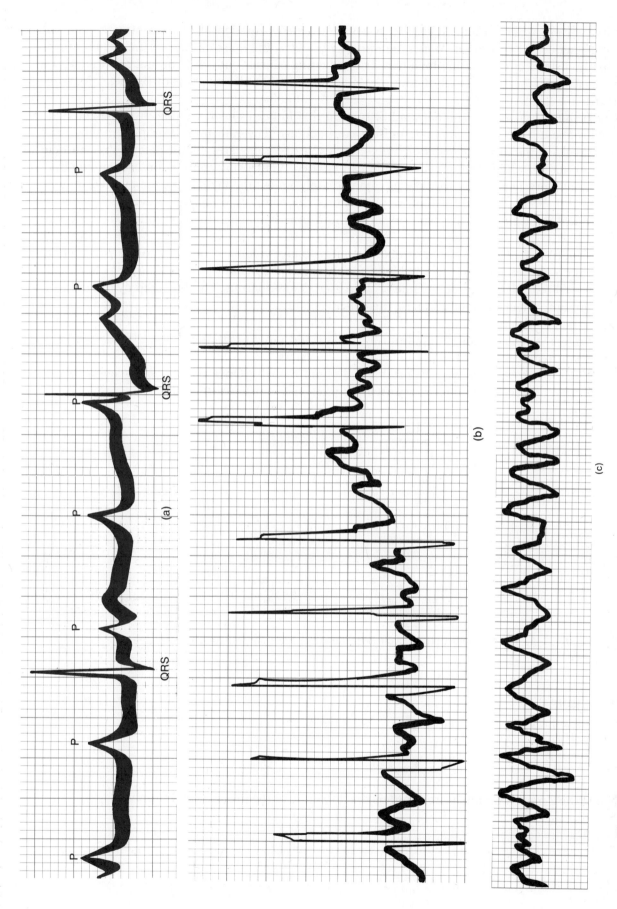

**Figure 32–4.** Abnormal electrocardiograms. (a) Complete heart block. There is no fixed ratio between atrial contractions (P waves) and ventricular contractions (QRS waves). (b) Atrial fibrillation. There is no regular atrial contraction and, therefore, no P wave. Since the ventricles contract irregularly and independently, the QRS wave appears at irregular intervals. (c) Ventricular fibrillation. In general, there is no rhythm of any kind.

hyperthyroidism. Atrial fibrillation results in complete uncoordination of atrial contraction so that atrial pumping ceases altogether. When the muscle fibrillates, the muscle fibers of the atrium quiver individually instead of contracting together. The quivering cancels out the pumping function of the atrium. In a strong heart, atrial fibrillation reduces the pumping effectiveness of the heart by 25 to 30 percent.

*Ventricular fibrillation* is another kind of rhythm that indicates heart trouble. It is characterized by asynchronous, irregular, ventricular muscle contractions. The rate may be rapid or slow. The impulse travels to the different parts of the ventricles at different rates. Thus, part of the ventricle may be contracting, while other parts are still unstimulated. Ventricular contraction becomes ineffective and circulatory failure and death occur immediately unless the arrhythmia is reversed quickly (Figure 32-4c). Ventricular fibrillation may be caused by coronary occlusion. It sometimes occurs during surgical procedures on the heart or pericardium. And it may be the cause of death in electrocution.

The condition that results when cardiac output or blood volume is reduced to the point where body tissues do not receive an adequate blood supply is called *circulatory shock*. The principal cause of circulatory shock is loss of blood volume or decreased cardiac output. This is caused by loss of blood volume through hemorrhage or through the release of histamine due to damage to body tissues (trauma). The characteristic symptoms of circulatory shock are a pale, clammy skin; cyanosis of the ears and fingers; a feeble, though rapid pulse; shallow and rapid breathing; lowered body temperature; and some degree of mental confusion or unconsciousness.

A *circulatory shock cycle* may be established (Figure 32-5). Once the shock reaches a certain level of severity, damage to the circulatory organs is so extensive that death ensues.

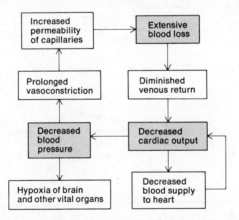

**Figure 32-5.** The circulatory shock cycle. Why does the cycle perpetuate itself until death results?

## CONCLUSIONS

1. What are the 3 main immediate causes of heart trouble?

2. Name the 2 most common heart disorders due to failure of the heart's blood supply.

3. What percentage of all sudden deaths is due to coronary heart disease?

4. Define ischemia, angina pectoris, infarction, and myocardial infarction.

5. What can be done to stop the pain of an angina pectoris attack?

6. Define arrhythmia.

7. Why are arrhythmias dangerous?

8. How many new babies are born with congenital or inborn heart defects every year in this country?

9. What is a septal defect?

10. Describe atrial septal defect and ventricular septal defect.

11. What is cyanosis?

12. Define valvular stenosis.

13. Describe tetralogy of Fallot.

14. What is rheumatic fever?

15. Which areas of the body are sometimes affected by rheumatic fever?

16. Define an atrioventricular (AV) block.

17. Describe atrial flutter, atrial fibrillation, and ventricular fibrillation.

18. What is circulatory shock?

## PART VIII   CARDIAC CATHETERIZATION

In this diagnostic procedure, the tip of a long plastic *catheter*, or tube, is introduced into a vein in the arm or leg. The catheter is radiopaque so that it can be seen with a fluoroscope. With the help of the fluoroscope, it is then threaded through the vena cava and into the right atrium, right ventricle, or pulmonary artery. The catheter can also be inserted into an artery of the arm or leg and worked up through the aorta to the left atrium and ventricle. When the catheter tip is in place, pressures may be recorded. This can tell the diagnostician about the functioning of the valves. An atrial or ventricular septal defect can be identified either by passing the tip of the catheter through the defect or by testing the oxygen content of the blood near the defect.

## PART IX   OPEN HEART SURGERY

Before surgeons can correct even the simplest heart defect, they have to be able to open up the walls of the heart and expose the chambers. This is what is meant by open-heart surgery. But before any open-heart surgery could be done, techniques had to be developed to capture the blood spurting out of the open chamber and pump it back into the vessels. One such life-support technique is extracorporeal circulation. Coupled with hypothermia, it has now made possible both heart surgery and heart transplants.

## PART X   HYPOTHERMIA

*Hypothermia,* or body cooling, slows metabolism and reduces the oxygen needs of the tissues. This allows the heart and brain to withstand short periods of interrupted or reduced blood flow. Lost blood is then replaced by transfusion during and after the operation.

## PART XI   EXTRACORPOREAL CIRCULATION

To sustain the patient for longer periods, surgeons turned to *extracorporeal circulation*. In this technique, blood bypasses the heart and lungs completely (Figure 32-6). It is pumped and oxygenated by a heart-lung machine located outside the body. Modern heart-lung machines may also chill the blood to produce hypothermia as well.

**322**

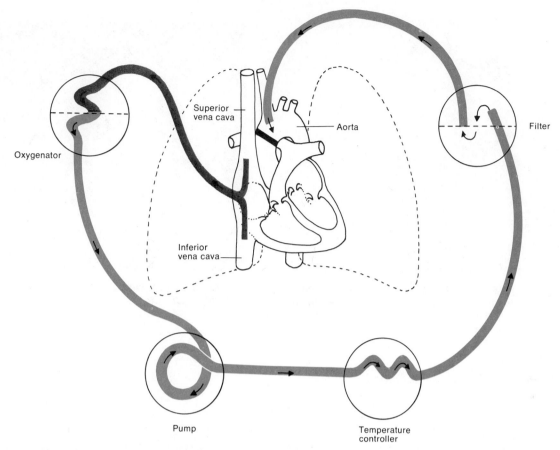

**Figure 32-6.** Diagram of the principle of the heart-lung bypass. Blood drawn from the vena cavae is oxygenated, rewarmed to body temperature or cooled, filtered to remove air and emboli, and returned to the aorta and coronary arteries at the proper pressure. If necessary, drugs, anesthetics, and transfusions may be added to the circuit.

## PART XII  ARTIFICIAL PARTS FOR THE HEART AND BLOOD VESSELS

The development of artificial blood vessels, heart valves, patches, and plugs made from synthetic materials has made it possible to repair many defects of the circulatory system. Replacing or bypassing diseased and damaged arteries with synthetic textile tubes has saved thousands of lives and limbs. The most important new development in treating coronary artery disease is the replacement of partially obstructed coronary vessels with synthetic tubes. This has saved many people from a critically ill, invalid life and allowed them to lead a normal one. Severely damaged aortic, mitral, or tricuspid valves have been replaced by artificial valves installed at the site of the natural valve in patients.

When one or more major elements of the conduction system are disrupted, heart block of varying degrees may result. Complete heart block results most commonly from a disturbance in AV conduction. The ventricles fail to receive any atrial impulses, causing the ventricles and atria to beat independently of each other. In patients with heart block, normal heart rate can be restored and maintained with an *artificial pacemaker* (Figure 32-7). Most pacemakers in current clinical use are compact devices, powered by long-lived batteries. Most of these devices are fixed-rate pacemakers. That is, they are set to pace the heart at a fixed rate—usually 80 beats per minute. Some more recent models make provisions for changing the firing rate of the artificial pacemaker. This allows for increased circulatory needs resulting from exertion or other factors.

**323**

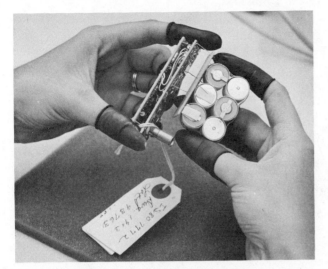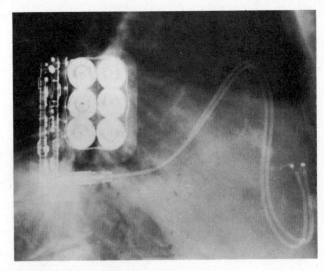

**Figure 32–7.** The cardiac pacemaker. (Left) Pacemaker with a six-cell powerpack prior to implantation. (Right) Roentgenogram of an implanted pacemaker. *(Photographs courtesy of General Electric.)*

## CONCLUSIONS

1. What is cardiac catheterization?

2. Describe hypothermia.

3. Define extracorporeal circulation.

4. Explain the artificial pacemaker.

**324**

## MEDICAL TERMINOLOGY

Define each of the following terms:

a. angiocardiography

b. arteriography

c. bradycardia

d. cardiac arrest

e. cardiomegaly

f. cyanosis

g. defibrillator

h. epistaxis

i. murmurs

j. occlusion

k. phlebitis

l. tachycardia

m. thrombophlebitis

## STUDENT ACTIVITY

Consider all the cardiovascular disorders that have been discussed in this module and separate them into 3 basic categories, listing the etiology (cause), if known, and the major symptoms of each of them: (1) disorders involving the blood vessels, (2) disorders involving the structure of the heart, including especially the valves, and (3) disorders that involve the conduction of the heartbeat or rhythmicity of the heart.

# MODULE 33
# STRUCTURE AND PHYSIOLOGY OF THE RESPIRATORY SYSTEM

**OBJECTIVES**

1. To identify the organs of the respiratory system

2. To compare the structure of the external and internal nose

3. To contrast the functions of the external and internal nose in filtering, warming, and moistening air

4. To differentiate between the 3 regions of the pharynx and describe their roles in respiration

5. To identify the anatomical features of the larynx related to respiration and voice production

6. To contrast tracheotomy and intubation as alternate methods of clearing air passageways

7. To describe the tubes that form the bronchial tree with regard to structure and location

8. To identify the coverings of the lungs and the gross anatomical features of the lungs

9. To describe the structure of a lobule of the lung

10. To describe the role of alveoli in the diffusion of respiratory gases

11. To compare the volumes and capacities of air exchanged in respiration

12. To demonstrate the spirograph, pneumograph, and vital capacity apparatus

Cells need a continuous supply of oxygen to carry out the activities that are vital to their survival. Many of these activities release quantities of carbon dioxide. Since an excessive amount of carbon dioxide is poisonous to cells, the gas must be eliminated quickly and efficiently. The 2 systems that are designed to supply oxygen and eliminate carbon dioxide are the circulatory system and the respiratory system. The *respiratory system* consists of organs that transmit gases between the atmosphere and blood. These organs are the nose, pharynx, larynx, trachea, bronchi, and lungs (Figure 33-1). In turn, the blood transports gases between the lungs and the cells. The overall exchange of gases between the atmosphere and the cells is called respiration. Both the respiratory and circulatory systems participate equally in respiration. Failure of either system has the same effect on the body: rapid death of cells from oxygen starvation and disruption of homeostasis.

Completely label Figure 33-1 indicating the various parts of the respiratory system.

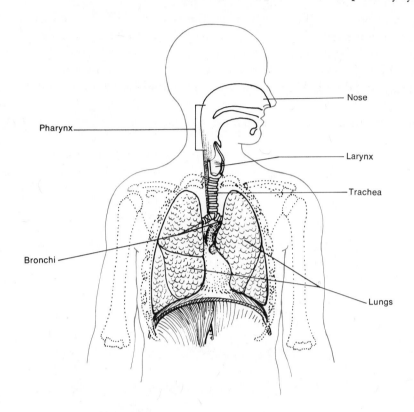

**Figure 33-1.** Organs of the respiratory system.

## PART I   INTERNAL AND EXTERNAL NOSE

The nose has an external portion jutting out from the face and an internal portion lying hidden inside the skull. On the undersurface of the external nose are 2 openings called the *nostrils*, or *external nares*. The internal region of the nose is a large cavity within the skull that lies below the cranium and above the mouth. Anteriorly, the internal nose merges with the external nose, and posteriorly it communicates with the throat (pharynx) through 2 openings called the *internal nares*. The inside of both the external and internal nose is divided into right and left nasal cavities by a vertical partition called the *nasal septum*.

The interior structures of the nose are specialized for 3 functions. First, incoming air is warmed, moistened, and filtered. Second, olfactory stimuli are received, and third, large hollow resonating chambers are provided for speech sounds. These 3 functions are accomplished in the following manner. When air enters the nostrils, it passes first through the vestibule. The vestibule is lined with mucous membrane covered with coarse hairs that filter out large dust particles. The air then passes into the rest of the cavity. Mucous membrane lines the cavity. The olfactory receptors lie in the membrane lining the upper portion of the cavity. Below the olfactory region, the membrane contains many goblet cells, cilia, and capillaries. As the air whirls around the turbinates (spiral-shaped bones) and meati (passageways), it is warmed by the capillaries. Mucus secreted by the goblet cells moistens the air and traps dust particles. The cilia move the resulting mucus-dust packages along to the throat so that they can be eliminated from the body. As the air passes through the top of the cavity, chemicals in the air may stimulate the olfactory receptors.

**327**

## PART II  PHARYNX

The *pharynx*, or throat, is a tube about 12.7 centimeters (5 inches) in length that starts at the internal nares and runs partway down the neck (Figure 33-2).

The uppermost portion of the pharynx is called the *nasopharynx*. This part lies behind the nose and extends down to the soft palate. There are 4 openings in its walls: 2 internal nares plus 2 openings that lead into the auditory tubes. The posterior wall of the nasopharynx also contains the pharyngeal tonsils, or adenoids. Through the internal nares the nasopharynx exchanges air with the nasal cavities and receives the packages of dust-laden mucus. Cilia in the walls of the nasopharynx move the mucus down toward the mouth. The nasopharynx also exchanges small amounts of air with the auditory canal so that the pressure inside the middle ear equals the pressure of the atmospheric air flowing through the nose and pharynx.

The second portion of the pharynx, the *oropharynx*, lies behind the mouth and extends from the soft palate down to the hyoid bone. It receives only one opening, the *fauces*, or opening from the

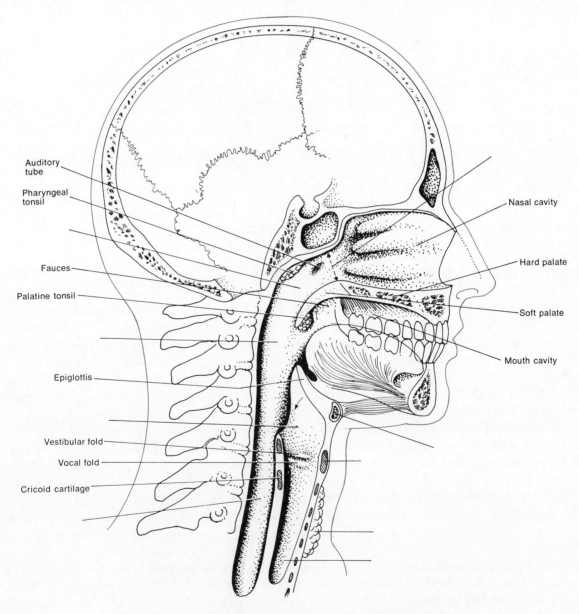

**Figure 33-2.** Sagittal section of the head, neck, and upper chest.

mouth. This portion of the pharynx is both respiratory and digestive in function since it is a common passageway for both air and food. Two pairs of tonsils, the palatine tonsils and the lingual tonsils, are found in the oropharynx.

The lowest portion of the pharynx is called the *laryngopharynx*. The laryngopharynx extends downward from the hyoid bone and empties into the esophagus (food tube) posteriorly and into the larynx (voice box) anteriorly. Like the oropharynx, the laryngopharynx is both respiratory and digestive in function.

## PART III  LARYNX

The *larynx*, or voice box, is a short passageway that connects the pharynx with the trachea. It lies in the midline of the neck. The walls of the larynx are supported by pieces of cartilage. The 3 most prominent pieces are the large thyroid cartilage and the smaller epiglottis and cricoid cartilage (See Figure 33-3a). The *thyroid cartilage*, or Adam's apple, consists of 2 fused plates that form the anterior wall of the larynx and give it its triangular shape (Figure 33-3a). In males the thyroid cartilage is bigger than it is in females.

The *epiglottis* is a large, leaf-shaped piece of cartilage lying on top of the larynx. The "stem" of the epiglottis is attached to the thyroid cartilage, but the "leaf" portion is unattached and free to move up and down like a door on a hinge. In fact, the epiglottis is sometimes called the trap door. As the larynx moves upward and forward during swallowing, the free edge of the epiglottis moves downward and forms a lid over the larynx (see Figure 33-3a). In this way, the larynx is closed off and liquids and foods are routed into the esophagus and kept out of the trachea. If anything but air passes into the larynx, a cough reflex attempts to expel the material.

The *cricoid cartilage* is a ring of cartilage forming the lower walls of the larynx. It is attached to the first ring of cartilage of the trachea.

The mucous membrane of the larynx is arranged into 2 pairs of folds, an upper pair called the *false vocal folds*, and a lower pair called simply the *vocal folds* or *true vocal cords* (Figure 33-3b). The air passageway between the folds is called the *glottis*.

Completely finish the labeling in Figure 33-2 and Figure 33-3.

## CONCLUSIONS

1. What are the openings to the respiratory system called?


2. Explain the expression "deviated septum."


3. What 3 functions are the interior structures of the nose specialized for?


4. What are the 3 parts of the pharynx?


5. Where are the pharyngeal tonsils or adenoids located?

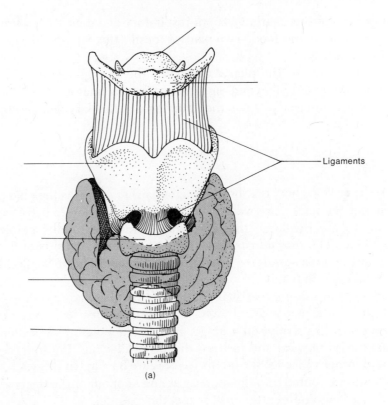

Ligaments

(a)

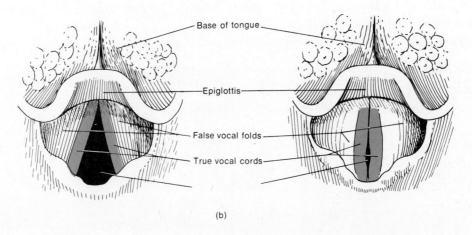

Base of tongue

Epiglottis

False vocal folds

True vocal cords

(b)

**Figure 33-3.** The larynx. (a) Anterior view. (b) Viewed from above. In the figure on the left the true vocal cords are relaxed; in the figure on the right the true vocal cords are pulled taut.

6.  Where are the palatine and lingual tonsils located?

7.  What is the more popular name for the larynx?

8. Name the 3 prominent pieces of cartilage that help make up the larynx.

9. What is the more popular name for the thyroid cartilage?

10. What is the main function of the epiglottis?

11. What is the glottis?

The trachea, or windpipe, is a tubular passageway for air, about 11.5 centimeters (4 and one half inches) in length, and 2.5 centimeters (one inch) in diameter. It is located in front of the esophagus, and it extends from the larynx to the fifth thoracic vertebra (see Figure 33-4 top), where it divides into right and left primary bronchi.

The trachea is lined with ciliated mucous membrane, providing the same protection against dust as the membrane lining the larynx. The walls of the trachea are composed of smooth muscle and elastic connective tissue. They are encircled by a series of horizontal rings of cartilage that look like a series of letter C's stacked one on top of the other. The open parts of the C's face the esophagus and permit the esophagus to expand into the trachea during swallowing. The solid parts of the C's provide a rigid support so that the tracheal walls do not collapse inward and obstruct the air passageway.

Occasionally the respiratory passageways are unable to protect themselves from obstruction. For instance, the rings of cartilage may be accidentally crushed, or the mucous membrane may become inflamed and swell so much that it closes off the air space. Inflamed membranes also secrete a great deal of mucus that may clog the lower respiratory passageways. Or a large object may be breathed in (aspirated) while the epiglottis is open. In any case, the passageways must be cleared quickly. If the obstruction is above the level of the chest, a *tracheotomy* may be performed. The first step in a tracheotomy is to make an incision through the neck and into the part of the trachea below the obstructed area. A metal tube is then inserted through the incision, and the patient breathes through the tube. Another method that may be employed is *intubation*. A tube is inserted into the mouth and passed down through the larynx and trachea. The firm walls of the tube push back any flexible obstruction, and the inside of the tube provides a passageway for air. If mucus is clogging the airways, it can be suctioned up through the tube.

## PART IV   BRONCHI

The trachea terminates in the chest by dividing into a *right primary bronchus*, which goes to the right lung, and a *left primary bronchus*, which goes to the left lung (Figure 33-4 top). The right primary bronchus is more vertical, shorter, and wider than the left. As a result, foreign objects that enter the air passageways frequently lodge in it. Like the trachea, the primary bronchi contain incomplete rings of cartilage and are lined by a ciliated columnar epithelium.

Upon entering the lungs, the primary bronchi divide to form smaller bronchi, the *secondary bronchi*, one for each lobe of the lung. (The right lung has 3 lobes, the left lung has 2.) The secondary bronchi continue to branch, forming still smaller tubes, the *bronchioles*. Bronchioles, in turn, branch into even smaller tubes, called the *terminal bronchioles*. The continuous branching of the trachea

**331**

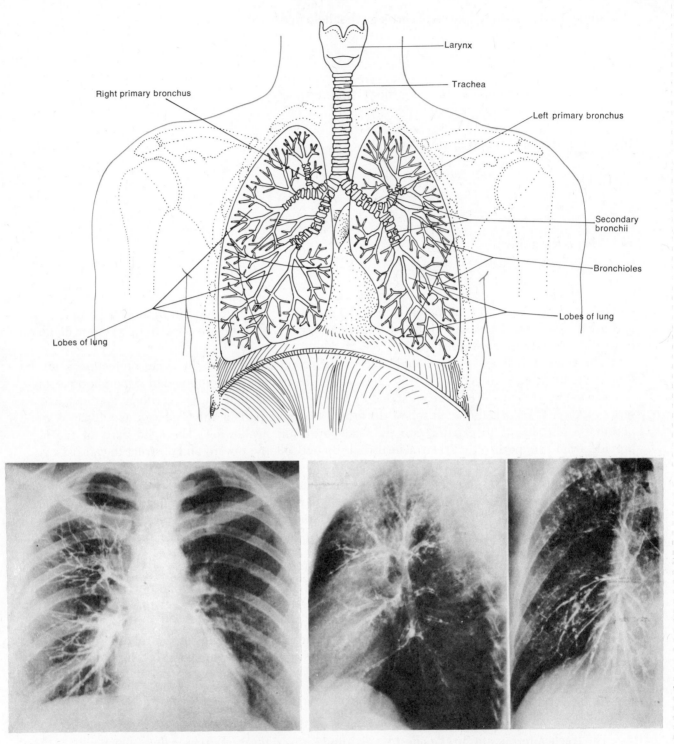

Right primary bronchus

Larynx

Trachea

Left primary bronchus

Secondary bronchii

Bronchioles

Lobes of lung

Lobes of lung

**Figure 33–4.** Air passageways to the lungs. (Top) Diagram of the bronchial tree in relationship to the lungs. (Bottom) Bronchograms of the lungs. *(Bronchograms from Lester W. Paul and John H. Juhl, The Essentials of Roentgen Interpretation, 3d ed., Harper & Row, Publishers, Inc., New York, N.Y., 1972.)*

into primary bronchi, secondary bronchi, bronchioles, and terminal bronchioles resembles a tree trunk with its branches and is commonly referred to as the *bronchial tree.*

*Bronchography* is a technique for examining the bronchial tree. The patient breathes in air that contains a safe dosage of a radioactive element. The element gives off rays that penetrate the chest walls and expose a film. The developed film, a *bronchogram,* provides a picture of the tree (see Figure 33-4 bottom).

## PART V  LUNGS

The lungs are paired, cone-shaped organs lying in the thoracic cavity (Figure 33-5a). They are separated from each other by the heart and other structures in the mediastinum. Two layers of serous membrane, collectively called the *pleural membrane,* enclose and protect each lung. The outer layer is attached to the walls of the pleural cavity and is called the *parietal pleura.* The inner layer, the *visceral pleura,* covers the lungs themselves. Between the visceral and parietal pleura is a small space, the *pleural cavity,* which contains a lubricating fluid secreted by the membranes. This fluid prevents friction between the membranes and allows them to move easily on one another during breathing. Inflammation of the pleural membranes, or *pleurisy,* causes friction during breathing that can be quite painful when the swollen membranes rub against each other.

The lungs extend from the diaphragm to a point just above the clavicles and lie against the ribs in front and back. The broad, inferior portion of the lung, the *base,* is concave and fits over the convex area of the diaphragm. The narrow, superior portion of the lung is referred to as the *apex.*

The surface of the lung lying against the ribs, the *costal surface,* is rounded to match the curvature of the ribs. Medially, the left lung contains a concavity, the *cardiac notch,* in which the heart lies.

Divided by 2 fissures into 3 *lobes,* the superior, middle, and inferior lobes, the right lung is thicker and broader than the left. It is also somewhat shorter than the left because the diaphragm is higher on the right side to accommodate the liver that lies below it. The left lung is thinner, narrower, and longer than the right and is divided into a superior and an inferior lobe.

Each lobe of the lungs is broken up into many small compartments called *lobules* (Figure 33-5b). Every lobule is wrapped in elastic connective tissue and contains a lymphatic, an arteriole, a venule, and a branch from a terminal bronchiole. Terminal bronchioles subdivide into microscopic branches called *respiratory bronchioles.* These, in turn, subdivide into several *alveolar ducts* that terminate in a cluster of *alveolar sacs.* The walls of the alveolar sacs, called *alveoli,* are composed of a single layer of squamous epithelium supported by an extremely thin elastic basement membrane. Over the alveoli, the arteriole and venule disperse into a network of capillaries. The exchange of gases between the lungs and blood takes place by diffusion across the alveoli and the walls of the capillaries. It has been estimated that each lung contains 150 million alveolar sacs, a situation that provides an immense surface area for the exchange of gases.

### CONCLUSIONS

1. Describe the branching of the trachea after it ends and divides into the structure called the "bronchial tree."

2. What is bronchography?

3. Explain the procedure.

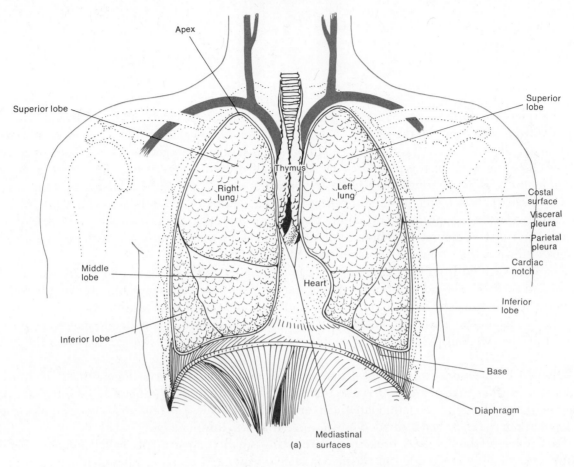

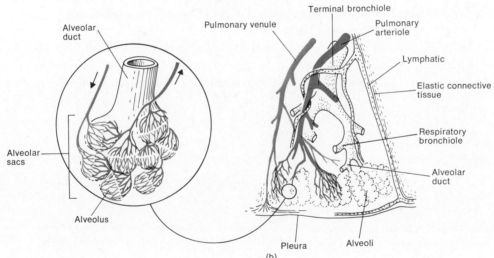

**Figure 33–5.** The lungs. (a) Coverings and external anatomy. (b) A lobule of the lung.

4. What is the film called?


5. Describe the different parts of the pleural membrane.


**334**

6. What is pleurisy?

7. Explain the cardiac notch.

8. Explain the subdivision of the terminal bronchioles until the end of the respiratory system.

9. What are alveoli?

10. What special structures at the alveolar level make gas diffusion very efficient?

## PART VI    AIR VOLUMES EXCHANGED IN RESPIRATION

In clinical practice the word *respiration* is used to mean one inspiration plus one expiration. The average healthy adult has 14 to 18 respirations a minute (that is, the individual inspires 14 to 18 times and expires 14 to 18 times).

During each respiration the lungs exchange given volumes of air with the atmosphere. A lower than normal exchange volume is usually a sign of pulmonary malfunction. The apparatus commonly used to measure the amount of air exchanged during breathing is referred to as a *spirometer* (Figure 33-6 top). During normal, quiet breathing, about 500 milliliters of air move into the respiratory passageways with each inspiration, and the same amount moves out with each expiration. This volume of air inspired (or expired) is called *tidal volume* (Figure 33-6 bottom). Actually, only about 350 milliliters of the tidal volume reach the alveolar sacs. The other 150 milliliters remain in the nose, pharynx, larynx, trachea, bronchi, and dead air space, and are known as *dead air.*

By taking a very deep breath, we can suck in a good deal more than 500 milliliters. This excess inhaled air, called the *inspiratory reserve volume*, averages 3000 milliliters above the 500 milliliters of tidal volume. Thus, the respiratory system can pull in as much as 3500 milliliters of air. If we inhale normally and then exhale as forcibly as possible, we should be able to push out 1100 milliliters of air in addition to the 500 milliliters tidal volume. This extra 1100 milliliters is called the *expiratory reserve volume.* Even after the expiratory reserve volume is expelled, a good deal of air still remains in the lungs because the lower intrathoracic pressure keeps the alveolar sacs slightly inflated. This air, the *residual volume*, amounts to about 1200 milliliters. Opening the thoracic cavity allows the intrathoracic pressure to equal the atmospheric pressure, forcing out the residual volume. The air still remaining is called the *minimal volume.*

Lung capacity can be calculated by combining various lung volumes. *Inspiratory capacity*, the total inspiratory ability of the lungs, is the sum of tidal volume plus inspiratory reserve volume.

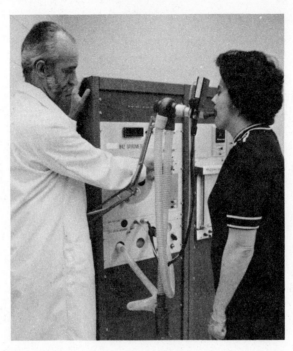

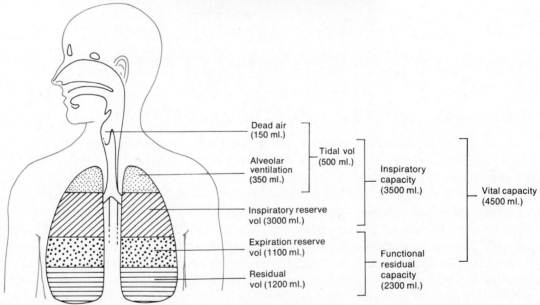

**Figure 33–6.** Air volumes exchanged in respiration. (Top) Determination of lung volumes using a spirometer. (Bottom) Respiratory air volumes. *(Photograph courtesy of Lenny Patti.)*

*Functional residual capacity* is the sum of residual volume plus expiratory reserve volume. *Vital capacity* is the sum of inspiratory reserve volume, tidal volume, and expiratory reserve volume. Finally, *total lung capacity* is the sum of all volumes.

A relatively easy way to measure pulmonary volumes is by using a spirometer.

## PART VII    USE OF THE SPIROMETER

The instructor demonstrates the use of the spirometer. The hose is then detached and rinsed with 70 percent alcohol. This procedure is repeated with every new student using the equipment. The students should work in pairs, with one student operating the spirometer while the other is tested. Before starting the recording, it is advisable to practice breathing into the spirometer hose. A little practice may be necessary to learn to inhale and exhale only through your mouth and into the hose.

Perform the following exercises:
1. Inspire normally, and then exhale as much air as possible. This volume is both your *tidal air* and *expiratory reserve* (supplemental) *volume*. Repeat 3 times and record the values on the chart.
2. Expire normally, and then exhale as much air as possible, recording this volume. This value is only your *expiratory reserve*. Repeat 3 times and record the values on the chart.
3. Subtract your expiratory reserve (number 2 above) from your expiratory reserve and tidal volume (number one above) and calculate your *tidal volume*.
4. After taking a deep breath, exhale as much air as possible. This volume is your *vital capacity*. Repeat 3 times and record the values on the chart.
5. Since your vital capacity consists of tidal, inspiratory reserve, and expiratory reserve, and since in number one above you measured tidal air plus expiratory reserve, subtracting the value of number 1 above from number 4 above provides you with the value for your *inspiratory reserve volume*.
6. Fill out the chart below:

| EXPIRATORY RESERVE AND TIDAL (1) | EXPIRATORY RESERVE (2) | TIDAL (3) | VITAL CAPACITY (4) | INSPIRATORY RESERVE (5) |
|---|---|---|---|---|
| First time _____ | _____ | _____ | _____ | _____ |
| Second time _____ | _____ | _____ | _____ | _____ |
| Third time _____ | _____ | _____ | _____ | _____ |
| Average _____ | _____ | _____ | _____ | _____ |

For an individual 5 feet 8 inches tall, the average vital capacity is 3600 milliliters. Your theoretical vital capacity can be approximated by adding or subtracting 150 milliliters for each inch above or below this average height. Compare with the value you actually obtained.

## PART VIII    USE OF THE PNEUMOGRAPH

The pneumograph can be used to measure respiratory movements. The chest pneumograph is a coiled rubber hose that fits around the chest. One end of this instrument is equipped with a thin tube that is attached to a recorder. As the student breathes, chest movements cause changes in the air pressure within the pneumograph that are transmitted to the recorder. Normal inspiration and expiration can thus be recorded, and the effects of a wide range of physical and chemical factors upon these movements can be studied.

The coiled rubber hose is placed around the student's chest at the level of the sixth rib and attached at the back. Connect the thin rubber tube to the recorder. As the student breathes normally, the needle should deflect. If this does not occur, the pneumograph should be adjusted.

Perform the following exercises:
1. Set the pneumograph at speed one and record normal, quiet breathing for about 30 seconds.
2. Have the student inhale deeply and hold his breath for as long as possible. Record his breathing after this time interval.
3. After normal breathing has returned to the student, have him hyperventilate (breathe deeply and rapidly) for approximately 30 seconds. Record his breathing after this period.
4. The student then breathes into a paper bag for 2 minutes, and his breath is recorded after this procedure. *Caution!* The instructor *must carefully watch* the student in this test for any untoward reactions while breathing into the paper bag.

5. Record the student's respiratory movements during the following activities:
   reading and talking
   swallowing water
   laughing
   yawning
   coughing
   sniffing
6. All records from the pneumograph experiments are given to the instructor who may wish to have the students interpret their results first.

## PART IX  MEASUREMENT OF THE CHEST AND ABDOMEN DURING RESPIRATION

Consult your textbook to see the relationship between inspiration and expiration as correlated with chest and abdomen expansion and contraction. With this relationship in mind, perform the following experiments and record your results in the table.

1. Determine the size of the chest during normal inspiration and expiration with a measuring tape, at about the level of the fifth rib.
2. Repeat the above procedure with a deep or forced inspiration and expiration.
3. Repeat using chest calipers.
4. Calculate the changes occurring at the abdomen during respiration with the tape and calipers.
5. Record all the results below:

|                    | THORAX | | ABDOMEN | |
|                    | INSPIRATION | EXPIRATION | INSPIRATION | EXPIRATION |
|--------------------|-------------|------------|-------------|------------|
| Tape (normal)      | _____ | _____ | _____ | _____ |
| Tape (forced)      | _____ | _____ | _____ | _____ |
| Calipers (normal)  | _____ | _____ | _____ | _____ |
| Calipers (forced)  | _____ | _____ | _____ | _____ |

## PART X  RESPIRATORY SOUNDS

As air flows through the respiratory tract there are characteristic sounds which can be detected with the use of a stethoscope.

Perform the following exercises:
1. Place the bell portion of the stethoscope just below the larynx and listen for bronchial sounds during both inspiration and expiration.
2. The stethoscope should then be slowly moved downward toward the bronchial tubes until the sounds are no longer heard.
3. Place the stethoscope over the chest at the following areas and listen for any sound during inspiration and expiration. (The sound may be a rumbling or rustling murmur.)
   different intercostal spaces
   beneath the scapula
   underneath the clavicle

## PART XI  VITAL CAPACITY APPARATUS

Using the vital capacity apparatus, allow the students to determine their own particular vital capacity values. Compare the values obtained with this apparatus to the values obtained using the spirometer (see Figure 33-7).

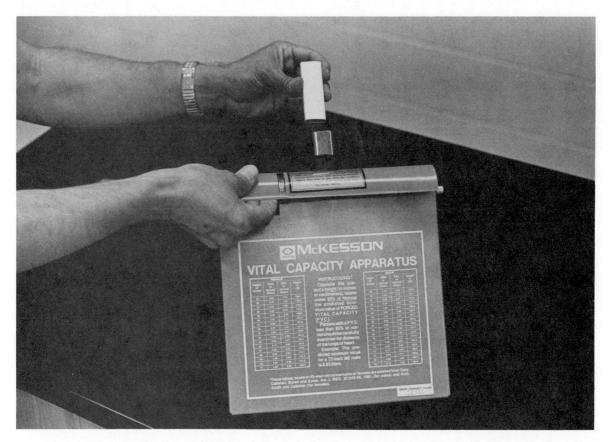

**Figure 33–7.** Placing a disposable air shield on a Vital Capacity Apparatus. *(Photograph courtesy of Lenni Patti.)*

## STUDENT ACTIVITY

Perform all the physiological experiments in this module with as many students participating as possible. Take the mean average values for each particular exercise and compare with previous mean average values. Notice whether or not any of the classes have above or below average values, and which particular physiological exercise shows the greatest deviation from normal.

# MODULE 34
# DISORDERS OF THE RESPIRATORY SYSTEM

**OBJECTIVES**

1. To list the basic steps involved in heart-lung resuscitation

2. To define hay fever, bronchial asthma, emphysema, pneumonia, tuberculosis, hyaline membrane disease, and neoplasm as disorders of the respiratory system

3. To describe the effects of pollutants on the epithelium of the respiratory system

4. To describe the administration of medication by nebulization

5. To define medical terminology associated with the respiratory system

This module presents the major disorders of the respiratory system. Before the disorders are discussed, the major aspects of heart and lung resuscitation will be described.

## PART I   HEART AND LUNG RESUSCITATION

A serious decrease in respiration or heart rate presents an urgent crisis because the body's cells cannot survive long if they are starved of oxygenated blood. In fact, if oxygen is withheld from the cells of the brain for 4 to 6 minutes, brain damage or death will result. Heart-lung resuscitation is the artificial reestablishment of normal or near normal respiration and circulation. The 2 simplest techniques for heart-lung resuscitation are exhaled air ventilation and external cardiac compression. Both techniques can be administered by a layman at the site of the emergency, and both are highly successful. They can be used for any sort of heart or respiratory failure, whether the cause is drowning, strangulation, carbon monoxide or insecticide poisoning, overdose of a drug or anesthesia, electrocution, or myocardial infarction. However, the success of heart-lung resuscitation is directly related to the speed and efficiency with which it is applied. Delay may be fatal.

## PART II   EXHALED AIR VENTILATION

A technique for reestablishing respiration is *exhaled air ventilation*. The first and most important step is immediate opening of the airway. This is accomplished easily and quickly by tilting the

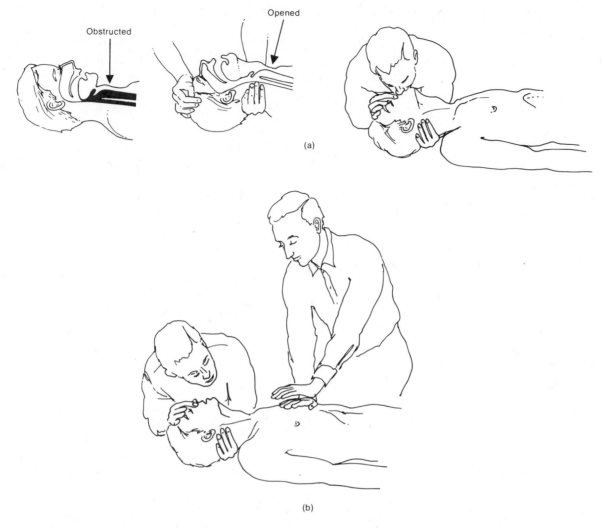

Obstructed

Opened

(a)

(b)

**Figure 34–1.** Heart-lung resuscitation. (a) Exhaled air ventilation. Shown on the left is the procedure for immediate opening of the airway. Shown on the right is the procedure for mouth-to-mouth respiration. (b) External cardiac compression technique in conjunction with exhaled air ventilation.

victim's head backward as far as it will go without being forced. The tilted position opens the upper air passageways to their maximum amount (Figure 34–1a). If the patient does not resume spontaneous breathing after his head has been tilted backward, immediately begin artificial ventilation by either the mouth-to-mouth or the mouth-to-nose method. In the more usual mouth-to-mouth method the nostrils are pinched together with the thumb and index finger of the hand. The rescuer then opens his mouth widely, takes a deep breath, makes a tight seal with his mouth around the patient's mouth, and blows in about twice the amount the patient normally breathes. He then removes his mouth and allows the patient to exhale passively. This cycle is repeated approximately 12 times per minute for adults. Atmospheric air contains about 21 percent $O_2$ and a trace of $CO_2$ Exhaled air still contains about 16 percent $O_2$ and 5 percent $CO_2$ This is more than adequate to maintain a victim's blood $pO_2$ and $pCO_2$ at normal levels if air is given at the prescribed rate and amount.

If the rescuer observes the following 3 signs, he knows that adequate ventilation is occurring:
The chest rises and falls with every breath.
He feels the resistance of the lungs as they expand
He hears the air escape during exhalation.

**341**

## PART III    EXTERNAL CARDIAC COMPRESSION

*External cardiac compression*, or closed-chest cardiac compression (CCCC), consists of the application of rhythmic pressure over the sternum (Figure 34-1b). The rescuer places the heels of his hands on the lower half of the sternum and presses down firmly and smoothly at least 60 times a minute. This action compresses the heart and produces an artificial circulation because the heart lies almost in the middle of the chest between the lower portion of the sternum and the spine. When properly done, external cardiac compression can produce systolic blood pressure peaks of over 100 millimeters Hg. It can also bring carotid arterial blood flow up to 35 percent of normal.

Compression of the sternum produces some artificial ventilation but not enough for adequate oxygenation of the blood. Therefore exhaled air ventilation must always be used with it. This combination constitutes *heart-lung resuscitation*. When there are 2 rescuers, the most physiologically sound and practical technique is to have one rescuer apply cardiac compression at a rate of at least 60 compressions per minute. The other rescuer should exhale into the patient's mouth between every fifth and sixth compression. The sequential steps in emergency heart-lung resuscitation must be continued uniformly and without interruption until the patient recovers or is pronounced dead.

### CONCLUSIONS

1.  If oxygen is withheld from brain cells, brain damage or death will result. How long is this period of time?

2.  How would you define heart-lung resuscitation?

3.  What are the 2 simplest techniques for heart-lung resuscitation?

4.  Name some different causes or conditions that would require these techniques at the site of a particular emergency.

5.  What are the 3 vital signs that indicate to the rescuer that adequate ventilation is occurring?

## PART IV    RESPIRATORY DISORDERS

### HAY FEVER

An allergic reaction to the proteins contained in foreign substances such as plant pollens, dust, and certain types of food is called *hay fever*. The response is localized in the respiratory membranes. The membranes become inflamed, and a watery fluid drains from the eyes and nose.

### BRONCHIAL ASTHMA

This usually allergic reaction is characterized by attacks of wheezing and difficult breathing. Attacks are brought on by spasms of the smooth muscles that lie in the walls of the smaller bronchi and bronchioles, causing the passageways to close partially. The patient has trouble exhaling, and the alveoli may remain somewhat inflated during expiration. Usually the mucuous membranes that line the respiratory passageways become irritated and secrete excessive amounts of mucus that may clog the bronchi and bronchioles and worsen the attack. About 3 out of 4 asthma victims are allergic to something they eat or to substances they breathe in, such as pollens, animal dander, house dust, or smog. Others are usually sensitive to the proteins of relatively harmless bacteria that inhabit the sinuses, nose, and throat.

## EMPHYSEMA

One lung disease that starts with the deterioration of some of the alveoli is *emphysema*. The alveolar walls lose their elasticity and remain filled with air during expiration. The name of the disease means "blown up" or "full of air." Reduced forced expiratory volume is the first symptom. Later, alveoli in other areas of the lungs are damaged. The lungs become permanently inflated. To adjust to the increased lung size, the size of the chest cage increases. The patient has to work to exhale. Oxygen diffusion does not occur as easily across the damaged alveoli, blood $pO_2$ is somewhat lowered, and any mild exercise that raises the oxygen requirements of the cells leaves the patient breathless. Carbon dioxide diffuses much more easily across the alveoli than does oxygen, so the $pCO_2$ is not affected initially. But as the disease progresses, the alveoli degenerate and are replaced with thick fibrous connective tissue. Even carbon dioxide does not diffuse easily through this fibrous tissue. If the blood cannot buffer all the carbonic acid that accumulates, the blood pH drops. Or, unusually high amounts of carbon dioxide may dissolve in the plasma. High carbon dioxide levels are toxic to the brain cells. Consequently, the inspiratory center becomes less active and the respiration rate slows down, further aggravating the problem. The capillaries that lie around the deteriorating alveoli are compressed and damaged and may no longer be able to receive blood. As a result, pressure increases in the pulmonary artery, and the right atrium overworks as it attempts to force blood through the remaining capillaries.

Emphysema is generally caused by any of a number of long-term irritations. Air pollution, occupational exposure to industrial dusts, and cigarette smoke are the most common irritants. Chronic bronchial asthma also may produce alveolar damage. Cases of emphysema are becoming more and more frequent in the United States. The irony is that the disease can be prevented and the progressive deterioration can be stopped by eliminating the harmful stimuli.

## PNEUMONIA

The term *pneumonia* means an acute infection or inflammation of the alveoli. In this disease the alveolar sacs fill up with fluid and dead white blood cells, reducing the amount of air space in the lungs. (Remember that one of the cardinal signs of inflammation is edema.) Oxygen has difficulty diffusing through the inflamed avleoli, and the blood $pO_2$ may be drastically reduced. Blood $pCO_2$ usually remains normal because carbon dioxide always diffuses through the alveoli more easily than oxygen does. If all the alveoli of a lobe are inflamed, the pneumonia is called *lobar pneumonia*. If only parts of the lobe are involved, it is called *lobular*, or *segmental*, *pneumonia*. If both the alveoli and the bronchial tubes are included, it is called *bronchopneumonia*.

The most common cause of pneumonia is the pneumococcus bacterium, but other bacteria or a fungus may be a source of the trouble. Viral pneumonia is caused by any of several viruses, including the influenza virus.

## TUBERCULOSIS

The bacterium called *Mycobacterium tuberculosis* produces an inflammation called tuberculosis. This disease is still one of the most serious of present-day illnesses and ranks as the number-one killer in the communicable disease category. Tuberculosis most often affects the lungs and the pleura. The bacterium destroys parts of the lung tissue, and the tissue is replaced by fibrous connective tissue. Because the connective tissue is inelastic and relatively thick, the affected areas of the lungs do not snap back during expiration, and larger amounts of air are retained. Gases no longer diffuse easily through the fibrous tissue.

Tuberculosis bacteria are spread by inhalation. Although they can withstand exposure to many disinfectants, they die quickly in sunlight. This is why tuberculosis is sometimes associated with crowded, poorly lit housing conditions. Many drugs are successful in treating tuberculosis. Rest, sunlight, and good diet are vital parts of treatment.

## HYALINE MEMBRANE DISEASE (HMD)

Sometimes called glassy-lung disease, *hyaline membrane disease* is responsible for approximately 20,000 newborn infant deaths per year. Autopsies reveal glassy membranes lining the alveoli and alveolar ducts. The membranes consist largely of fibrin deposits that prevent oxygen-carbon dioxide exchange and prevent the opening of the alveoli. Asphyxiation occurs for most infants within 72 hours of birth.

A new treatment currently being developed called PEEP—positive end expiratory pressure—could reverse the mortality rate from 90 percent deaths to 90 percent survival. This treatment consists of passing a tube through the air passage to the top of the lungs to provide needed oxygen-rich air at continuous pressures of up to 14 millimeters Hg. Continuous pressure keeps the baby's lungs open and available for gas exchange.

## SMOKING AND THE RESPIRATORY SYSTEM

As part of ordinary breathing, many irritating substances are inhaled. Almost all pollutants, including inhaled smoke, have an irritating effect on the bronchial tubes and lungs and may be regarded as stresses.

Close examination of the epithelium of a bronchial tube reveals that it consists of 3 kinds of cells (Figure 34-2a). The uppermost cells are columnar cells that contain the cilia on their surfaces. At intervals between the ciliated columnar cells are the mucus-secreting goblet cells. The bottom of the epithelium normally contains 2 rows of basal cells above the basement membrane. The bronchial epithelium is important clinically because researchers have learned that one of the most common types of lung cancer, *bronchogenic carcinoma,* starts in the walls of the bronchi.

The stress of constant irritation by inhaled smoke and pollutants causes an enlargement of the goblet cells of the bronchial epithelium (Figure 34-2b). They respond by secreting excessive amounts of mucus. The basal cells respond to the stress by undergoing cell division so fast that the basal cells push into the area occupied by the goblet and columnar cells. As many as 20 rows of basal cells may be produced. Many researchers believe that if the stress is removed at this point, the epithelium can return to normal.

If the stress persists, more and more mucus is secreted and the cilia become less effective. As a result, mucus is not carried toward the throat; instead, it remains trapped in the bronchial tubes. The individual then develops a "smoker's cough." Moreover, the constant irritation from the pollutant slowly destroys the alveoli, which are replaced with thick, inelastic connective tissue. Mucus that has accumulated becomes trapped in the air sacs. Millions of the sacs rupture. This results in a loss of diffusion surface for the exchange of oxygen and carbon dioxide. The individual has now developed emphysema. If the stress is removed at this point, there is little chance for improvement. Any alveolar tissue that has been destroyed cannot be repaired. But removal of the stress can stop further destruction of lung tissue.

Assuming that stress continues, the emphysema gets progressively worse, and the basal cells of the bronchial tubes continue to divide and break through the basement membrane. At this point the stage is set for bronchogenic carcinoma. Columnar and goblet cells disappear and may be replaced with squamous cancer cells (Figure 34-2c). If this happens, the malignant growth spreads throughout the lung and may block a bronchial tube. If the obstruction occurs in a large bronchial tube, very little oxygen enters the lung, and disease-producing bacteria thrive on the mucoid secretions. In the end, the patient may develop emphysema, carcinoma, and a host of infectious diseases. Treatment involves surgical removal of the diseased lung. However, metastasis of the growth through the lymphatic or blood system may result in new growths in other parts of the body such as the brain and liver.

**344**

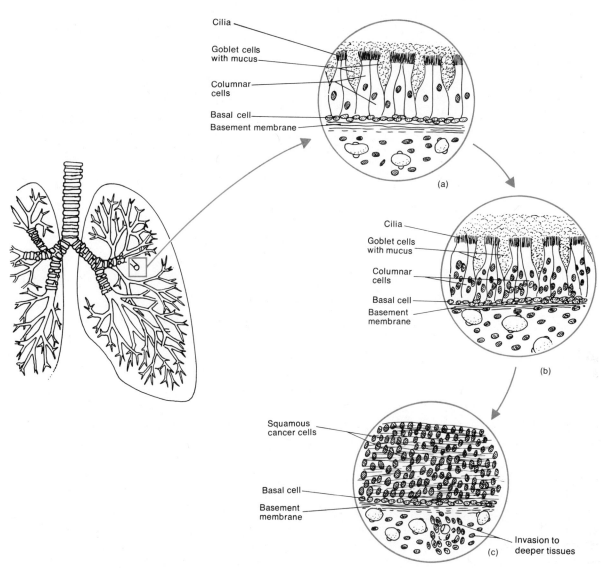

**Figure 34–2.** Effects of smoking on the respiratory epithelium. (a) Microscopic view of the normal epithelium of a bronchial tube. (b) Initial response of the bronchial epithelium to irritation by pollutants. (c) Advanced response of the bronchial epithelium.

The processes that we have just described have been observed in some heavy smokers. One should realize, though, that there are other causes of emphysema, such as chronic asthma. There are also other factors that may be associated with lung cancer. For instance, breast, stomach, and prostate malignancies can metastasize to the lungs. People who apparently have not been exposed to pollutants do occasionally develop bronchogenic carcinoma. However, the occurrence of bronchogenic carcinoma is probably over 20 times higher in heavy cigarette smokers than it is in nonsmokers.

## PART V    THERAPY BY NEBULIZATION

Many of the previously mentioned respiratory disorders are treated by means of a comparatively new method of treatment called *nebulization*. This procedure is the administering of medication, in the form of droplets that are suspended in air, to selected areas of the respiratory tract. The patient inhales the medication as a fine mist (Figure 34-3). Droplet size is directly related to the number of droplets suspended in the mist. Smaller droplets (approximately $2 \mu m$ in diameter) can be suspended in

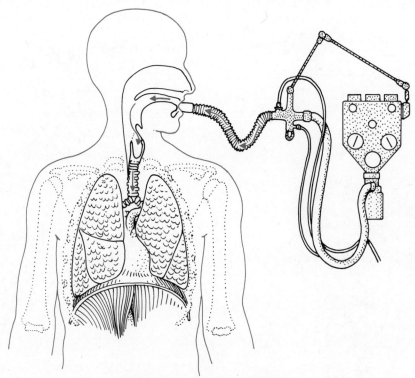

**Figure 34–3.** Use of a nebulizer.

greater numbers than can large droplets and will reach the alveolar ducts and sacs. The larger droplets (approximately 7 to 16μm in diameter) will be deposited mostly in the bronchi and bronchioles. Droplets of 40μm and larger will be deposited in the upper respiratory tract—the mouth, pharynx, trachea, and main bronchi. Nebulization therapy can be used with many different types of drugs, such as chemicals that relax the smooth muscle of the respiratory passageways, chemicals which reduce the thickness of mucus, and antibiotics.

## CONCLUSIONS

1. Define hay fever.



2. Describe bronchial asthma.



3. Connect bronchial asthmatic patients with allergies.

4. What is emphysema?

5. What is the relationship between emphysema and the right atrium of the heart?

6. What are possible causes of emphysema?

7. Define pneumonia.

8. Why is pneumonia dangerous?

9. Differentiate between the 3 types of pneumonia.

10. What are some causes of pneumonia?

11. Despite their resistance to many disinfectants, what kills the tuberculosis bacteria very quickly?

12. What is the more popular name for glassy-lung disease?

13. How many newborn infants die from this disorder every year?

14. Describe this condition.

15. What is the new treatment for HMD?

16. How does it work?

17. What is bronchogenic carcinoma?

18. What basically causes "smoker's cough"?

19. Describe metastasis of lung cancer.

20. What is "therapy by nebulization"?

21. What is the significance of droplet size in this therapy?

22. Of what real value is nebulization therapy?

## MEDICAL TERMINOLOGY

Define each of the following terms:

a. apnea

b. asphyxia

c. atelectasis

d. bronchitis

e. diphtheria

f. dyspnea

g. eupnea

h. hypoxia

i. influenza

j. orthopnea

k. pneumothorax

l. pulmonary edema

m. pulmonary embolism

n. rales

o. respirator

## STUDENT ACTIVITY

Conduct the various physiological tests as outlined in the previous module, dividing the class into 2 major groups: smokers, and nonsmokers. Take the mean average of each group for every test and compare. Determine if, at least in these limited sampling and testing procedures, smokers are handicapped in any way.

Study the different pathology slides of the respiratory system and make sure you understand why each of them is considered pathological.

# MODULE 35
# STRUCTURE AND PHYSIOLOGY OF THE DIGESTIVE SYSTEM

**OBJECTIVES**

1. To define digestion as a chemical and mechanical process

2. To identify the organs of the alimentary canal and the accessory organs of digestion

3. To define the mesentary, lesser omentum, and greater omentum as extensions of the peritoneum

4. To describe the role of the mouth in mechanical digestion

5. To identify the location of the salivary glands

6. To define the function of saliva in digestion

7. To define the action of salivary amylase

8. To describe the mechanisms that regulate the secretion of saliva

9. To describe the structural features of the stomach and the relationship between these features and digestion

10. To compare mechanical and chemical digestion in the stomach

11. To describe the structural features of the small intestine that adapt it for digestion and absorption

12. To describe the mechanisms involved in the hormonal control of digestion in the stomach and small intestine

13. To describe the anatomy of the large intestine

We all know that food is vital to life. Food is required for the chemical reactions that occur in every cell—both those that synthesize new enzymes, cell structures, bone, and all the other components of the body, and those that release the energy needed for the building processes. However, the vast majority of foods we eat are simply too large to pass through the plasma membranes of the cells.

Therefore, chemical and mechanical *digestion* must occur first, and this module will discuss those processes.

## PART I   DIGESTION

*Chemical digestion* is a series of hydrolytic reactions that break down the large carbohydrate, lipid, and protein molecules which we eat into monosaccharides, glycerol and fatty acids, and amino acids, respectively. These products of digestion are small enough to pass through the walls of the digestive organs, into the blood and lymph capillaries, and eventually into the cells of the body. *Mechanical digestion* consists of various movements that aid chemical digestion. Food must be pulverized by the teeth before it is small and flexible enough to be swallowed. After it has been swallowed, the smooth muscles of the stomach and small intestine churn the food so it is thoroughly mixed with the enzymes that catalyze the hydrolysis reactions.

## PART II   GENERAL ORGANIZATION

The organs of digestion are traditionally divided into 2 main groups. First is the *alimentary canal* or *gastrointestinal tract*, a continuous tube running through the ventral body cavity and extending

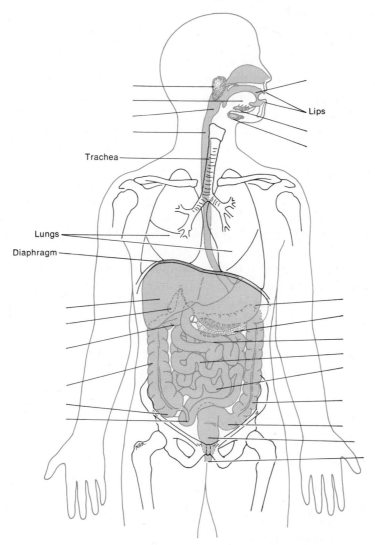

Trachea

Lungs

Diaphragm

Lips

**Figure 35–1.** The organs of the digestive system.

from the mouth to the anus (Figure 35-1). The length of a tract taken from a cadaver is about 35 meters (30 to 40 feet). In a living person it is somewhat shorter because the muscles lying in its walls are in a state of tonic contraction. Organs comprising the alimentary canal include the mouth, pharynx, esophagus, stomach, small intestine, and large intestine. The alimentary canal contains the food while it is being eaten, digested, and eliminated. Muscular contractions in the walls of the alimentary canal break down the food physically by churning it. Secretions produced by cells in the alimentary canal break down the food chemically.

The second group of organs comprising the digestive system are the *accessory organs*—the teeth, tongue, salivary glands, liver, gallbladder, and pancreas. Teeth are cemented to bone, protrude into the alimentary canal, and aid in the physical breakdown of food. The other accessory organs lie totally outside the canal and produce or store secretions—which aid in the chemical breakdown of the food—that are dumped into the canal through ducts.

Fill in the missing labels in Figure 35-1.

## PART III   THE PERITONEUM

The *peritoneum* is the largest serous membrane of the body. It consists of a layer of simple squamous epithelium and an underlying supporting layer of connective tissue. The *visceral peritoneum* covers some of the organs and constitutes their serosa; the *parietal peritoneum* lines the walls of the abdominopelvic cavity. The space between the parietal and visceral peritonea is called the *peritoneal cavity.*

Unlike the 2 other serous membranes of the body, the pericardium and pleura, the peritoneum contains large folds that weave in between the viscera. The folds bind the organs to each other and to the walls of the cavity and contain the blood and lymph vessels and the nerves that supply the abdominal organs. One extension of the peritoneum is called the *mesentery* and is an outward fold of the serous coat of the intestines (Figure 35-2a). Attached to the posterior abdominal wall is the tip of the fold. The mesentery binds the small and most of the large intestines to the wall. It also carries blood vessels and lymphatics to the intestines.

Two other important peritoneal folds are the lesser omentum and the greater omentum. The *lesser omentum* arises as 2 folds in the serosa of the stomach and duodenum (Figure 35-2b). They extend anteriorly and connect with the visceral peritoneum of the liver. The lesser omentum suspends the stomach and duodenum from the liver (Figure 35-2d). An extension of the visceral peritoneum of the liver ties it, in turn, to the diaphragm and the upper abdominal wall. The *greater omentum* is a large fold in the serosa of the stomach that hangs down like an apron over the front of the intestines (Figure 35-2c). It then passes up to a part of the large intestine (the transverse colon), wraps itself around it, and finally attaches to the parietal peritoneum of the posterior wall of the abdominal cavity. Because the greater omentum contains large quantities of adipose tissue, it commonly is called the "fatty apron." The greater omentum contains numerous lymph nodes. If an infection occurs in the intestine, plasma cells formed in the nodes combat the infection and help to prevent it from spreading to the peritoneum. Inflammation of the peritoneum, *peritonitis*, is a serious condition because the peritoneal membranes are continuous with each other, enabling the infection to spread to all the organs in the cavity (Figure 35-2d).

Fill in the missing labels in Figure 35-2.

## PART IV   THE MOUTH

The *mouth*, also referred to as the *oral* or *buccal cavity*, is formed by the cheeks, hard and soft palates, and tongue (Figure 35-3). During chewing the cheeks and lips help to keep food between the upper and lower teeth. The tongue muscles move the tongue from side to side and in and out, and maneuver food for chewing and swallowing.

Finish the labeling of the mouth in Figure 35-3.

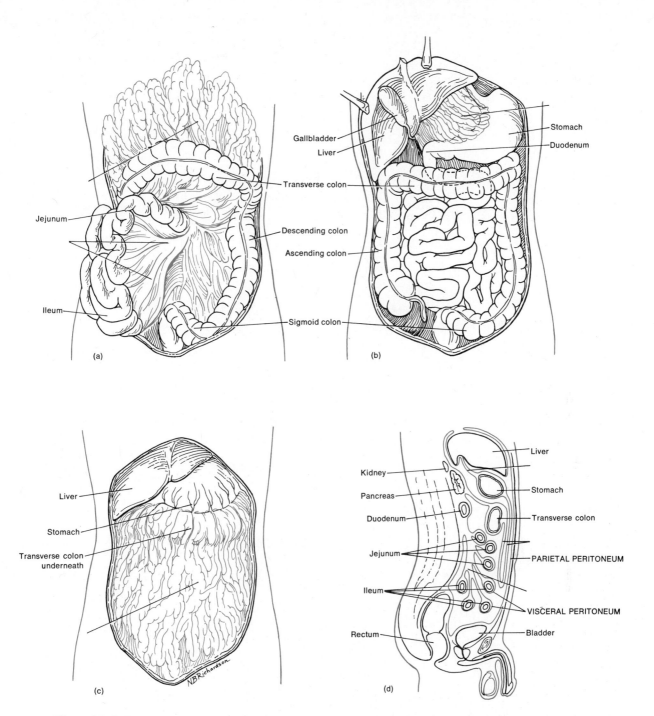

**Figure 35-2.** Extensions of the peritoneum. (a) Mesentery. The greater omentum has been lifted. (b) Lesser omentum. The liver and gallbladder have been lifted. (c) Greater omentum. (d) Sagittal section through the abdomen and pelvis indicating the relationship of the peritoneal extensions to each other.

## PART V SALIVARY GLANDS

Saliva is a fluid that is continuously secreted by glands lying in or near the mouth. Ordinarily, just enough saliva is secreted to keep the mucous membranes of the mouth moist, but when food enters the mouth, secretion increases so the saliva can lubricate, dissolve, and chemically break down the food. The mucous membrane lining the mouth contains many small glands, the *buccal glands*,

**353**

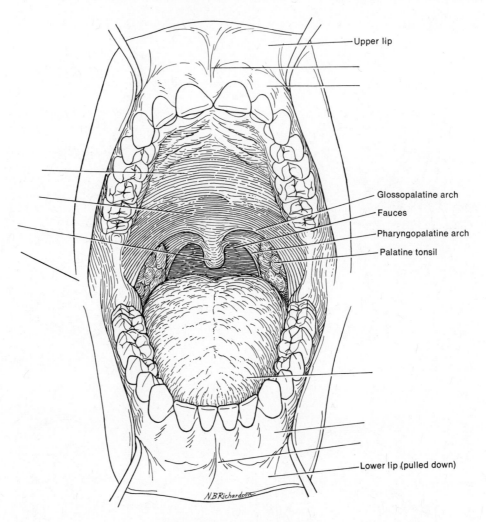

**Figure 35–3.** The mouth.

Upper lip

Glossopalatine arch

Fauces

Pharyngopalatine arch

Palatine tonsil

Lower lip (pulled down)

N B Richardson

that secrete small amounts of saliva. However, the major portion of saliva is secreted by the *salivary glands*, which lie outside the mouth and pour their contents into ducts that empty into the oral cavity. There are 3 pairs of salivary glands: the parotid, submandibular, and sublingual glands (Figure 35–4). The *parotid glands* are located under and in front of the ears. Each secretes into the oral cavity via a duct that opens into the inside of the cheek opposite the upper second molar tooth. The *submandibular glands* are found beneath the base of the tongue in the posterior part of the floor of the mouth, and their ducts are situated on either side of the frenulum. The *sublingual glands* are anterior to the submandibular glands, and their ducts open into the floor of the mouth.

Label completely Figure 35–4.

The enzyme *salivary amylase* initiates the breakdown of carbohydrates, which is the only chemical digestion that occurs in the mouth. The function of salivary amylase is to break the chemical bonds between some of the monosaccharides that make up the polysaccharides, which constitute the vast majority of the carbohydrates that we eat. In this way, the enzyme catalyzes the breaking of the long-chain polysaccharides into shorter polysaccharides called *dextrins*.

Food stimulates the glands to secrete heavily. When food is taken into the mouth, chemicals in the food stimulate the taste receptors. Rolling a dry, indigestible object over the tongue produces friction, which also may stimulate the receptors. Impulses are conveyed from the receptors to 2 salivary centers in the brain stem. Returning autonomic impulses from one of the centers activate the

**354**

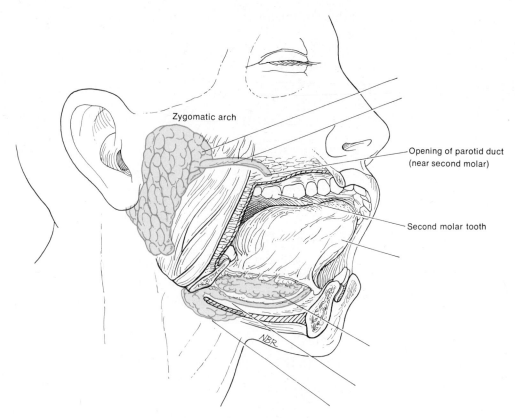

**Figure 35-4.** The salivary glands.

secretion of saliva from the parotid glands, while returning autonomic impulses from the other center activate the submandibular and sublingual glands.

The smell or sight of food also serves as a stimulus for increased saliva secretion.

## CONCLUSIONS

1. Define chemical and mechanical digestion.

2. What is another name for the alimentary canal?

3. List the organs that constitute the alimentary canal.

4. What are the accessory organs of the digestive system?

5. What are the 2 main layers of the peritoneum?

6. Name the 3 important folds of the peritoneum.

7. Give each of their locations in the body.

8. Which structure is nicknamed the "fatty apron"?

9. What is peritonitis and why is it a dangerous condition?

10. What are other names for the mouth?

11. Name and locate the 3 pairs of salivary glands.

12. Which enzyme is found in saliva?

13. Give its function.

14. Which mechanisms are involved in stimulating the salivary glands to secrete heavily?

## PART VI   THE STOMACH

The stomach is divided into 4 areas: the cardia, fundus, body, and pylorus (Figure 35-5 left). The *cardia* surrounds the cardiac sphincter, and the rounded portion above and to the left of the cardia is the *fundus*. Below the fundus, the large central portion of the stomach is called the *body*, and the more narrow, inferior region is the *pylorus*. The concave medial border of the stomach is called the *lesser curvature*, and the convex lateral border is referred to as the *greater curvature*. The pylorus communicates with the duodenum via a sphincter called the *pyloric valve*.

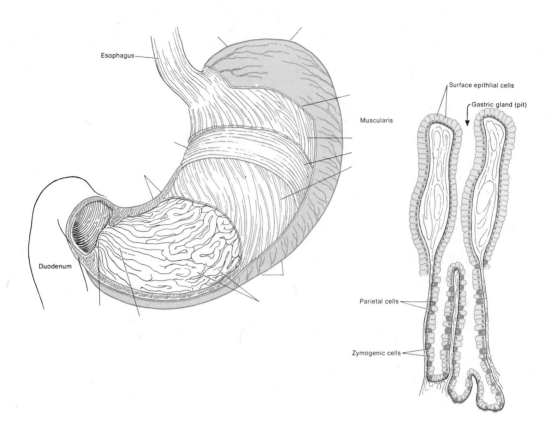

Esophagus

Muscularis

Duodenum

Surface epithlial cells

Gastric gland (pit)

Parietal cells

Zymogenic cells

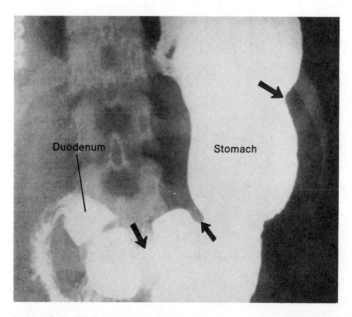

Duodenum

Stomach

**Figure 35–5.** The stomach. (Left) External and internal anatomy. (Right) Microscopic view of gastric glands from the fundus region. (Bottom) Roentgenogram of a normal stomach. Note the peristaltic waves. *(Bottom: from Lester W. Paul and John H. Juhl, The Essentials of Roentgen Interpretation, 3d. ed., Harper & Row, Publishers, Inc., New York, N.Y., 1972.)*

Two abnormalities of the pyloric valve sometimes are found in infants. One of these abnormalities, *pylorospasm*, is characterized by failure of the muscle fibers encircling the opening to relax normally. As a result, ingested food does not pass easily from the stomach to the small intestine. The stomach becomes overly full, and the infant vomits frequently to relieve the pressure. Pylorospasm is treated by drugs that relax the muscle fibers of the valve. The other abnormality, called *pyloric stenosis*, is a narrowing of the pyloric valve caused by a tumorlike mass that apparently is formed by enlargement of the circular muscle fibers. The mass obstructs the passage of food and must be surgically corrected.

Label the different areas of the stomach in Figure 35-5.

The wall of the stomach is composed of the same 4 basic layers as the rest of the alimentary canal, with certain modifications. When the stomach is empty, the mucosa lies in large folds that can be seen with the naked eye. These folds are called *rugae* (Figure 35-5 left). As the stomach fills and distends, the rugae gradually smooth out and disappear. Microscopic inspection of the mucosa reveals a layer of simple columnar epithelium containing many narrow openings that extend down into the lamina propria. These pits are called *gastric glands,* and they are lined with 3 kinds of secreting cells: zymogenic, parietal, and mucous (Figure 35-5 right). The *zymogenic*, or *chief cells*, secrete the digestive enzymes of the stomach. Hydrochloric acid, which stimulates one of the digestive enzymes, is produced by the *parietal cells*. The *mucous cells* secrete mucus and the intrinsic factor, a substance involved in the absorption of vitamin $B_{12}$. Secretions of the gastric glands are collectively called *gastric juice*.

Stomach contractions churn food, break it into small particles, mix it with gastric juice, and pass it to the duodenum. The principal chemical activity of the stomach is to begin the digestion of proteins.

The pancreas, liver, and gallbladder are accessory organs that play important roles in the chemical digestive process.

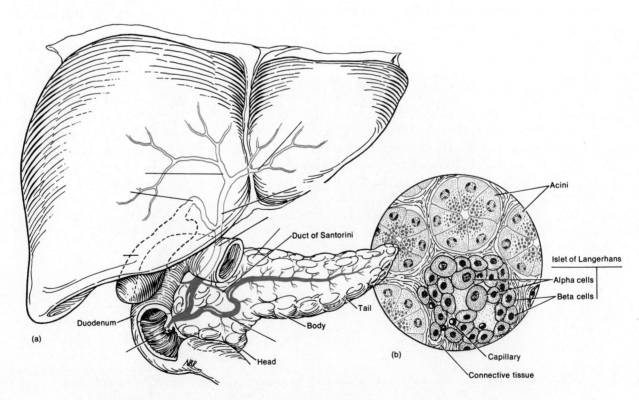

**Figure 35–6.** The pancreas. (a) The pancreas, liver, and gallbladder in relation to the duodenum. (b) Microscopic view of pancreatic cells.

Consult your textbook and label the important parts of these organs that are not labeled in Figure 35-6.

## PART VII    THE SMALL INTESTINE

The small intestine is divided into 3 segments: duodenum, jejunum, and ileum. The *duodenum*, the broadest part of the small intestine, originates at the pyloric valve of the stomach and extends about 25 to 30 centimeters (10 to 12 inches) until it merges with the jejunum. The *jejunum* is about one meter (3 to 4 feet) long and extends to the ileum. The final portion of the small intestine, the *ileum*, measures about 2 meters (6 to 7 feet) and joins the large intestine at the *ileocecal valve*. A roentgenogram of the normal small intestine is shown in Figure 35-7 top.

The walls of the intestine are composed of the same 4 coats that make up most of the alimentary canal. However, both the mucosa and submucosa are modified to allow the small intestine to complete the processes of digestion and absorption. The mucosa contains many pits lined with glandular epithelium. These pits—the *intestinal glands*, or *crypts of Lieberkühn*—secrete the intestinal digestive enzymes (Figure 35-7 bottom). The submucosa contains *Brunner's glands*, which secrete mucus to protect the walls of the small intestine from the action of the enzymes. *Succus entericus* is the name for the composite of all the intestinal secretions.

Since almost all the absorption of nutrients occurs in the small intestine, its walls need to be specially equipped to do this job. The epithelium covering and lining the mucosa consists of simple columnar epithelium. Some of the epithelial cells have been transformed to goblet cells, which secrete additional mucus. The rest contain microvilli, fingerlike projections of the plasma membrane. Digested nutrients diffuse more quickly into the intestinal wall because the microvilli increase the surface area of the plasma membrane.

The mucosa lies in a series of 2.5 centimeters (one inch) high projections called *villi*, giving the intestinal mucosa its velvety appearance (Figure 35-7 bottom). The enormous number of villi (4 to 5 million) vastly increases the surface area of the epithelium available for the epithelial cells specializing in absorption. Each villus is lined with the lamina propria, the connective tissue layer of the mucosa. Embedded in this connective tissue are an artery, a venule, a capillary, and a *lacteal* (lymphatic vessel). Nutrients that diffuse through the adjacent epithelial cells are able to pass through the walls of the capillary and lacteal and enter the blood.

A third set of projections called *plicae circulares* further increases the surface area for absorption. The plicae are deep folds in the mucosa and submucosa (Figure 35-7 left). Some of the folds extend all the way around the circumference of the intestine, and others extend only part-way around.

The muscularis of the small intestine consists of 2 layers of smooth muscle. The outer, thinner layer contains longitudinally arranged fibers, and the inner, thicker layer contains circularly arranged fibers. Except for a major portion of the duodenum, the serosa, or visceral peritoneum, completely covers the small intestine.

The hormonal control of digestion is summarized in Table 35-1.

Consult your textbook and complete Table 35-1 under the columns "where produced" and "action."

## PART VIII    THE LARGE INTESTINE

The overall functions of the large intestine are the completion of absorption, the manufacture of some vitamins, the formation of feces, and the expulsion of feces from the body.

The *large intestine* is about 1.5 meters (5 feet) in length and averages 6.5 centimeters (2.5 inches) in diameter. It extends from the ileum to the anus and is attached to the posterior abdominal wall by its mesentery. Structurally, the large intestine is divided into 4 principal regions: the cecum, colon, rectum, and anal canal (Figure 35-8). Let us now look at the parts of the large intestine in the order in which food passes through them.

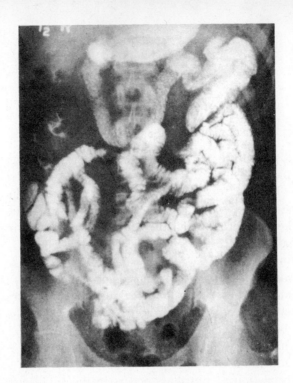

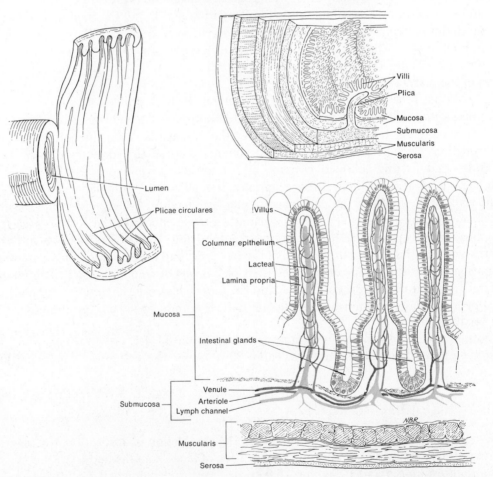

**Figure 35–7.** The small intestine. (Top) Roentgenegram of the normal small intestine one half hour after taking a barium "meal." (Left) Section of small intestine cut open to expose plicae circulares. (Middle) Villi in relation to the coats of the small intestine. (Bottom) Enlarged aspect of several villi. *(Top: from Lester W. Paul and John H. Juhl, The Essentials of Roentgen Interpretation, 3d ed., Harper & Row, Publishers, Inc., New York, N.Y., 1972.)*

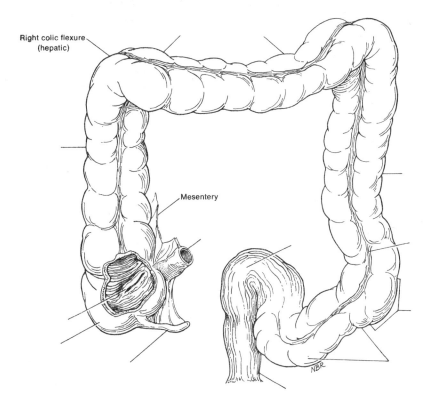

Right colic flexure
(hepatic)

Mesentery

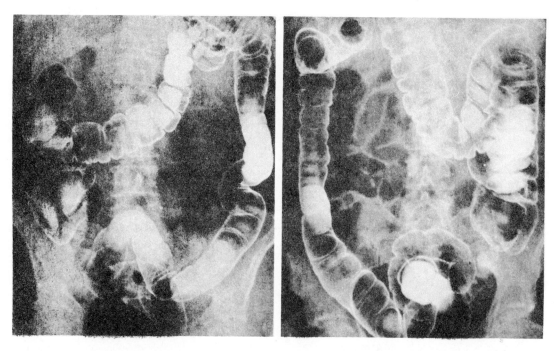

NBR

**Figure 35–8.** The large intestine. (Top) Anatomy of the large intestine. (Bottom) Roentgenogram of the large intestine in which several haustra are clearly visible. *(Bottom): from Lester W. Paul and John H. Juhl, The Essentials of Roentgen Interpretation, 3d ed., Harper & Row, Publishers, Inc., New York, N.Y., 1972.)*

## Table 35–1.
## SUMMARY OF THE HORMONAL CONTROL OF DIGESTION IN THE STOMACH AND SMALL INTESTINE

| HORMONE | WHERE PRODUCED | STIMULANT | ACTION |
|---|---|---|---|
| Gastrin | | Partially digested proteins | |
| Gastrinlike hormone | | Partially digested proteins | |
| Secretin | | Acidity of chyme | |
| Pancreozymin | | Acidity of chyme | |
| Enterocrinin | | Acidity of chyme | |
| Cholecystokinin | | Combination of acid and fat | |
| Enterogastrone | | Fats | |

The opening from the ileum into the large intestine is guarded by a fold of mucous membrane called the *ileocecal valve*. This structure allows materials from the small intestine to pass into the large intestine but prevents them from moving in the opposite direction. Hanging below the ileocecal valve is the *cecum*, a blind pouch about 6 centimeters (2 to 3 inches) long. Attached to the cecum is a twisted, coiled tube, called the *vermiform appendix* (vermis=worm). Inflammation of the appendix is called *appendicitis*.

The open end of the cecum merges with a long tube called the *colon*. Based on location, the colon is divided into ascending, transverse, descending, and sigmoid portions. The *ascending colon* ascends on the right side of the abdomen, reaches the undersurface of the liver, and turns abruptly to the left. The colon continues across the abdomen to the left side as the *transverse colon*. It curves beneath the lower end of the spleen on the left side and passes downward to the level of the iliac crest as the *descending colon*. The *sigmoid colon* is the S-shaped portion that begins at the iliac crest, projects inward to the midline, and terminates as the *rectum* at about the level of the third sacral vertebra.

The rectum, the last 20 centimeters (7 to 8 inches) of gastrointestinal tract, lies anterior to the sacrum and coccyx. The terminal part of the rectum is referred to as the *anal canal*. Internally, the mucous membrane of the anal canal is arranged in longitudinal folds called *anal columns* that contain a network of arteries and veins. Inflammation and enlargement of the anal veins is known as *hemorrhoids* or *piles*. The opening of the anal canal to the exterior is called the *anus*. It is guarded by an internal sphincter of smooth muscle and an external sphincter of skeletal muscle. Normally the anus is closed except during the elimination of the wastes of digestion.

Completely label Figure 35–8, the large intestine.

## CONCLUSIONS

1. What are the 4 main areas of the stomach?

2. Explain the 2 abnormalities of the pyloric valve.

3. Name the 3 types of secreting cells of the gastric glands in the stomach and their secretions.

4. What are the 3 main accessory glands of digestion?

5. List the 3 segments of the small intestine and their respective sizes.

6. How does the small intestine join the large intestine?

7. Which structures secrete the intestinal digestive enzymes?

8. How are the walls of the small intestine protected from digestive juices?

9. What is the name for the composite of all intestinal secretions?

10. Name the 3 distinct projections that help the small intestine absorb nutrients by increasing its surface area considerably.

11. List the overall functions of the large intestine.

12. What are the 4 principal regions of the large intestine?

13. What is the significance of the cecum in man?

14. Name the 4 regions of the colon.

15. What is the termination of the sigmoid colon called.

16. What is the significance of the veins contained in the anal columns?

17. How is the anus protected?

**STUDENT ACTIVITY**

Divide the class into 3 groups; one group will take the digestive accessory organs; the second group will take the small intestine; and the third group will take the large intestine. Each group will list all the details of their particular subject from memory, mentioning various regions, special structures, and basic functions of all the parts. The 3 groups will then combine their information and determine how accurate their concepts of the digestive system were.

# MODULE 36
# NUTRITION: THE UTILIZATION OF FOODS

**OBJECTIVES**

1. To define a nutrient and list the functions of the 6 classes of nutrients

2. To define enzymes as catalysts that speed up many of the reactions in the body

3. To define activation energy and enzyme-substrate complexes

4. To perform tests using enzymes as catalysts to demonstrate digestion of carbohydrates (starch), proteins, and lipids

5. To demonstrate the effects of temperature and pH on starch digestion

6. To demonstrate the effect of bile on fats

In the last module you learned how food is digested and absorbed. Now you will learn what happens to the food after it reaches the cells of the body. You also will learn what nutrients are needed for survival and why they are needed.

*Nutrients* are chemical substances in food that provide energy, act as building blocks in forming new body components, or assist body processes. There are 6 major classes of nutrients: carbohydrates, lipids, proteins, minerals, vitamins, and water. Carbohydrates, proteins, and lipids are the raw materials for reactions occurring inside cells. The cells either break them down to release energy or use them to build new structures and new regulatory substances, such as hormones and enzymes. Some minerals and many vitamins are used by enzyme systems that catalyze the reactions undergone by carbohydrates, proteins and lipids. Many minerals have other functions. Water has 4 major functions. It acts as a reactant in hydrolysis reactions, as a solvent and suspending medium, as a lubricant, and as a coolant.

Proteins that are produced by living cells to catalyze or speed up many of the reactions in the body are *enzymes*. Protein in the form of an enzyme acts as a *catalyst*, which is any chemical substance that speeds up a reaction without being permanently altered by the reaction.

For any chemical or biochemical reaction to occur, a certain amount of energy is required. The amount of energy needed is called the *activation energy*. One way to cause most of the molecules to obtain an activation-level energy is to heat them. Unfortunately, heat happens to destroy the body's proteins. The role of an enzyme, then, is to decrease the amount of energy needed to start the reaction.

Exactly how enzymes lower activation energies is not fully understood. However, we do know that an enzyme attaches itself to one of the reacting molecules, called a *substrate,* and the two form a temporary *enzyme-substrate complex.* Many kinds of enzymes exist, but each kind can attach to only one kind of substrate. Apparently, the enzyme molecule must fit perfectly with the substrate molecule like pieces of a jigsaw puzzle. If the enzyme and substrate do not fit properly, no reaction occurs.

Since digestive enzymes function outside of the cells that produce them, they are capable of also reacting within a test tube, and therefore provide an excellent means of studying enzyme activity.

## PART I   POSITIVE TESTS FOR SUGAR AND STARCH

Salivary amylase (or ptyalin) is produced by the salivary glands which empty their secretions by way of ducts into the mouth. This enzyme starts *starch* digestion and *hydrolyzes* (converts) it into maltose. We measure the amount of starch and sugar present before and after enzymatic activity. It is expected that the amount of starch should *decrease* while the sugar level should *increase* as a result of salivary amylase activity.

## PART II   TEST FOR SUGAR

Benedict's test is a commonly used test for detecting reducing sugars. Glucose, maltose, or any other reducing sugar reacts with *Benedict's solution,* forming insoluble red cuprous oxide. The *precipitate* of cuprous oxide can usually be seen in the bottom of the tube when standing. Benedict's solution turns green, yellow, orange, or red depending on the amount of reducing sugar present.

Test for the presence of sugar (maltose) with the following procedure. Place 2.0 milliliters of maltose and 2.0 milliliters of Benedict's solution in a test tube. Place the test tube in a boiling water bath and heat it for 5 minutes; note the color change from blue to red. Repeat the same procedure using starch solution instead of maltose. Notice that there is no color change because of the absence of sugar. Now you have a method for detecting the presence of sugar.

## PART III   TEST FOR STARCH

*Lugol's solution* is a special iodine solution that is used to test certain polysaccharides, especially starch. Starch, for example, gives a *deep blue* to *"black"* color with Lugol's solution (the black is really a concentrated blue color). Cellulose, monosaccharides, and disaccharides do not react. A negative test is indicated by a yellow to brown color of the solution itself, or possibly some other color (other than blue to black) resulting from pigments present in the substance being tested.

Place a drop of starch solution on a spot plate, and then add a drop of Lugol's solution to it. Notice the black color that forms as the starch-iodine complex develops. Repeat the test using a maltose solution in place of the starch solution. Note that there is no color change, since Lugol's solution and maltose do not combine. Now you have a method for detecting the presence of starch.

## PART IV   DIGESTION OF STARCH

Salivary amylase from freshly collected saliva is used in this procedure to show digestion of starch. Collect some saliva by rinsing the mouth thoroughly with distilled water two or three times. Then chew a piece of paraffin wax or sugarless gum to stimulate saliva flow, collecting 4 to 5 milliliters of saliva in a small container. Add an equal amount of tap water carefully, so that you have diluted saliva which contains the enzyme to be studied. You may also use commercially prepared salivary amylase.

**366**

Transfer 5.0 milliliters of a starch solution to a small beaker, add 5.0 milliliters of the saliva solution (or a pinch of anylase powder) and mix thoroughly. Wait one minute, record the time, remove one drop of the mixture with a glass rod to the depression of a spot plate, and then test for starch with Lugol's solution. At one-minute intervals test one-drop samples of the mixture until there is no longer a positive test for starch. Keep the glass rod in the mixture, stirring it from time to time.

After the starch test is seen to be negative, test the remaining mixture for the presence of glucose. Do this by adding 2.0 milliliters of the mixture to 2.0 milliliters of the Benedict's solution and heat in a boiling water bath.

How long did it take for the starch to be digested?

Did your observation indicate the presence of maltose?

What is the meaning of this result?

## PART V    EFFECT OF TEMPERATURE ON STARCH DIGESTION

In the following procedure, we will test starch digestion at 5 different temperatures, to determine how temperature influences enzyme activity. Lugol's solution is again used for presence or absence of starch.

A fresh enzyme solution (salivary amylase) should be prepared as before. Five constant-temperature water baths should be available, starting with the lowest temperature; an ice bath at 1°C or lower, 10°C, 40°C, 60°C, and boiling.

Prepare 10 test tubes so that 5 of them each contain 1.0 milliliters of the enzyme solution, while the other 5 each contain 1.0 milliliters of the starch solution. Place one pair of tubes (i.e., a tube containing 1.0 milliliters enzyme, and a tube with 1.0 milliliters starch) into the ice bath, the 10°C, 40°C, 60°C, and boiling water baths respectively, and permit the tubes to adapt to the bath temperatures for about 5 minutes. After this time combine and mix the enzyme and starch solutions together, replacing the single test tube in the respective water bath. After 30 seconds, test all 5 tubes for starch on a spot plate. Repeat every 30 seconds until you have determined the time for the disappearance of the starch. Record your results on the following lines and notice what the optimum temperature is for starch digestion.

0°–1°_____    40°_____        Boiling _____

10°  _____    60°_____

## PART VI    EFFECT OF pH ON STARCH DIGESTION

In this procedure the effectiveness of salivary amylase digestion at different pH levels is demonstrated. Three buffer solutions are prepared and should be made up as follows:

solution A: pH—4.0
solution B: pH—7.0
solution C: pH—9.0

Once again a fresh enzyme solution (salivary amylase) should be prepared. Mix 4.0 milliliters of a starch solution with a 2.0 milliliters of buffer solution A in a test tube. Repeat this procedure with buffer solutions B and C, and you end up with 3 test tubes of a starch-buffer solution, each at a different pH (4.0, 7.0 and 9.0).

Place one drop of starch-buffer solution A on a spot plate, and immediately add one drop of the saliva to it. Test for starch disappearance as in the previous experiment with Lugol's solution, and record the time when starch first disappears completely. Repeat this test for the other 2 starch buffers (B and C) and record the time when starch is no longer present at each pH. Explain your results.

A: pH 4.0 _____
B: pH 7.0 _____
C: pH 9.0 _____

## PART VII    DIGESTION OF FATS

In this experiment the effect of pancreatic juice on fat is demonstrated. Pancreatin contains all the enzymes present in pancreatic juice and is the substance used. Since the optimum pH of the pancreatic enzymes ranges from 7.0 to 8.8, the pancreatin is prepared in sodium carbonate. The enzyme activity in this test is lipase, which digests fat to fatty acids and glycerol. The fatty acid produced changes the color of *blue* litmus to *red*.

Place 5.0 milliliters of litmus cream (heavy cream to which powdered litmus has been added to give it a blue color) in a test tube, and put it in a 40°C water bath. Repeat the procedure with another 5.0 milliliter portion in a second tube, but put it in an ice bath. When the tubes have adapted to their respective temperatures (around 5 minutes), add 5.0 milliliters of pancreatin to each tube and replace them in their water baths until a color change occurs in one tube.

Complete the following table:

| TUBE NO. | TEMPERATURE OF WATER BATH | CHANGE IN pH (COLOR) | EXPLANATION OF RESULTS |
|---|---|---|---|
| 1 | _____ | _____ | _____ |
| 2 | _____ | _____ | _____ |

## PART VIII    ACTION OF BILE ON FATS

Bile is important in the process of digestion because of its emulsifying (breaking down of large fat globules to smaller, uniformly-distributed particles) effect on fats and oils. *Bile does not contain any enzymes.* Emulsification of lipids by means of bile serves to increase the surface area of the lipid which will be exposed to the action of lipase.

Place 5.0 milliliters of water into one test tube, and 5.0 milliliters of a bile solution into a second test tube. Add one drop of vegetable oil which has been colored with a fat-soluble dye, such as Sudan B, into each tube. Shake both tubes *vigorously*, then let them stand in a test-tube rack undisturbed for 10 minutes. If a fat or oil is broken into small enough droplets it will remain suspended in water in the form of an *emulsion*.

1. What difference can you detect in the appearance of the mixtures in the 2 vials?

2. How does emulsification of lipids by means of bile aid the digestion of fat?

## PART IX    DIGESTION OF PROTEIN

Here we demonstrate the effect of pepsin on protein and the factors affecting the rate of action of pepsin. Pepsin is a proteolytic enzyme secreted by the chief cells in the lining of the stomach, and it digests proteins (fibrin in this experiment) to proteoses and peptones. The efficiency of pepsin activity depends upon the pH of the solution, the optimum being 1.5 to 2.5. Pepsin is almost completely inactive in neutral or alkaline solutions.

Prepare the following 5 test tubes, numbering them from one to 5. In this test it is important to measure the quantity of each solution carefully.

Tube 1—5.0 ml. of 0.5% pepsin; 5.0 ml. of 0.8% HCL
Tube 2—5.0 ml. of pepsin; 5.0 ml. of water
Tube 3—5.0 ml. of pepsin, boiled for 10 minutes in a water bath; 5.0 ml. of 0.8% HCL
Tube 4—5.0 ml. of pepsin; 5.0 ml. of 0.5% NaOH
Tube 5—5.0 ml. of water; 5.0 ml. of 0.8% HCL

First determine the approximate pH of each test tube with Hydrion paper (range one to 11) and put the values in the table at the end of this test.

Place the same amount of fibrin (the protein) in each test tube. An amount near the size of a pea will be sufficient. Put the tubes in a 40°C water bath and shake them once in a while. Maintain 40°C temperature closely. The tubes should remain in the water bath for *at least one and a half hours*. Watch the changes that the fibrin undergoes. The swelling that occurs in some tubes should not be confused with digestion. When the fibrin is digested, it becomes transparent and *disappears* (dissolves) as the protein is digested to soluble proteoses and peptones.

Finish the experiment when the fibrin is digested in one of the 5 tubes.

Record all your observations in the following Table:

| TUBE NO. | TUBE CONTENTS | pH | DIGESTION (+ OR —) | EXPLANATION OF RESULTS |
|---|---|---|---|---|
| 1 | | | | |
| 2 | | | | |
| 3 | | | | |
| 4 | | | | |
| 5 | | | | |

## CONCLUSIONS

1. Define a nutrient.

2. Name the 6 major classes of nutrients.

3. What are the 4 major functions of water?

4. Define enzyme.

5. What is a catalyst?

6. A chemical or biochemical reaction needs a certain amount of energy. What is this energy called?

7. What is your interpretation of a substrate?

8. Which enzyme is present in saliva?

9. What is its function in the mouth?

10. What solution is used in testing for reducing sugars such as glucose and maltose, and what is the color change?

11. Name the solution used in testing for starch and its color change when positive.

12. What is the purpose of chewing on paraffin wax or sugarless gum?

13. How do you explain the fact that in the temperature-effect-on-starch-digestion experiment, neither the 60°C or boiling temperatures worked?

14. In the test for the digestion of fats why is pancreatin used?

15. What is the role of bile in fat digestion?

16. What is the correlation between pepsin and hydrochloric acid?

**STUDENT ACTIVITY**

If available, conduct the same digestive experiments from this module on various other carbohydrates, lipids, and proteins. Compare the results to determine if there are individual differences between them in each group of nutrients.

# MODULE 37
# DISORDERS OF THE DIGESTIVE SYSTEM

**OBJECTIVES**

1. To list the causes and symptoms of dental caries and periodontal disease

2. To contrast between the location and effects of gastric and duodenal ulcers

3. To compare pancreatitis and cirrhosis as disorders of the accessory organs of digestion

4. To describe the location of tumors of the gastrointestinal tract

5. To define phenylketonuria (PKU), cystic fibrosis, and celiac disease as disorders related to faulty metabolism

6. To define the causes and treatment of obesity

7. To define medical terminology associated with the digestive system

Now that we have discussed the structure, physiology, and nutrition of the digestive system, this module looks at some disorders that are related to it.

## PART I    DENTAL CARIES

*Dental caries*, or tooth decay, involve a gradual disintegration of the enamel and dentin. If this condition remains untreated, various microorganisms may invade the pulp cavity, causing infection and inflammation of the living tissue. If the pulp is destroyed, the tooth is pronounced "dead."

No individual microbe is responsible for dental caries, but oral bacteria that create a pH of 5.5 or lower start the process. Acids can come directly from foods, such as the ascorbic acid of citrus fruits, or they may be breakdown products of carbohydrates. Microbes that digest carbohydrates include 2 bacteria, *Lactobacillus acidophilus* and streptococci, as well as some yeasts. Research suggests that the streptococci break down carbohydrates into *dental plaque*, a polysaccharide which adheres to the tooth surface. When other bacteria digest the plaque, acid is produced. Saliva cannot reach the tooth surface to buffer the acid because the plaque covers the teeth.

Certain measures can be taken to prevent dental caries. First, the diet of the mother during pregnancy is very important in forestalling tooth decay of the newborn. Simple, balanced meals are the best diet during pregnancy. Supplementation with multivitamins, with emphasis on vitamin D and the minerals calcium and phosphorus, is customary because they are responsible for normal bone and teeth development.

Other preventive measures have centered around fluoride treatment because teeth are less susceptible to acids when they are permeated with fluoride. Fluoride may be incorporated into the drinking water or applied topically to erupted teeth. Maximum benefit often occurs when fluoride is used in drinking water during the period when teeth are being calcified. Excessive fluoride may cause a light brown to brownish-black discoloration of the enamel of the permanent teeth called "mottling."

Brushing the teeth immediately after eating removes the plaque from flat surfaces before the bacteria have a chance to go to work. Dentists also suggest that the plaque between the teeth be removed every 24 hours with dental floss.

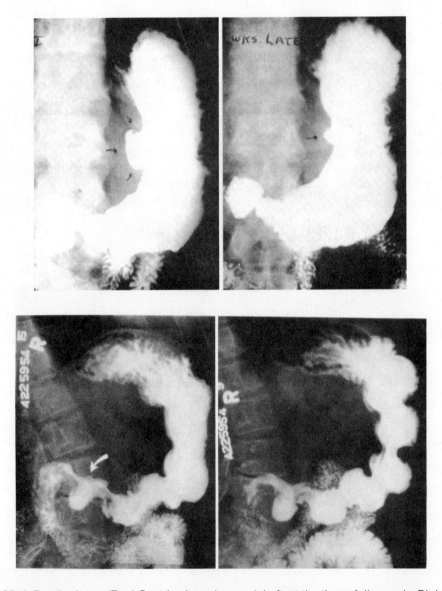

**Figure 37–1.** Peptic ulcers. (Top) Gastric ulcers (arrows). Left, at the time of diagnosis. Right, 3 weeks after treatment. (Bottom) Duodenal ulcer (arrow). Left, prior to treatment. Right, after treatment. *(Courtesy of Lester W. Paul and John H. Juhl, The Essentials of Roentgen Interpretation, 3d ed., Harper & Row, Publishers, Inc., New York, N.Y., 1972.)*

## PART II    PERIODONTAL DISEASES

*Periodontal disease* is a collective term for a variety of conditions characterized by inflammation and/or degeneration of the gingivae, alveolar bone, periodontal membrane, and cementum. The initial symptoms are enlargement and inflammation of the soft tissue. Without treatment, the soft tissue may deteriorate and the alveolar bone may be resorbed, causing loosening of the teeth and receding of the gums.

Periodontal diseases are frequently caused by local irritants, such as bacteria, impacted food, cigarette smoke, or by a poor "bite." The latter may put a strain on the tissues supporting the teeth. Methods of prevention and treatment include good mouth care to remove plaque and other sources of irritation. Periodontal disease may also be caused by allergies, vitamin deficiencies, and a number of systemic disorders, especially those that affect bone, connective tissue, or circulation. In these cases, the systemic disorder must be treated as well.

## PART III    ULCERS

An *ulcer* is a craterlike lesion in a membrane. Ulcers that develop in areas of the alimentary canal exposed to acid gastric juice are called *peptic ulcers.* Peptic ulcers occasionally develop in the lower end of the esophagus. However, most of them occur on the lesser curvature of the stomach, in which case they are called *gastric ulcers,* or in the first part of the duodenum where they are called *duodenal ulcers* (Figure 37-1).

The cause of ulcers is obscure. However, hypersecretion of acid gastric juice seems to be the immediate cause in the production of duodenal ulcers and in the reactivation of healed ulcers. Hypersecretion of acid gastric juice is not implicated as much in gastric ulcer patients because the stomach walls are highly adapted to resist gastric juice through their secretion of mucus. A possible cause of gastric ulcers is hyposecretion of mucus. Hypersecretion of pepsin also may contribute to ulcer formation.

Among the factors believed to stimulate an increase in acid secretion are certain foods or medications, such as alcohol, coffee, or aspirin, and overstimulation of the vagus nerve. Normally, the mucous membrane lining the stomach and duodenal walls resists the secretions of hydrochloric acid and pepsin, and no ulcer develops. In some people, however, this resistance breaks down, and an ulcer develops.

The danger inherent in ulcers is the erosion of the muscular portion of the wall of the stomach or duodenum. This could damage blood vessels and possibly produce fatal hemorrhage. If an ulcer erodes all the way through the wall, the condition is called *perforation.* Perforation allows bacteria and partially digested food to pass into the peritoneal cavity, producing peritonitis.

## PART IV    CIRRHOSIS

*Cirrhosis* is a chronic disease of the liver in which the parenchymal liver cells are replaced by fibrous connective tissue, a process called *stromal repair.* Often there is a lot of replacement by adipose connective tissue as well. The liver has a high ability for parenchymal regeneration, so stromal repair occurs whenever any parenchymal cell is killed or when damage to the cells occurs continuously over a long time. These conditions could be caused by *hepatitis* (inflammation of the liver), certain chemicals that may destroy liver cells, parasites that sometimes infect the liver, and alcoholism. Malnutrition, particularly deficiencies of essential amino acids, is common among alcoholics. It is not known whether degeneration of the cells is caused by alcohol, malnutrition, or both.

## PART V    TUMORS

Both benign and malignant *tumors* occur in all parts of the gastrointestinal tract. The benign growths are much more common, but the malignant tumors are responsible for 30 percent of all deaths from cancer in the United States. To achieve relatively early diagnosis, complete periodic routine examinations are necessary. Cancers of the mouth usually are detected through routine dental checkups.

A regular physical checkup should include rectal examination. Fifty percent of all rectal carcinomas are within reach of the finger, and 75 percent of all colonic carcinomas can be seen with the *sigmoidoscope* (Figure 37-2). Both the fiberoptic sigmoidoscope and the more recent fiberoptic *endoscope* are flexible tubular instruments composed of a light and many tiny glass fibers. They allow visualization, magnification, and even photography of almost the entire length of the gastrointestinal tract and have been invaluable in the correct diagnosis of a wide range of gastrointestinal disorders without surgery

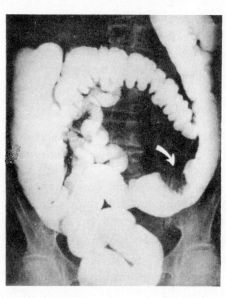

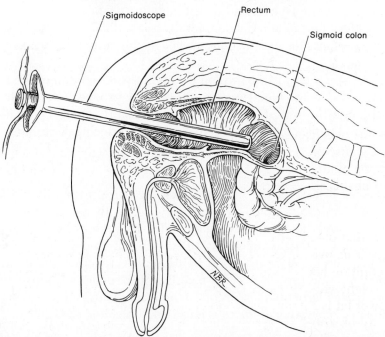

**Figure 37-2.** Tumors of the alimentary canal. (Top) Carcinoma of the sigmoid colon is indicated at the arrow. (Bottom) Detection of carcinomas by use of the sigmoidoscope. *(Top from Lester W. Paul and John H. Juhl, The Essentials of Roentgen Interpretation, 3d ed., Harper & Row, Publishers, Inc., New York, N.Y., 1972.)*

Another test in a routine examination for intestinal disorders is the filling of the gastrointestinal tract with barium, which is either swallowed or given in an enema. Barium, a mineral, shows up on x-rays the same way that calcium appears in bones. Tumors as well as ulcers can be diagnosed this way. The only definitive treatment for gastrointestinal carcinomas is surgery.

## PART VI   NUTRITIONAL AND METABOLIC DISORDERS

### PHENYLKETONURIA

By definition, *phenylketonuria* is an inborn eror of metabolism characterized by an elevation of the amino acid phenylalanine in the blood and is frequently associated with mental retardation. The DNA of people with phenylketonuria lacks the gene that normally programs the manufacture of the enzyme phenylalanine hydroxylase. This enzyme is necessary for the conversion of phenylalanine into the amino acid tyrosine, an amino acid that enters the Krebs cycle. As a result, phenylalanine cannot be metabolized, and what is not used in protein synthesis builds up in the blood. High levels of phenylalanine are toxic to the brain during the early years of life when the brain is developing, and mental retardation is produced. Mental retardation can be prevented, when the condition is detected early, by restricting the child to a diet that supplies only the amount of phenylalanine necessary for growth.

### CYSTIC FIBROSIS

*Cystic fibrosis* is an inherited disease of the exocrine glands, affecting the pancreas, respiratory system, and salivary and sweat glands. It is characterized by the production of thick exocrine secretions that do not drain easily from the passageways. The buildup of the secretions leads to inflammation and replacement of injured cells with connective tissue that blocks the passageways. One of the prominent features is blockage of the pancreatic ducts so that the digestive enzymes cannot reach the intestine. Since pancreatic juice contains the only fat-digesting enzyme in the adult body, the person fails to absorb fats or fat-soluble vitamins and thus suffers from vitamin A, D, and K deficiency diseases. Calcium also needs fat to be absorbed, so tetany also may result.

A child suffering from cystic fibrosis is given pancreatic extract and large doses of vitamins A, D, and K. The therapeutic diet is low, but not lacking, in fats and high in carbohydrates and proteins that can be used for energy and can also be converted by the liver into the lipids essential for life processes.

### CELIAC DISEASE

Cystic fibrosis was once confused with an allergy to gluten—the protein in wheat, rye, barley, and oats—called *celiac disease*. The allergy causes changes in the muscosa of the small intestine that decrease the absorption of all nutrients. The condition is easily remedied by administering a diet that excludes all cereal grains except rice and corn.

### OBESITY

Despite the fact that obesity is a disorder that usually can be diagnosed by inspection, it is not easily defined. There is no overall agreement as to the degree of overweight that divides obesity from nonobesity. One definition holds that an obese individual is one who is 20 percent or more over his so-called desirable weight when the extra weight is in the form of stored fat.

Because storage of fat is a normal function of adipose tissue, it is difficult to determine the point at which the quantity of stored fat becomes excessive. It is usually assumed that the "normal" or "best" weight for an individual is achieved between the ages of 18 and 25 years. Therefore, any significant positive deviation from this norm could be considered "obesity." Fatness, however, cannot always be predicted on the basis of weight. Height-weight relationships do not necessarily take into consideration an individual's build.

Much obesity, particularly when it is severe, has its roots in childhood. Children may become obese at any time, but this disorder is said to develop most commonly in 3 phases of childhood: in the latter part of infancy, at the time of starting school, and during adolescence. Obese children are likely to become obese adults. The more severe the obesity in childhood, the more likely it is to persist into adult life.

Causes of obesity can be classified under 2 broad headings: regulatory and metabolic. People with *regulatory obesity* have no apparent metabolic abnormality that can account for the obesity. They just seem to ingest more high-energy-releasing foods than their bodies need. Causes include neurotic overeating, cultural dietary habits in otherwise normal people, and inactivity. Occasionally, regulatory obesity is caused by a disorder in the hypothalamus that destroys or reduces sensations of satiety. *Metabolic obesity* results primarily from a disorder that reduces the catabolism of carbohydrates and/or fats. An example is hyposecretion of thyroxin. It is not yet certain to what degree these metabolic disorders may be caused by changes in diet or physical activity.

Regulatory obesity seems to be far more common than metabolic obesity. However, in the clinical setting, obesity seems more usefully categorized according to factors such as age of onset, degree of severity, presence of an associated pertinent disorder such as diabetes, hypertension, osteoarthritis, or hyperlipidemia.

*Treatment of Obesity.* Reduction of body weight involves restricting calorie intake to a level well below that of energy expenditure. The goals during weight decrease are:

Loss of body fat with a minimal accompanying breakdown of lean tissue
Maintenance of physical and emotional fitness during the reducing period
Establishment of eating and exercise habits that will help the formerly obese individual maintain his weight at the recommended level

A number of factors must be considered in the formulation of a reducing diet. These include the individual's degree of overweight, age, state of physical fitness, normal level of physical activity, and the presence of related illness such as hypertension, coronary heart disease, diabetes mellitus, or gastrointestinal disorders.

In recent years, "unbalanced" diets, particularly diets that are very low in carbohydrate, have achieved both popularity and notoriety and have been praised by some as being remarkably effective and condemned by others for promoting side effects. Two of the latest unbalanced diets are the *Dr. Stillman* "quick weight loss" diet—high on water and low on food—and *Dr. Robert Atkins'* method—high protein, high fat, and extremely low carbohydrate. Both diets produce ketosis because neither contains enough carbohydrate. The Dr. Stillman diet can also produce hypercholesterolemia (an excess of cholesterol in the blood). This diet is essentially limited to fat and protein. However, the fat is almost entirely of animal origin and thus is highly saturated. The average amount of cholesterol consumed per day in this diet is more than twice the amount in the average American diet.

The following points are essential for proper and effective dieting:

1. Have a complete physical examination to determine whether or not a special diet is needed.
2. Get a list of the nutrients that are required daily and foods containing them. One such booklet is *Nutritive Value of Foods*, Home and Garden Bulletin No. 72, United States Department of Agriculture.
3. Use the calorie chart contained in nutrition booklets and devise a high nutrition diet that has less calories than you have been taking in.

Medical authorities agree that a sensible diet results in the loss of approximately 2 pounds a week and not much more.

Surgery may be the answer to gross obesity that does not respond to dieting. Surgical procedures are considered when a person weighs twice as much as he or she should and has maintained this weight for at least 5 years. The surgical procedures are all drastic measures and primarily involve removing portions of the stomach or intestinal tract.

## CONCLUSIONS

1. Define dental caries.

2. What is dental plaque?

3. List some preventative measures to combat dental caries.

4. What is periodontal disease?

5. What is an ulcer?

6. Name the 2 main types and describe their location.

7. What are some possible causes of ulcers?

8. What is the inherent danger of ulcers?

9. What is this condition called?
10. Define cirrhosis.

11. What are some causes of cirrhosis?

12. Which 2 instruments are used to detect and diagnose tumors of the gastrointestinal tract?

13. How do they work?




14. Define PKU.




15. What is cystic fibrosis?




16. What does it do to the body?




17. Name and define the 2 types of obesity.




18. Which type is more common?
19. What are the 3 goals of weight decrease?

20. What is wrong with "unbalanced" diets, specifically Dr. Stillman's and Dr. Atkins'?

21. According to medical authorities, how much should you lose per week on a sensible diet?

## MEDICAL TERMINOLOGY

Define each of the following terms:

a. calculus

b. cholecystitis

c. colitis

d. colostomy

e. constipation

f. diarrhea

g. diverticulosis

h. flatus

i. heartburn

j. hepatitis

k. hernia

l. mumps

m. pancreatitis

n. vomiting

## STUDENT ACTIVITY

Divide the class into 3 groups. One group starts with the mouth and goes as far as, but not including, the small intestine. The second group takes the small intestine, the liver and gallbladder, the third group the pancreas and large intestine. Each of the groups compiles a list of diseases and/or disorders of their particular regions of the digestive system. They are asked to include causes, symptoms, and treatment, if possible.

# MODULE 38
# STRUCTURE AND PHYSIOLOGY
# OF THE URINARY SYSTEM

**OBJECTIVES**

1. To identify the gross anatomical features of the urinary system and the kidneys

2. To define the structural adaptations of a nephron for urine formation

3. To describe the blood and nerve supply to the kidneys

4. To describe the process of urine formation

5. To define glomerular filtration, tubular reabsorption, and tubular secretion

6. To compare the lungs, integument, and alimentary canal as organs of excretion that help maintain body pH

7. To describe the structure and physiology of the ureters, urinary bladder, and urethra

8. To describe the physiology of micturition

9. To compare the causes of incontinence, retention, and suppression

The metabolism of nutrients results in the production of wastes by body cells, including carbon dioxide and excesses of water and heat. Protein catabolism produces toxic nitrogenous wastes, such as ammonia and urea. In addition, too many of the essential ions such as sodium, chloride, sulfate, phosphate, and hydrogen tend to be accumulated in the body. All the toxic materials and excess essential materials must be eliminated by the body.

The module deals with the structure and physiology of the urinary system. The primary function of the urinary system is to keep the body in hemeostasis by controlling the concentration and volume of blood by removing and restoring selected amounts of water and solutes. It also excretes selected amounts of various wastes.

Two kidneys, 2 ureters, one urinary bladder, and a single urethra comprise the system (Figure 38-1). The kidneys control the concentration and volume of the blood and remove wastes from the blood, manufacturing urine in the process. Urine drains out of each kidney through its ureter and is stored in the urinary bladder until it is expelled from the body through the urethra.

Finish labeling Figure 38-1.

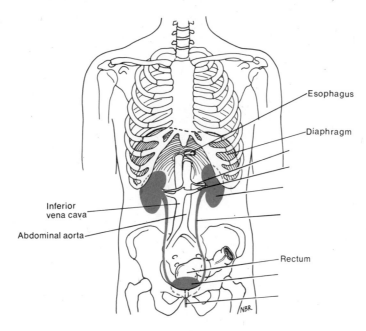

**Figure 38-1.** Organs of the urinary system.

## PART I  THE KIDNEYS

The paired kidneys are reddish organs that resemble lima beans in shape (see Figure 38-1). They are found just above the waist, between the parietal peritoneum and the posterior wall of the abdomen. Since they are external to the peritoneal lining of the abdominal cavity, their placement is described as *retroperitoneal*.

The average adult kidney measures about 11.25 centimeters (4 inches) long, 5.0 to 7.5 centimeters (2 to 3 inches) wide, and 2.5 centimeters (one inch) thick. Its concave medial border faces the vertebral column. Near the center of the concave border is a notch called the *hilum* through which the ureter leaves the kidney. Blood, lymph vessels, and nerves also enter and exit the kidney through the hilum.

Three layers of tissue surround each kidney. The innermost layer, the *renal capsule*, is a smooth, transparent, fibrous membrane that adheres to the kidney and is continuous with the outer coat of the ureter at the hilum. It serves as a barrier against trauma and the spread of infection to the kidney. The second layer, the *adipose capsule*, is a mass of fatty tissue surrounding the renal capsule. It also protects the kidney from trauma and holds it firmly in place in the abdominal cavity. The outermost layer, the *renal fascia*, is a thin layer of fibrous connective tissue that anchors the kidneys to their surrounding structures and to the abdominal wall. Some individuals, especially thin ones in whom either the adipose capsule or renal fascia is deficient, may develop a condition called *ptosis* (dropping) of one or both kidneys. Ptosis is dangerous because it may cause kinking of the ureter with reflux of urine and back pressure. Ptosis of the kidneys below the rib cage also makes the individual susceptible to blows and penetrating injuries.

If you make a longitudinal section through the kidney, you will see an outer, reddish area called the *cortex*, and an inner, reddish-brown region called the *medulla* (Figure 38-2). Within the medulla are 8 to 18 striated, triangular-shaped structures termed *renal*, or *medullary pyramids*. The bases of the pyramids face the cortical area, and their apices, called *renal papillae*, are directed toward the center of the kidney. The cortex is the smooth-textured area extending from the renal capsule to the bases of the pyramids and into the spaces between the pyramids. Together, the cortex and renal pyramids constitute the parenchyma of the kidney. Structurally, the parenchyma of each kidney consists of approximately one million microscopic units called *nephrons*, collecting ducts, and their

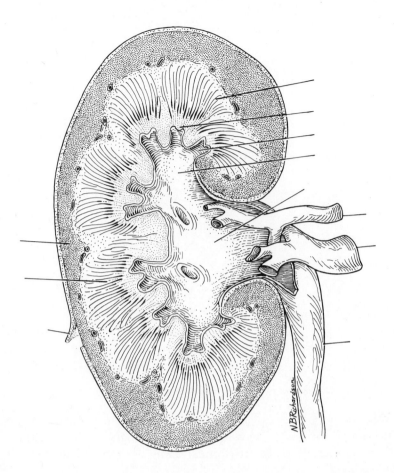

**Figure 38–2.** Longitudinal section through the right kidney illustrating gross internal anatomy.

associated vascular supply. Nephrons are the functional units of the kidney. They form the urine and regulate the blood composition.

In the center of the kidney is a large cavity called the *renal pelvis* of the kidney. The edge of the pelvis is divided into cuplike extensions called the *major* and *minor calyces*. Each minor calyx collects urine from collecting ducts. From the calyx, the urine drains into the body of the pelvis and out through the ureter.

Completely label Figure 38–2.

## PART II   THE NEPHRON

The physiological unit of the kidney is referred to as a *nephron* (Figure 38–3). Essentially, each nephron is a *renal tubule* plus its associated blood supply. The parts of a nephron are as follows: glomerular capsule, proximal convoluted tubule, descending limb of Henle, loop of Henle, ascending limb of Henle, and distal convoluted tubule. Let us examine each of these parts of a nephron in detail. It begins as a double-walled globe called the *glomerular,* or *Bowman's capsule,* lying in the cortex of the kidney. The inner wall of the capsule consists of simple squamous epithelium surrounding a capillary network called the *glomerulus.* A space separates the inner wall from the outer one, which is also composed of simple squamous epithelium. Collectively, the Bowman's capsule and the enclosed glomerulus are called a *renal corpuscle.*

The first section of the renal tubule, the *proximal convoluted tubule,* also lies in the cortex. Convoluted means that the tubule is highly coiled rather than straight, and the word "proximal" refers to the fact that the tubule is nearest its point of origin at the Bowman's capsule. The wall of the proxi-

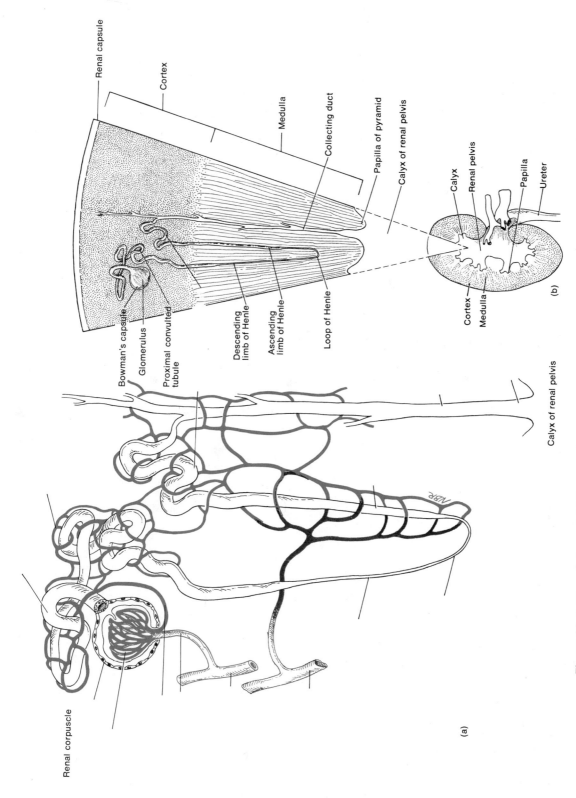

Renal capsule

Cortex

Medulla

Collecting duct

Papilla of pyramid

Calyx of renal pelvis

Bowman's capsule

Glomerulus

Proximal convulted tubule

Descending limb of Henle

Ascending limb of Henle

Loop of Henle

Calyx

Renal pelvis

Papilla

Ureter

Cortex

Medulla

(b)

Renal corpuscle

NBR

Calyx of renal pelvis

(a)

**Figure 38–3.** The nephron. (a) Microscopic appearance of an isolated nephron. (b) Position of a nephron in relation to the cortex and medulla.

mal convoluted tubule consists of cuboidal epithelium with microvilli. These cytoplasmic extensions, like those of the small intestine, increase the surface area for reabsorption and secretion.

The second section of the renal tubule, the *descending limb of Henle*, dips into the medulla. It consists of squamous epithelium. The tubule then bends into a C-shaped structure called the *loop of Henle*. As the tubule straightens out, it increases in diameter and ascends toward the cortex as the *ascending limb of Henle*, which consists of cuboidal and columnar epithelium. In the cortex, the tubule again becomes convoluted. Because of its distance from the point of origin at the Bowman's capsule, this section is referred to as the *distal convoluted tubule*. Like those of the proximal tubule, the cells of the distal tubule are cuboidal with microvilli. The distal tubule terminates by merging with a straight *collecting duct*. In the medulla, the collecting ducts receive the distal tubules of several nephrons, pass through the renal pyramids, and open into the calyces of the pelvis through a series of *papillary ducts*.

Completely label the missing parts in Figure 38-3.

## CONCLUSIONS

1. What is the main function of the urinary system?

2. List the organs that make up the urinary system.

3. Explain why the position of the kidneys in the body is called retroperitoneal.

4. Everything that enters and leaves the kidney goes through a notch in the concave border. What is this called?

5. What is another name for a "dropped kidney"?

6. What name is given to the outer area of the kidney? The inner area?

7. What is a renal corpuscle?

8. Name the physiological units of the kidneys.

## PART III   BLOOD AND NERVE SUPPLY

Because the nephrons are responsible for removing wastes from the blood and regulating its fluid and electrolyte content, it should not seem surprising that they are abundantly supplied with blood vessels. The 2 *renal arteries* transport about one-fourth the total cardiac output to the kidneys (see Figure 38-1). Thus, approximately 1,200 milliliters of blood pass through the kidneys each minute.

**384**

Before or immediately after entering through the hilum, the renal artery divides into several branches that enter the parenchyma and pass between the renal pyramids. Further divisions of the branches produce a series of interlobular arteries (see Figure 38-3a). The interlobular arteries enter the cortex and divide into *afferent arterioles*. One afferent arteriole is distributed to each glomerular capsule, where the arteriole breaks up into the capillary network termed the *glomerulus*. The glomerular capillaries then reunite to form an *efferent arteriole*, leading away from the capsule, that is smaller in diameter than the afferent arteriole. This situation is unique because blood usually flows out of capillaries into venules and not into other arterioles. Each efferent arteriole divides to form a second network of capillaries, called the *peritubular capillaries*. The peritubular capillaries supply the renal tubule and then eventually reunite to form *intralobular veins*. The blood then drains through veins running between the pyramids and leaves the kidney through a single *renal vein* that exits at the hilum.

The nerve supply to the kidneys is derived from the *renal plexus* of the autonomic system. Nerves from the plexus accompany the renal arteries and their branches and are distributed to the vessels. Because the nerves are vasomotor, they regulate the circulation of blood in the kidney by regulating the diameters of the small blood vessels.

## PART IV  URINE FORMATION

Urine formation requires 3 principal processes: glomerular filtration, tubular reabsorption, and tubular secretion.

The first step in urine formation is *glomerular filtration*. Filtration occurs in the renal corpuscle of the kidneys. When blood enters the glomerulus, the blood pressure forces water and dissolved blood components through the walls of the capillaries and on through the adjoining inner wall of the Bowman's capsule. The resulting fluid is called the *filtrate*. In a healthy person, the filtrate consists of all the materials present in the blood except for the formed elements and proteins, which are too large to pass through the capillary walls.

The second step in urine formation is *tubular reabsorption*, which is the movement of the filtrate back into the blood. It is a very discriminating process with only specific amounts of specific substances reabsorbed, depending on the needs of the body at a particular moment. Materials that are reabsorbed include water, glucose, amino acids, any proteins that managed to get into the filtrate, and ions such as $Na^+$, $K^+$, $Ca^{++}$, $Cl^-$, and $HCO_3^-$. Tubular reabsorption is an important process because it allows the body to retain most of its nutrients.

The third process involved in urine formation is *tubular secretion* or *tubular excretion*. Whereas tubular reabsorption removes substances from the filtrate, tubular secretion adds materials to the filtrate. In man, these secreted substances include $K^+$, $H^+$, ammonia, creatine, and the drugs penicillin and para-aminohippuric acid, among others. Tubular secretion has 2 principal effects, it rids the body of certain materials, and it controls the blood pH.

Table 38-1 summarizes filtration, reabsorption, and secretion in the nephrons.

## PART V  OTHER EXCRETORY ORGANS

The lungs, integument, and alimentary canal all perform specialized excretory functions (Table 38-2). Primary responsibility for regulating body temperature through the excretion of water is assumed by the skin. The lungs maintain blood-gas homeostasis through the controlled excretion of $CO_2$. One way in which the kidneys maintain homeostasis is by coordinating their activities with the activities of the other excretory organs. For example, when the integument increases its excretion of water, the renal tubules increase their reabsorption of water, and blood volume is maintained. Or, if the lungs fail to eliminate enough carbon dioxide, the kidneys attempt to compensate. They change some of the $CO_2$ into hydrogen ions, which are excreted, and into sodium bicarbonate, which becomes a part of the blood buffer systems.

Table 38–1.
SUMMARY OF FILTRATION, REABSORPTION, AND SECRETION.

| REGION OF NEPHRON | ACTIVITY |
|---|---|
| Renal corpuscle | Filtration of glomerular blood under hydrostatic pressure results in formation of a filtrate free of plasma proteins and cellular elements of blood. |
| Proximal convoluted tubule and descending and ascending limbs of Henle | Reabsorption of physiologically important solutes such as $Na^+$, $K^+$, $Cl^-$, $HCO_3^-$, and glucose. Obligatory reabsorption of water by osmosis. |
| Distal convoluted tubule | Reabsorption of $Na^+$. Secretion of $H^+$, $NH_3$, $K^+$, creatine, and certain drugs. Conservation of sodium bicarbonate. |
| Collecting duct | Facultative reabsorption of water under control of vasopressin-ADH. |

Table 38–2.
EXCRETORY ORGANS OF THE BODY AND PRODUCTS ELIMINATED.

| EXCRETORY ORGANS | PRODUCTS ELIMINATED PRIMARY | SECONDARY |
|---|---|---|
| Kidneys | Water, soluble salts from protein catabolism, and inorganic salts. | Heat and carbon dioxide |
| Lungs | Carbon dioxide | Heat and water |
| Skin | Heat | Carbon dioxide, water, and salts |
| Alimentary tract | Solid wastes and secretions | Carbon dioxide, water, salts, and heat |

## PART VI   THE URETERS

Once urine is formed by the nephrons, it drains through the collecting ducts into the calyces surrounding the renal papillae. The minor calyces join with the major calyces that unite to become the renal pelvis (see Figure 38–2). From the pelvis, the urine drains into the ureters and is carried down to the urinary bladder. From the bladder, the urine is discharged from the body through the single urethra.

The body has 2 *ureters*, one for each kidney. Each ureter is an extension of the pelvis of the kidney and runs 25 to 30 centimeters (10 to 12 inches) to the bladder (see Figure 38–1). As the ureters descend, their thickened walls progressively increase in diameter, but at their widest point they measure less than 1.7 centimeters (½ inch) in diameter. Like the kidneys, the ureters are retroperitoneal in placement.

The principal function of the ureters is to carry urine from the renal pelvis into the urinary bladder. Urine is carried through the ureters primarily by peristaltic contractions of the muscular walls of the ureters, but hydrostatic pressure and gravity also contribute.

## PART VII   THE URINARY BLADDER

The *urinary bladder* (Figure 38-4) is a hollow muscular organ situated in the pelvic cavity posterior to the symphysis pubis. In the male it is directly in front of the rectum, whereas in the female it is also in front, under the uterus and in front of the vagina.

At the base of the bladder is a small triangular area with its apex pointing anteriorly. The opening to the urethra is found in the apex of the triangle. At the 2 points that form the base of the triangle, the ureters drain into the bladder. This triangular area is called the *trigone*.

Four coats comprise the walls of the bladder. The mucosa, the innermost coat, is a mucous membrane containing transitional epithelium. Recall that transitional epithelium is able to stretch, a marked advantage for an organ that must continually inflate and deflate. Stretchability is further enhanced by the rugae, or folds in the mucosa that appear when the bladder is empty. The second coat, the submucosa, is a layer of dense connective tissue that connects the mucous and muscular coats. The third coat—a muscular one called the *detrusor muscle*—consists of 3 layers: inner longitudinal, middle circular, and outer longitudinal muscles. In the area around the opening to the urethra, the circular fibers form an *internal sphincter* muscle. Below the internal sphincter is the *external sphincter* that is composed of skeletal muscle. The outermost coat, the serous, is formed by the peritoneum and covers only the superior surface of the organ.

Label the indicated structures in Figure 38-4.

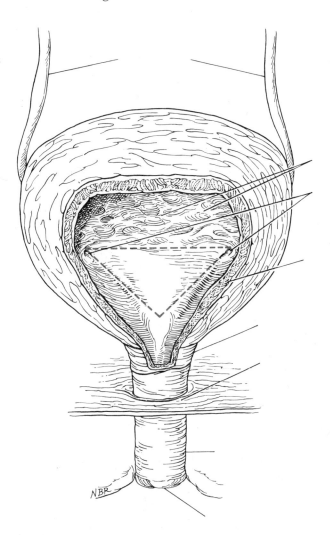

**Figure 38-4.** The urinary bladder and urethra.

## PART VIII   THE URETHRA

The *urethra* is a small tube leading from the floor of the bladder to the exterior of the body (see Figure 38-4). In females, it lies directly behind the symphysis pubis and is embedded in the anterior wall of the vagina. Its undilated diameter is about 6 millimeters (one-fourth inch) and its length is approximately 3.8 centimeters (one and one-half inches). The female urethra is directed obliquely downward and forward, and the opening of the urethra to the exterior, the *urinary meatus*, is located between the clitoris and vaginal opening.

In males, the urethra is about 20 centimeters (8 inches) long, and it follows a different course from that of the female urethra. Immediately below the bladder it runs vertically through the prostate gland. It then penetrates the penis and takes a curved course through its body. Unlike the female urethra, the male urethra serves as a common tube for urinary and reproductive systems.

Label completely Figure 38-4.

## PART IX   PHYSIOLOGY OF MICTURITION

Urine is expelled from the bladder by an act called *micturition*. This response is brought about by a combination of involuntary and voluntary nervous impulses. The average capacity of the bladder is 700 to 800 milliliters. When the amount of urine in the bladder exceeds about 200 to 400 milliliters, stretch receptors in the bladder wall transmit impulses to the lower portion of the spinal cord. These impulses initiate a conscious desire to expel urine and an unconscious reflex referred to as the *micturition reflex*. In the micturition reflex, parasympathetic impulses transmitted from the spinal cord reach the bladder wall and internal urethral sphincter, bringing about contraction of the bladder and relaxation of the internal sphincter. Then the conscious portion of the brain sends impulses to the external sphincter, the sphincter relaxes, and urination takes place. Although emptying of the bladder is controlled by reflex, it may be initiated voluntarily and may be started or stopped at will because of cerebral control of the external sphincter.

A lack of voluntary control over micturition is referred to as *incontinence*. In infants about 2 years and under, incontinence of urine is normal because they have not developed voluntary control over the external sphincter muscle. Infants void whenever the bladder is sufficiently distended to arouse a reflex stimulus. Proper training overcomes incontinence if the latter is not caused by emotional stress or irritation of the bladder.

Involuntary micturition in the adult may occur as a result of unconsciousness, injury to the spinal nerves controlling the bladder, irritation due to abnormal constituents in urine, disease of the bladder, and emotional stress due to failure of the detrusor muscle to relax.

*Retention* is a term used to describe a condition in which there is a failure to void urine. Retention may be due to an obstruction in the urethra or neck of the bladder, nervous contraction of the urethra, or lack of sensation to urinate.

A condition far more serious than retention is *suppression*, or *anuria*—the failure of the kidneys to secrete urine. It usually occurs when blood plasma is prevented from reaching the glomerulus as a result of inflammation of the glomeruli. Anuria also may be caused by a low filtration pressure.

## CONCLUSIONS

1. Which main blood vessels enter and exit from the kidneys?


2. Which part of the autonomic system supplies nerves to the kidneys?

3. What are the 3 processes that participate in urine formation?

4. Name the other areas of the body, besides the kidneys, that are involved in excretory functions.

5. What part of the kidney becomes the ureter as it leaves the kidney?

6. What is the trigone? How is it formed?

7. What is the significance of the detrusor muscle?

8. Compare the length of the urethra in females and males.

9. What is micturition?

10. What is the average capacity of urine (in millimeters) of the urinary bladder?

11. Contrast incontinence, retention, and suppression.

## STUDENT ACTIVITY

Trace a drop of blood from its entrance into the renal artery to its exit through the renal vein. In tracing the drop of blood, make sure that you do the following: (1) name, in sequence, each blood vessel through which it passes, and (2) explain what happens during glomerular filtration, tubular reabsorption, and tubular secretion.

# MODULE 39
# URINALYSIS AND DISORDERS OF THE URINARY SYSTEM

**OBJECTIVES**

1. To list the physical characteristics of urine

2. To define albuminuria, glycosuria, hematuria, pyuria, ketosis, casts, and calculi

3. To perform various procedures involved in urinalysis

4. To discuss the causes of ptosis, kidney stones, gout, glomerulonephritis, pyelitis, and cystitis

5. To discuss the operational principle of hemodialysis

6. To define medical terminology associated with the urinary system

The kidneys perform their homeostatic functions through the manufacture of *urine*. Urine is a fluid that contains a very high concentration of solutes. In a healthy person its volume, pH, and solute concentration vary with the needs of the internal environment. During certain pathological conditions, the characteristics of urine may change drastically. An analysis of the volume and physical and chemical properties of urine tells us much about the state of the body.

## PART I    PHYSICAL CHARACTERISTICS OF URINE

Normal urine is usually a yellowish or amber-colored, transparent liquid with a characteristic and aromatic odor. The color is caused by the presence of *urochrome*, a pigment derived from the destruction of hemoglobin by reticuloendothelial cells. The color varies considerably with the ratio of solutes to water in the urine. For example, the less water there is, the darker the color of the urine. Freshly voided urine is usually transparent, but a turbid (cloudy) urine does not necessarily indicate a pathological condition since turbidity may result from mucin secreted by the lining of the urinary tract.

The reaction of normal urine is usually slightly acid, averaging about 6.0 in pH. It ranges between 5.0 and 7.8 and rarely becomes more acid than 4.5 or more alkaline than 8. Variations in urine pH are closely related to diet. Moreover, these variations are due to differences in the end products of metabolism that appear in the urine. Whereas a high-protein diet increases acidity, a diet composed largely of vegetables increases alkalinity.

*Specific gravity* is the ratio of the weight of a volume of a substance to the weight of an equal volume of distilled water. Water has a specific gravity of 1.000. The specific gravity of urine depends on the amount of solid materials in solution and ranges from 1.008 to 1.030 in normal urine. The greater the concentration of solutes, the higher the specific gravity. The lower the concentration, the lower the specific gravity. Table 39-1 summarizes the physical characteristics of normal urine.

## PART II   ABNORMAL CONSTITUENTS OF URINE

If some of the chemical processes of the body are not operating efficiently, traces of particular substances that are not normally present may appear in the urine, Or, normal constituents may appear in abnormal amounts. Analyzing the physical and chemical properties of a patient's urine often provides information that aids diagnosis. Such an analysis is called a urinalysis.

Table 39-1.
PHYSICAL CHARACTERISTICS OF NORMAL URINE.

| | |
|---|---|
| Volume | 1000-1800 ml. in 24 hours, but varies considerably with many factors |
| Color | Yellow or amber colored, but varies with quantity voided and diet |
| Turbidity | Transparent when freshly voided, becomes turbid upon standing |
| Odor | Aromatic, becomes ammonialike upon standing |
| Reaction | Ranges in pH from 5.0-7.8, average 6.0, varies considerably with diet |
| Specific gravity | Ranges from 1.008-1.030 |

## ALBUMIN

Protein albumin is one of the things a technician looks for when he does a urinalysis. Albumin is a normal constituent of plasma, but it usually does not appear in urine because the particles are too large to pass through the pores in the capillary walls. The presence of albumin in the urine is called *albuminuria*. Albuminuria indicates an increase in the permeability of the glomerular membrane. Conditions that lead to albuminuria include injury to the glomerular membrane as a result of disease, increased blood pressure, and irritation of kidney cells by substances such as bacterial toxins, ether, or heavy metals. Other proteins, such as globulin and fibrinogen, may also appear in the urine under certain conditions.

## GLUCOSE

The presence of sugar in the urine is termed *glycosuria*. Normal urine contains such small amounts of glucose that clinically it may be considered absent. The most common cause of glycosuria is a high blood sugar level. Remember that glucose is filtered into the Bowman's capsule. Later, in the renal tubules, the tubule cells actively transport the glucose back into the blood. However, the number of glucose carrier molecules is limited. If a person ingests more carbohydrates than his body can convert to glycogen or fat, more sugar is filtered into the Bowman's capsule than can be removed by the carriers. This condition, called *temporary* or *alimentary glycosuria*, is not considered pathological.

## RED BLOOD CELLS

The appearance of red blood cells in the urine is called *hematuria*. Hematuria generally indicates a pathological condition. One possible cause is acute inflammation of the urinary organs as a result of disease or of irritation from stones in the organs. Whenever blood is found in the urine, additional tests are performed to ascertain the part of the urinary tract that is bleeding. One should also make sure that the sample was not contaminated with blood from the vagina.

## WHITE BLOOD CELLS

The presence of leucocytes and other components of pus in the urine, referred to as *pyuria*, is evidence that there is some kind of infection in the kidney or other urinary organs. Again, the source of the pus must be located, and care should be taken to make sure the urine was not contaminated.

## KETONE BODIES

Ketone, or acetone, bodies appear in normal urine in very small amounts. However, their appearance in urine in unusually high quantities, a condition called *ketosis*, or *acetonuria*, may indicate a number of abnormalities. For example, it may be caused by diabetes mellitus, starvation, or simply too little carbohydrate in the diet. Whatever the cause, excessive quantities of fatty acids are oxidized in the liver, and the ketone bodies are filtered from the plasma into the Bowman's capsule.

## CASTS

Microscopic examination of urine may reveal the presence of *casts*—tiny masses of material that have hardened within and assumed the shape of the lumens of the tubules, later flushed out of the tubules by a buildup of filtrate behind them. Casts are named on the basis of either the substances that compose them or their appearance. Accordingly, there are white-blood-cell casts, red-blood-cell casts, epithelial casts that contain cells from the walls of the tubes, granular casts which contain decomposed cells that form granules, and fatty casts from cells which have become fatty.

## CALCULI

Occasionally, the salts found in urine may solidify into insoluble stones called *calculi*. They may be formed in any portion of the urinary tract from the kidney tubules to the external opening. Conditions leading to calculi formation include the ingestion of excessive amounts of mineral salts, a decrease in the amount of water, and abnormally alkaline or acid urine.

## PART III   URINALYSIS

In this part the student will determine some of the characteristics of urine, and perform tests for some abnormal constituents which may be present in urine. Some of these tests may be used in determining unknowns in urine specimens.

## PHYSICAL ANALYSIS

Obtain a fresh specimen of urine. Prior to testing *always* mix urine by swirling, inverting the container, or stirring with a wooden swab stick. *Keep all containers clean!* Wrap and discard all papers, sticks, etc., in the garbage pail. Rinse all the test tubes and glass containers carefully with *cold water* after they have cooled. Flush sinks well with cold water.

Obtain your own freshly-voided urine sample, or you will be provided with a sample to be analyzed.

The following tests can be performed immediately and quickly using the "Hema-Combistix" or "Bili-Labstix." These are special paper strips impregnated with certain chemicals in designated areas, so that a color reaction occurs when they are dipped into a urine sample.

Depending on what type of paper strip is used, perform the tests and record your results in the following area:

| TEST | NORMAL | YOUR SAMPLE |
|---|---|---|
| pH | 5.0–7.8 | _____ |
| Protein | none | _____ |
| Glucose | none | _____ |
| Ketones | none–trace | _____ |

| TEST | NORMAL | YOUR SAMPLE |
|------|--------|-------------|
| Blood | none | _____ |
| Transparency | clear–slightly cloudy | _____ |
| Odor | faintly aromatic | _____ |
| Hemoglobin | none | _____ |
| Color | straw–amber | _____ |
| Sediment | none | _____ |

The color of the urine may show some possible variations as shown below:

| COLOR | POSSIBLE CAUSE |
|-------|----------------|
| Straw-amber | urochrome, pigment found in normal urine |
| Colorless | reduced concentration |
| Silvery, milky | pus, bacteria, epithelial cells |
| Smoky brown | blood |
| Port wine | porphyrins |
| Yellow foam | bile or medications |
| Orange, green, blue, red | medications |
| Dark brown or black | increase of melanin pigment |

## SPECIFIC GRAVITY

This test is easily performed using a urinometer (hydrometer). The urinometer is a float with a numbered scale near the top that indicates directly the specific gravity of urine into which it is placed.

Mix the urine, and pour it into the urinometer cylinder. Place the urinometer in urine, spinning it slightly to make sure that it floats free. Record your reading when the urinometer float is at rest.

NORMAL                          YOUR SAMPLE

.008–1.030                      _____

## PART IV  CHEMICAL ANALYSIS

### GLUCOSE

Glycosuria occurs in patients with diabetes mellitus, as well as in certain other disorders. Traces of this sugar may occur in normal urine, but to detect these small amounts requires special tests. *Benedict's solution* is commonly used to detect the presence of reducing sugars in urine, and is not specific for the presence of glucose.

A. In a test tube (Pyrex) combine 10 drops of urine with 5.0 milliliters of Benedict's solution. Mix and place it in a boiling water bath for 5 minutes. Remove it from the heat and read the results according to the chart below. Compare your results with the Combistix data.

| COLOR | RESULT |
|-------|--------|
| Blue | negative |
| Greenish-blue | trace |
| Greenish-yellow | 1+ |
| Greenish-brown | 2+ |
| Orange-yellow | 3+ |
| Brick-red (with precipitate) | 4+ |

A high concentration of phosphates in the urine sample may produce a white precipitate, whereas the precipitate of cuprous oxide in a positive Benedict's test is red.

B. An alternate method is to use Clinitest Reagent tablets. Place 10 drops of water and 5 drops of urine in a test tube. Add one Clinitest tablet. The concentrated sodium hydroxide in the tablet generates enough heat so that the liquid will boil in the test tube. The color of the solution is graded as in Benedict's test in Part A.

## PROTEIN

Normal urine contains a trace of proteins which are hard to detect under regular laboratory procedure. Albumin is the most abundant of serum proteins and is the protein usually detected. Since tests for albumin are determined by precipitating the protein either by heat (coagulation) or by adding a reagent to it, the urine sample should either be filtered or centrifuged (see Figure 39-1). This test is done by the *Sulfosalicylic Method.*

Sulfosalicylic acid will precipitate protein in urine with a turbidity (cloudiness) that is approximately proportional to the concentration of protein present.

A. Place 3.0 milliliters of clear urine (supernatant of filtered or centrifuged urine) in a test tube and add 10% acetic acid drop by drop until the specimen is just acid. Check pH after each drop with litmus paper. Add 3.0 milliliters of 20% sulfosalicylic acid. If proteins are present, a cloudy precipitate will appear at the junction of the 2 fluids. Record the results according to the chart below. Compare your results with the Combistix data.

| APPEARANCE | RESULT |
|---|---|
| Clear (no cloudiness) | negative |
| Cloudiness or ring barely preceptible | very faint trace |
| Faint cloudiness: ring is very fine | trace |
| Dense or granular turbidity | 1+ |
| Flocculated cloudiness: thick heavy ring | 2+ |
| Curdy cloud: curdy ring | 3+ |
| Solid coagulation: solid ring | 4+ |

B. An alternate method is by using Albutest Reagent tablets containing bromophenol blue. Place the tablet on a clean surface and add one drop of urine. After the drop has been absorbed, add 2 drops of water, and allow these to penetrate before reading.

Compare the color (in daylight or fluorescent light) on top of the tablet with the color chart provided. If negative, the original color of the tablet will not be changed much at the completion of the test. If protein is present in the urine, a *blue-green* spot will remain on the tablet surface after the addition of water. The amount of protein present is indicated by the intensity of the blue-green spot.

PROTEIN RESULTS

A. Sulfosalicylic method    _____
B. Albutest reagent tablets    _____

## KETONE ACETONE

The presence of ketone or acetone bodies in urine is a result of abnormal fat catabolism. Reagents such as sodium nitroprusside, ammonium sulfate, and ammonium hydroxide are available in the form of a tablet. Ketones turn purple when added to these chemicals.

A. Place an Acetest tablet on a piece of white paper, and add one drop of urine on the tablet. If acetone or ketone is present, a *lavender-purple* color develops within 30 seconds. Compare results with the color chart that comes with the reagent.

B. An alternate method is the use of ketostix, paper strips specially prepared for absence or presence of ketones.

Dip the test end of the strip in urine (do not touch the test end with your fingers) and wait one minute. Compare the color of the dipped end with the color chart provided. A negative test shows the original color unchanged or cream colored. A positive test will produce a color from *pink* to *purple,* depending on the amount of ketone bodies present.

C. Dissolve a crystal of sodium nitroprusside in 5.0 milliliters urine (Caution: Do not handle the sodium nitroprusside with your fingers). Then add 5 drops of acetic acid to this mixture. With an eye dropper, *carefully* place one drop of sodium hydroxide on the side of the tube permitting it to run down to the mixture. A *reddish-purple ring* indicates the presence of acetone.

KETONE RESULTS

A. Acetest tablet       _____
B. Ketostix       _____
C. Sodium nitroprusside       _____

## BILE PIGMENTS

Bile pigments, biliverdin, and bilirubin are not normally present in urine. The presence of large quantities of bilirubin in the extracellular fluids produces jaundice. This is a yellowish tint to the body tissues, including yellowness of the skin and also of the deep tissues.

A. Fill a test tube halfway with urine and shake it vigorously. A yellow color of the foam is an indication of the presence of bile pigments.
B. *Rosenbach test for biliverdin.* Place some filter paper in a funnel and moisten the paper with a small amount of urine. Add a drop of *concentrated nitric acid* to the tip of the paper cone. A *greenish* color indicates a positive test for biliverdin. *Note*: Use caution in dispensing the concentrated nitric acid.
C. *Ictotest for bilirubin.* Place a drop of urine on one square of the special mat provided in the ictotest kit. Next, place one ictotest reagent tablet in the center of the moistened area. Now add 2 drops of water directly to the tablet and observe the color. The presence of bilirubin will turn the mat *blue* or *purple*.

BILE PIGMENT RESULTS

A. Shaken tube       _____
B. Rosenbach test (Biliverdin)       _____
C. Ictotest (Bilirubin)       _____

## HOMOGLOBIN

Hemoglobin is not normally found in urine.

*Weber test.* Combine and mix 10.0 milliliters of urine, 1.0 milliliter of glacial acetic acid, and 10.0 milliliters of ether in a test tube and shake vigorously. *Note*: Use caution with ether since it is very flammable. Allow the liquids to separate and remove 5.0 milliliters of the upper ether layer with a pipette. To this 5.0 milliliters portion add 1.0 milliliter of a saturated solution of benzidene in alcohol, and 1.0 milliliter of 3% hydrogen peroxide. A positive test for hemoglobin will produce a *greenish-blue* color. Compare these results with your Combistix strip.

HEMOGLOBIN RESULTS

Weber test       _____

## UREA

Place 2 drops of urine on a clean slide and carefully add 2 drops of concentrated nitric acid. *Note:* Use caution in dispensing the concentrated nitric acid. Gently place the slide on a hot plate and *slowly* warm the mixture. Examine the slide under a microscope, after the mixture has dried. Look for crystals of urea nitrate which should be quite abundant since urea is a major portion of urine.

## CREATINE

Place 5.0 milliliters of urine in a test tube and add 2.0 milliliters (use a syringe) of Picric acid solution. Warm this mixture very *slowly* and *gently* in a water bath. A positive test for creatine is a *reddish* color. Creatine is nearly as abundant as urea and should be present.

**Figure 39–1.** Tabletop centrifuge used for spinning urine samples to obtain sediment for microscopic analysis. *(Photograph courtesy of Lenni Patti.)*

## PART V  MICROSCOPIC ANALYSIS

If you allow a urine specimen to stand undisturbed for a length of time, many suspended materials will settle to the bottom. A much faster method is to centrifuge a urine sample for approximately 10 minutes (see Figure 39-1). Pour off the supernatant fluid, leaving only the sediment at the bottom for microscopic examination. Place a drop of the sediment on a clean glass slide, add one drop of Sedi-Stain, and place a cover glass over the specimen.

Starting with a low power, examine the urinary sediment for any of the following: red cells, white cells, epithelial cells, bacteria, vegetable fibers, and crystals of many types. Crystals can be identified as follows (see Figure 39-2):

Triple phosphates—exist as prisms or feathery forms
Calcium oxalate—dumbbell and octahedral shapes
Calcium carbonate—granules, spheres or dumbbells
Calcium phosphate—very pointed, wedge-shaped formations which may occur as individual crystals or grouped together to form rosettes.
Calcium Sulfate–long, thin needles or prisms. Look like calcium phosphate
Uric acid—rhombic prisms, wedges, dumbbells, rosettes, irregular crystals. Are pigmented in sediment and the color varies from yellow to dark reddish brown

*Note*: Crystals are usually much smaller than the other materials found in the urinary sediment, therefore it may be necessary to examine them under high power or oil immersion magnification.

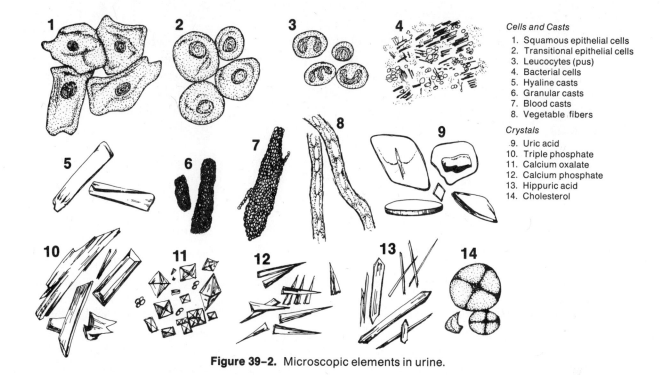

Cells and Casts
1. Squamous epithelial cells
2. Transitional epithelial cells
3. Leucocytes (pus)
4. Bacterial cells
5. Hyaline casts
6. Granular casts
7. Blood casts
8. Vegetable fibers

Crystals
9. Uric acid
10. Triple phosphate
11. Calcium oxalate
12. Calcium phosphate
13. Hippuric acid
14. Cholesterol

**Figure 39-2.** Microscopic elements in urine.

## PART VI  UNKNOWN SPECIMENS

In chemistry, when the composition of a substance is not known, it is called an *unknown*. In this part, the unknowns will be urine specimens to which glucose, albumin, or anything else can be added by the instructor, without the student knowing about it. Any and all of the tests previously outlined can then be performed on these unknowns.

Record your results, give them to the instructor, and you will be graded on your unknown results.

## PART VII  DISORDERS OF THE URINARY SYSTEM

### FLOATING KIDNEY

*Floating kidney* (ptosis) occurs when the kidney no longer is held in place securely by the adjacent organs or by its covering of fat and drifts or slips from its normal position. Pain occurs if the ureter is twisted or bent. Such an abnormal orientation also may obstruct the flow of urine.

### KIDNEY STONES

If urine becomes too concentrated, some of the chemicals that are normally dissolved in it may crystallize out, forming *kidney stones* (renal calculi). Some common constituents of stones are uric acid, calcium oxalate, and calcium phosphate. The stones usually form in the pelvis of the kidney, where they cause pain, hematuria, and pyuria. Severe pain occurs when a stone passes through a ureter and stretches its walls. Ureteral stones are seldom completely obstructive because they are usually needle-shaped and urine can flow around them.

### GOUT

*Gout*, as you may recall, is a hereditary condition associated with a high level of uric acid in the blood. When the purine-type nucleic acids are catabolized, a certain amount of uric acid is produced as a waste. Some people, however, seem to produce excessive amounts of uric acid, and others seem to have trouble excreting normal amounts. In either case, uric acid accumulates in the body and tends to solidify into crystals that are deposited in the joints and kidney tissue. Gout is further aggravated by excessive use of diuretics, dehydration, and starvation.

**397**

## INFECTIOUS DISORDERS

*Glomerulonephritis* is an inflammation of the kidney that involves the glomeruli. One of the most common causes of glomerulonephritis is an allergic reaction to the toxins given off by strepto-cocci bacteria that have recently infected another part of the body, especially the throat. The glomeruli become so inflamed, swollen, and engorged with blood that the glomerular membranes become highly permeable and allow blood cells and proteins to enter the filtrate. Thus the urine contains many erythrocytes and much protein. The glomeruli may be permanently changed, leading to chronic renal disease and renal failure.

*Pyelitis* is an inflammation of the kidney pelvis and its calyces, and *pyelonephritis* is the interstitial inflammation of one or both kidneys. The latter usually involves both the parenchyma and the renal pelvis, due to bacterial invasion from the middle and lower urinary tracts or the bloodstream.

*Cystitis* is an inflammation of the urinary bladder involving principally the mucosa and submucosa. It may be caused by bacterial infection, chemicals, or mechanical injury.

## HEMODIALYSIS

If the kidneys are impaired so severely by disease or injury that they are unable to excrete nitrogenous wastes and regulate pH and electrolyte concentration of the plasma, then the blood must be filtered by an artificial device. Such filtering of the blood is called *hemodialysis*. As you remember, *dialysis* means using a semipermeable membrane to separate large particles from smaller ones. One of the most well-known devices for accomplishing dialysis is the kidney machine (Figure 39-3). When the machine is in operation, a tube connects it with a much smaller tube implanted in the patient's radial artery. The blood is pumped from the artery and through the tubes to one side of a semiporous cellophane membrane. The other side of the membrane is continually washed with an artificial solution called the "dialyzing" solution.

All substances (including wastes) in the blood except protein molecules and erythrocytes can diffuse back and forth across the semipermeable membrane. The electrolyte level of the blood is controlled by keeping the dialyzing solution electrolytes at the same concentration as that found in normal plasma. Any excess blood electrolytes move down the concentration gradient and into the dialyzing solution. If the blood electrolyte level is normal, it is in equilibrium with the dialyzing solution, and no electrolytes are gained or lost. Since the dialyzing solution contains no wastes, substances such as urea move down the concentration gradient and into the dialyzing solution. Thus, wastes are removed, and normal electrolyte balance is maintained.

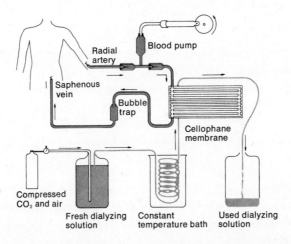

**Figure 39-3.** Diagrammatic representation of the operation of an artificial kidney. The blood route is indicated in color. The route of the dialyzing solution is indicated in gray.

**CONCLUSIONS**

1. What causes the normal color of urine?

2. What is the relationship between either a high-protein or high-vegetable diet and pH?

3. What is temporary or alimentary glycosuria?

4. Define a "floating kidney."

5. What is the more common name for renal calculi?

6. What is gout?

7. Define hemodialysis.

8. What materials go back and forth across the semipermeable membrane?

**MEDICAL TERMINOLOGY**

Define each of the following terms:

a. cystoscope

b. dysuria

c. nephrosis

d. oliguria

e. polyuria

f. stricture

g. uremia

**STUDENT ACTIVITY**

The class should be divided into 2 groups. Each group adds certain materials to normal, freshly voided urine, keeping accurate records as to what was added to which samples. The 2 groups then exchange urine samples, and each group does the basic chemical tests on the urine trying to detect accurately which substances where added.

# MODULE 40
# FLUID AND ELECTROLYTE DYNAMICS

**OBJECTIVES**

1. To identify the routes of fluid intake and fluid output

2. To explain the regulation of fluid intake on the basis of the thirst cycle

3. To explain the regulation of fluid output by describing the actions of vasopressin–ADH and aldosterone

4. To define an electrolyte and explain its functions in the body

5. To calculate the concentration of an electrolyte in milliequivalents per liter

6. To explain how fluids move between body compartments

7. To describe the role of buffers, respirations, and kidney excretion in maintaining pH

8. To define acidosis and alkalosis and distinguish between respiratory and metabolic types

In this module, you will be introduced to some of the major principles governing fluid and electrolyte dynamics. We would like you to keep in mind that water comprises the bulk of body fluids. Therefore, any reference to fluid is also a reference to water.

About two-thirds of body fluid is located inside cells and is referred to as **intracellular fluid**, or **ICF**. The remaining third of body fluid is called **extracellular fluid**, or **ECF**, and is found outside of body cells. Examples of extracellular fluid include interstitial fluid, plasma, lymph, cerebrospinal fluid, synovial fluid, and the fluids in the eyes and ears.

Body fluids contain a variety of dissolved chemicals. Some are compounds with covalent bonds called *nonelectrolytes*. Others contain ionic bonds and are referred to as *electrolytes*. When electrolytes dissolve in body fluids, they dissociate into positive and negative ions.

## PART I   WATER

*Water* comprises 45-75 percent of the total body weight. The major sources of intake include ingested liquids and foods and water produced by catabolism. Refer to Figure 40-1 and indicate the amount of water intake per day next to the label for each source. The avenues of fluid output include the kidneys, skin, lungs, and gastrointestinal tract. Refer to Figure 40–1 again and indicate the amount of water output per day next to the label for each avenue of exit.

Fluid intake is regulated by the presence or absence of thirst. When water loss is greater than intake, the condition is called dehydration. Dehydration stimulates thirst through both local and general responses. Figure 40–2 shows a cycle of changes in the body resulting from thirst. Fill in the key words that have been deleted from some of the boxes.

Under normal circumstances, fluid output is regulated directly by vasopressin-ADH and indirectly by aldosterone. Both hormones regulate urine production. A cycle of the vasopressin-ADH regulatory mechanism is presented in Figure 40–3. Fill in the key words that have been deleted from some of the boxes.

Under abnormal conditions, vomiting and diarrhea lead to fluid loss via the gastrointestinal tract; fever and skin burns lead to water loss through the skin; and hyperventilation leads to water loss by the lungs.

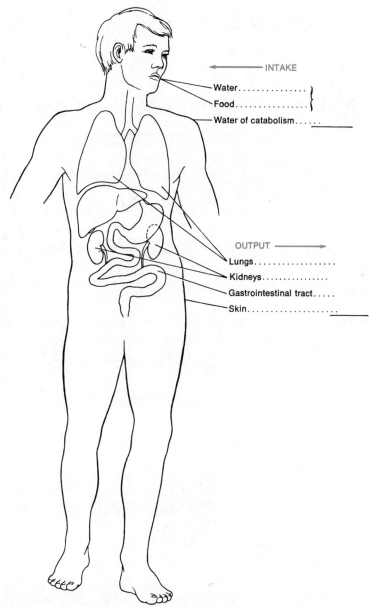

**Figure 40–1.** Routes of fluid intake and output.

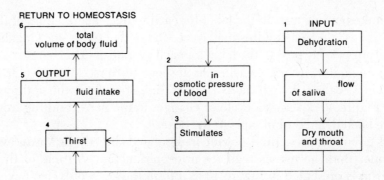

**Figure 40-2.** Thirst cycle.

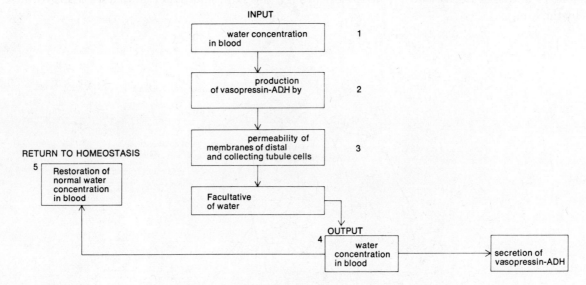

**Figure 40-3.** Control of fluid output by vasopressin–ADH.

## CONCLUSIONS

1. Distinguish between intracellular fluid and extracellular fluid.

2. What is meant by fluid balance?

3. How are the various fluid compartments isolated from each other?

4. Based upon the data you recorded in Figure 40-1, what relationship exists between fluid intake and fluid output?

## PART II  ELECTROLYTES

Body fluids contain a variety of dissolved chemicals. Some, called **nonelectrolytes**, do not form ions. Examples are most organic compounds, such as glucose, urea, and creatine. Other dissolved chemicals, called **electrolytes**, dissociate into positive ions (cations) and negative ions (anions). Most electrolytes are inorganic compounds, such as acids, bases, and salts.

In general, electrolytes serve 3 major functions in the body: (1) many are minerals needed to carry on living processes, (2) some control the movement of water between body compartments, and (3) others are needed to maintain acid-base balance.

An electrolyte exerts a far greater effect on osmosis than a nonelectrolyte does, because an electrolyte dissociates into at least 2 particles, both of which are charged. Consider the following:

$C_6H_{12}O_6$ Nonelectrolyte (glucose) — does not dissociate → $C_6H_{12}O_6$ Only one particle

NaCl Electrolyte (sodium chloride) — dissociates → $Na^+ + Cl^-$ 2 particles

Notice that glucose, a nonelectrolyte, does not dissociate and contributes only one particle to the solution. Sodium chloride, an electrolyte, dissociates into 2 particles. Thus, it has twice the effect on solute concentration as glucose.

The concentration of an ion in a solution is commonly expressed in *milliequivalents per liter* (*mEq./L.*). It is the number of electrical charges in a liter of solution, that is, the number of charges the ion carries times the number of ions in solution. The number of milliequivalents of an ion in a liter of solution is expressed as follows:

$$mEq./L. = \frac{milligrams\ (mg.)\ of\ ion\ per\ liter\ X\ number\ of\ charges}{atomic\ weight}$$

As an example, let's calculate the mEq./L. for calcium. Its atomic weight is 40, its number of charges is 2 ($Ca^{++}$) and there are 100 mg. of calcium in a liter of plasma. By substituting these values we arrive at:

$$mEq./L. = \frac{100\ x\ 2}{40}$$

$$mEq./L. = \frac{200}{40}$$

$$mEq./L. = 5$$

## CONCLUSIONS

1. Distinguish between an anion and cation.


2. After dissociation, how many particles would each of the following contribute to a solution?

   a. KCl

   b. $CaCl_2$

   c. $CaCO_3$

   d. $MgCl_2$

3. Why do electrolytes have a greater effect on a solution than nonelectrolytes?

4. Calculate the mEq./L. of sodium in plasma given the following data: mg./L.=3300, number of charges=1, atomic weight=23.

5. Compare the principal differences in electrolytic composition between plasma and interstitial fluid.

6. What is the most abundant cation in extracellular fluid?
   The most abundant anion?

7. What is the most abundant cation in intracellular fluid?
   The most abundant anion?

## PART III   MOVEMENT OF BODY FLUIDS

The movement of water between plasma and interstitial fluid occurs across capillary membranes (Figure 40-4). This movement depends on 4 forces: (1) *blood hydrostatic pressure (BHP)*, the blood pressure in a capillary, (2) *interstitial fluid hydrostatic pressure (IFHP)*, the pressure of the interstitial fluid against the cells of a tissue, (3) *blood osmotic pressure (BOP)*, the pull of water into plasma, and (4) *interstitial fluid osmotic pressure (IFOP)*, the pull of water into the interstitial fluid.

The difference between the 2 forces that move fluid out of the blood (blood hydrostatic pressure and interstitial fluid osmotic pressure) and the 2 forces that push it into the blood (interstitial fluid hydrostatic pressure and blood osmotic pressure) is called *effective filtration pressure (P$_{eff}$)*. P$_{eff}$ determines the direction of fluid movement and is represented by the following equation:

$$P_{eff} = (BHP+IFCP)-(IFHP+BOP)$$

### CONCLUSIONS

1. Refer to Figure 40-4 and, by substituting the values given, calculate the P$_{eff}$ at the arterial end of a capillary.

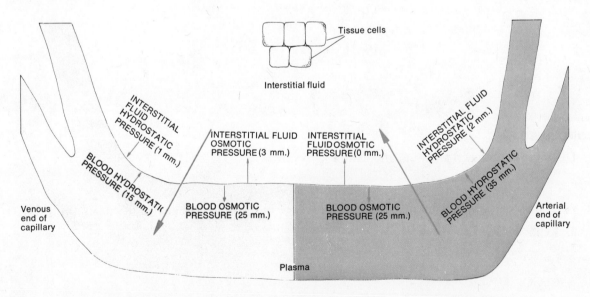

**Figure 40–4.** Forces that move water between plasma and interstitial fluid.

404

2. In which direction does fluid move at the arterial end of a capillary?
3. Refer to Figure 40-4 and, by substituting the values given, calculate the $P_{eff}$ at the venous end of a capillary.

4. In which direction does fluid move at the venous end of a capillary?

5. What is meant by Starling's law of the capillaries?

6. Define edema.

7. What are some causes of edema?

Water movement between interstitial fluid and intracellular fluid results from the same types of pressures that exist between plasma and interstitial fluid. When a fluid imbalance exists between interstitial and intracellular fluid, it is usually caused by a change in the $Na^+$ or $K^+$ concentration. As an example, we will consider how a sodium electrolyte imbalance leads to a fluid imbalance.

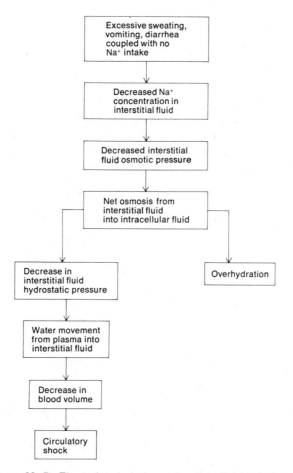

**Figure 40-5.** Electrolyte imbalance leads to fluid imbalance.

Sodium balance is normally controlled by: (1) vasopressin-ADH, which regulates the electrolyte concentration of extracellular fluid by regulating the amount of water reabsorbed into the blood by the kidney tubules, and (2) aldosterone, which regulates extracellular fluid volume by regulating the amount of sodium reabsorbed by the blood from the kidney tubules. A decrease in sodium concentration may be brought about by excessive sweating, vomiting, and diarrhea. If these occur and there is little or no sodium intake, the sodium deficit in interstitial fluid lowers the IFOP. As a result, water moves from interstitial fluid into intracellular fluid (see Figure 40–5). The movement of excess water into cells may produce overhydration which results in neurological symptoms including convulsions, coma, and even death. A second effect of the water movement is a loss of interstitial fluid that leads to a decrease in IFHP. As a result, water moves out of plasma, there is a decrease in blood volume, and shock may result.

## PART IV   ACID-BASE BALANCE

Electrolytes help to regulate acid-base balance. This overall balance is maintained by controlling the $H^+$ concentration of body fluids, particularly extracellular fluid. Normally, the pH of the extracellular fluid ranges from 7.35-7.45. This very narrow range of pH is necessary for maintaining homeostasis and depends upon 3 major mechanisms: (1) buffer systems, (2) respirations, and (3) kidney excretion.

### BUFFERS

A *buffer system* consists of a weak acid and a salt of that acid. Its function is to prevent drastic changes in the pH of a body fluid by changing strong acids and bases into weak acids and bases. This is important because strong acids dissociate into $H^+$ ions more easily than weak acids and therefore lower pH more. Similarly, strong bases raise pH more than weak ones because strong bases dissociate more easily into $OH^-$ ions.

The carbonic acid-bicarbonate buffer system is an important regulator of the pH of blood. In this system, the weak acid is carbonic acid ($H_2CO_3$) and the salt of the acid is usually sodium bicarbonate ($NaHCO_3$). If a strong base, such as NaOH, is introduced into blood, the weak acid of the buffer system reacts with it and converts it to a salt of a weak acid as follows:

$$H_2CO_3 + NaOH \longrightarrow H_2O + NaHCO_3)$$

The important thing to notice is that when carbonic acid reacts with sodium hydroxide, the products are water and sodium bicarbonate. Neither of the products has any change on pH.

If, however, a strong acid, such as HCl, is introduced into blood, the salt of the weak acid of the buffer system reacts with it and converts it to a weak acid. This reaction is as follows:

$$NaHCO_3 + HCl \longrightarrow NaHCl + H_2CO_3$$

Again, the important thing to notice is that when sodium bicarbonate reacts with the strong acid, the products are salt and a weak acid. Neither of these products has any significant change on pH.

### CONCLUSIONS

1.  Define a buffer system.


2.  Write an equation for the phosphate buffer system.


3.  Where does the phosphate buffer system operate in the body?

4. Explain how and where the hemoglobin-oxyhemoglobin buffer system operates.

5. What is the protein buffer system?

## RESPIRATIONS

The role of respirations in helping to maintain pH is as follows. An increase in the $CO_2$ concentration of body fluids lowers pH according to the following equation:

$$CO_2 + H_2O \rightleftharpoons H_2CO_3 \rightleftharpoons H^+ + HCO_3^-$$

Conversely, a decrease in the $CO_2$ concentration of body fluids raises pH.

The pH of body fluids may be adjusted by a change in the rate of breathing. If the rate of breathing is increased, more $CO_2$ is exhaled, and blood pH rises. Slowing down the respiration rate means less $CO_2$ is exhaled, and the blood pH falls.

The pH of body fluids, in turn, affects the rate of breathing. This relationship is shown in Figure 40-6. Fill in the key words that have been deleted from some of the boxes.

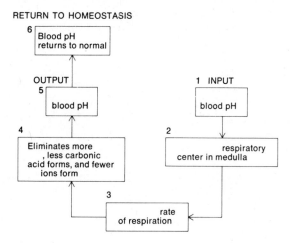

RETURN TO HOMEOSTASIS

6 Blood pH returns to normal

OUTPUT
5 blood pH

1 INPUT
blood pH

4 Eliminates more ____, less carbonic acid forms, and fewer ____ ions form

2 respiratory center in medulla

3 ____ rate of respiration

**Figure 40-6.** Relationship between pH and respirations.

## KIDNEY EXCRETION

Refer to module 38 for a description of how the kidneys help to maintain pH.

## CONCLUSIONS

1. Define acidosis.

2. Define alkalosis.

3. Distinguish between respiratory and metabolic acidosis and respiratory and metabolic alkalosis.

4. What are the effects of acidosis and alkalosis on the body?

5. Discuss how acidosis and alkalosis are treated.

**STUDENT ACTIVITY**
1. Explain briefly how edema is produced.
2. Describe how an electrolyte imbalance may lead to a fluid imbalance.

# MODULE 41
# STRUCTURE AND PHYSIOLOGY OF THE REPRODUCTIVE SYSTEMS

## OBJECTIVES

1. To identify the male organs of reproduction and describe the function of each

2. To explain the control of testosterone secretion and its effects on the body

3. To identify the female organs of reproduction and describe the functions of each

4. To explain the control of estrogen and progesterone secretion and their effects on the body

5. To correlate the events of the ovarian and menstrual cycles

6. To describe the hormonal interactions of the female sexual cycles

In this module, you will study the structure and physiology of the male and female **reproductive systems.** In addition, emphasis will be placed on the endocrine relations between the menstrual and ovarian cycles.

## PART I   MALE REPRODUCTIVE SYSTEM

The **testes,** or male gonads, are paired oval-shaped glands supported by the scrotum. Each testis is covered by *white fibrous tissue* that extends inward and divides it into internal compartments called *lobules.* The lobules contain one to three tightly coiled tubules, the *seminiferous tubules,* that produce sperm. Label these structures in Figure 41-1a.

Sperm production is called *spermatogenesis.* Immature sperm cells called *spermatogonia* develop into mature sperm called *spermatozoa.* Between the developing cells in the seminiferous tubules are *Sertoli cells* that supply nutrients for the spermatozoa. Between the tubules are clusters of *interstitial cells of Leydig* that secrete the male hormone testosterone. Label the structures shown in the cross section of the seminiferous tubule in Figure 41-1b.

**Testosterone** controls the development, growth, and maintenance of the male sex organs; the development of male secondary sex characteristics; bone growth; protein anabolism; and normal sexual behavior. Testosterone is secreted at the time of puberty under the influence of 2 anterior pituitary hormones, FSH and ICSH. This interaction of hormones represents another feedback

**409**

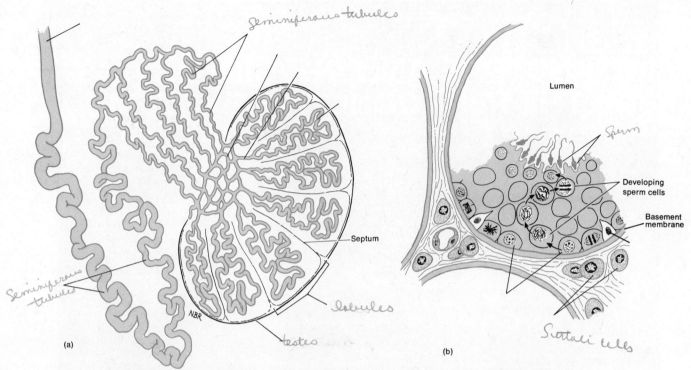

**Figure 41-1.** The testes. (a) Sagittal section. (b) Cross section of a seminiferous tubule.

Handwritten labels: Seminiferous tubules; Seminiferous tubules; testes; lobules; NBR; Lumen; Sperm; Developing sperm cells; Basement membrane; Sertoli cells; Septum; (a); (b)

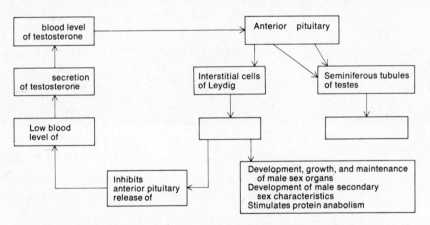

**Figure 41-2.** Interaction of anterior pituitary hormones and testosterone.

Boxes: blood level of testosterone; Anterior pituitary; secretion of testosterone; Interstitial cells of Leydig; Seminiferous tubules of testes; Low blood level of; Development, growth, and maintenance of male sex organs / Development of male secondary sex characteristics / Stimulates protein anabolism; Inhibits anterior pituitary release of

system and is illustrated in Figure 41-2. Fill in the key words that have been deleted from some of the boxes.

Mature sperm move from the seminiferous tubules to the *straight tubules, rete testes, efferent ducts, epididymis, ductus deferens, ejaculatory duct,* and *urethra*. This system of ducts enables the spermatozoa to move from the testes to the exterior. Label as many of these parts as you can in Figure 41-1a. Label the remainder in Figure 41-3.

Whereas the ducts of the system store or transport sperm, a series of glands secrete the liquid portion of semen. These include the *seminal vesicles, prostate gland,* and *bulbourethral glands.* Label these glands in Figure 41-3. Also label the other structures shown.

**Semen,** or **seminal fluid,** is a mixture of sperm and the secretions of the seminal vesicles, prostate, and bulbourethral glands. An average ejaculation of 3-4ml. contains about 400,000,000 spermatozoa. If the number falls below about 100,000,000, the male is likely to be physiologically sterile.

**410**

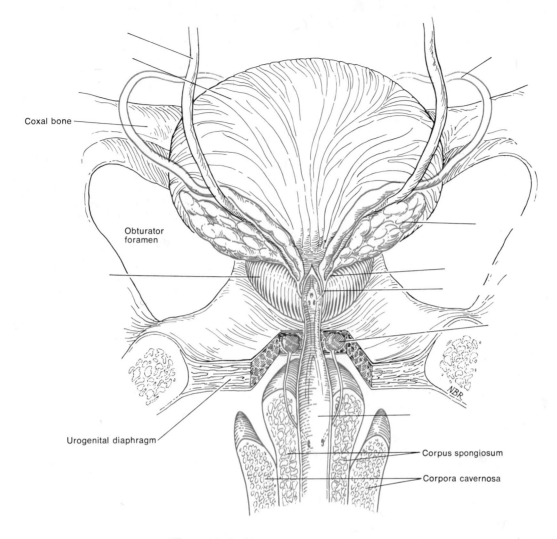

Coxal bone

Obturator
foramen

Urogenital diaphragm

Corpus spongiosum

Corpora cavernosa

NBR

**Figure 41–3.** Glands of the male reproductive system.

The **penis** introduces spermatozoa into the vagina. It contains an enlarged distal end, the *glans*, that is covered by the *prepuce* or foreskin, the part removed during circumcision. Internally, the penis consists of 3 masses of spongelike tissue that contain venous sinuses. The 2 dorsal masses are the *corpora cavernosa penis* and the ventral mass which contains the urethra is the *corpus spongiosum penis*. Label the penis and associated structures in Figure 41–4.

## CONCLUSIONS

1. Why are the testes located outside of the abdominal cavity?

2. What is the function of the scrotum?     Supports the testes

3. Define cryptorchidism.

   How is the condition corrected?

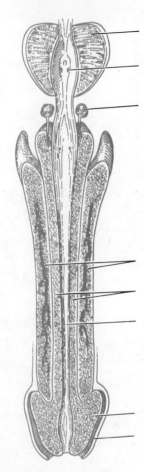

**Figure 41–4.** The penis.

4. What are the 2 principal functions of the testes?

5. Describe the parts of a spermatozoan.

6. In what ways does the anterior pituitary gland affect puberty in the male?

7. Explain the functions of the epididymis.

8. What are the functions of the ductus deferens and ejaculatory duct?

9. Describe the secretions of each of the following in producing semen:

    a. seminal vesicles

    b. prostate gland

    c. bulbourethral glands

10. What is the importance of hyaluronidase secreted by spermatozoa?


11. Why is semen slightly alkaline?


12. Describe how an erection is brought about.


## PART II   FEMALE REPRODUCTIVE SYSTEM

The **ovaries, or female gonads**, are paired glands in the upper pelvic cavity, one on each side of the uterus. They are maintained in position by a series of ligaments. These are the *mesovarium, ovarian ligament,* and *suspensory ligament.* Label these structures in Figure 41-5.

In sectional view it can be seen that each ovary has an outer layer of simple epithelium, the *germinal epithelium;* an inner area filled with *connective tissue;* and *ovarian follicles,* ova and their surrounding epithelial cells in various stages of development. Label these structures in Figure 41-6.

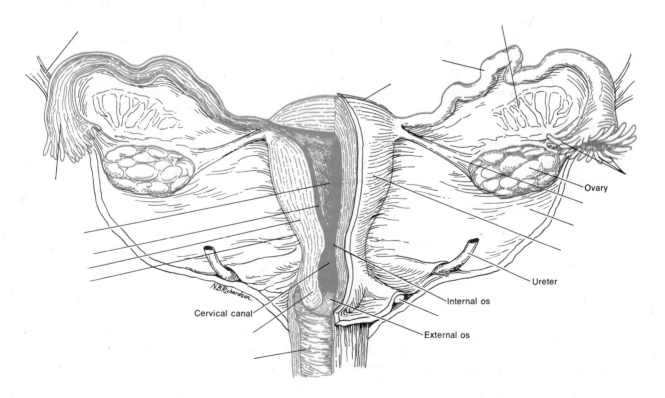

N.B.Richardson

Cervical canal

Internal os

External os

Ureter

Ovary

**Figure 41-5.** Female reproductive structures.

**413**

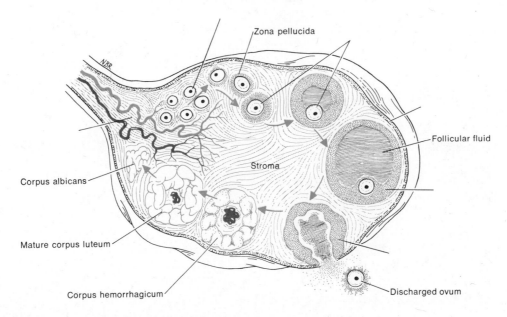

Figure 41-6. Sectional view of an ovary.

Two **uterine** or **fallopian tubes transport ova** from the ovaries to the uterus. The open end of each tube is called the *infundibulum* and is surrounded by fingerlike projections termed *fimbriae*. Label these structures in Figure 41-5.

The **uterus** is an inverted, pear-shaped organ between the bladder and rectum. Its major parts include the *fundus*, the dome-shaped portion above the uterine tubes; the *body*, the major central portion; the *cervix*, the inferior narrow portion opening into the vagina; and the *uterine cavity*, the interior of the body. Label these parts in Figure 41-5.

In sectional view, it can be seen that the uterus consists of 3 layers: the outer *serous layer* or *peritoneum*, the *myometrium* or middle muscular layer, and the inner mucous membrane called the *endometrium*. The 2 layers of the endometrium are the *functionalis*, which is shed during menstruation, and the *basalis*, which produces a new functionalis. Both are highly vascular. The normal position of the uterus is maintained by the *broad ligaments, uterosacral ligaments, cardinal ligament*, and *round ligaments*. Label the uterine layers and ligaments shown in Figure 41-5.

The **vagina** serves as a passageway for the menstrual flow, the receptacle for the penis, and the lower portion of the birth canal. At the lower end of the vagina there is an opening, the *vaginal orifice*. The orifice is bordered by a vascularized mucous membrane called the *hymen*. Label these parts in Figure 41-7.

The **vulva** or **pudendum** is a collective term for the female external genitals. Its parts include the **mons pubis**, a fat pad over the symphysis pubis covered by hair; the *labia majora*, 2 longitudinal folds of skin covered by hair; the *labia minora*, 2 folds of skin medial to the majora devoid of hair; the *clitoris*, a cylindrical mass of erectile tissue behind the junction of the labia minora; the *vestibule*, the cleft between the labia; the *vaginal orifice*; the *urethral orifice* in front of the vaginal orifice and behind the clitoris; the *lesser vestibular glands* on either side of the urethral orifice; and the *greater vestibular glands*, on either side of the vaginal orifice. Label the parts of the vulva in Figure 41-7.

The **perineum** is the diamond-shaped area at the lower end of the trunk between the thighs and buttocks in both males and females. It is surrounded anteriorly by the symphysis pubis, laterally by the ischial tuberosities, and posteriorly by the coccyx. A transverse line drawn between the ischial tuberosities divides the perineum into an anterior urogenital triangle that contains the external genitalia and a posterior anal triangle that contains the anus. If the vagina is too small to accommodate the head of an emerging fetus, the skin and underlying tissue of the perineum tears. To avoid this, a

**414**

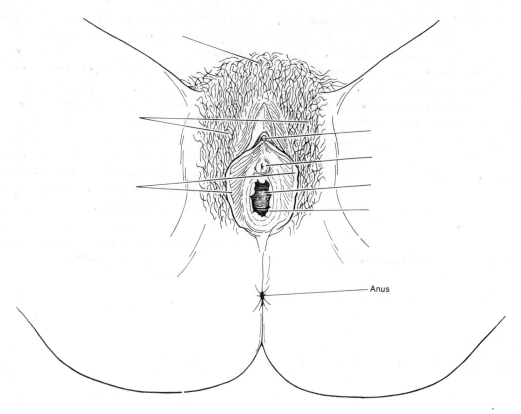

**Figure 41-7.** Female external genitals.

small incision, called an *episiotomy*, is made in the perineal skin just prior to delivery. Label the borders and triangles of the female perineum in Figure 41-8.

The **mammary glands**, or *breasts*, lie over the pectoralis major muscles and are attached to them by a layer of connective tissue. Internally, each mammary gland consists of *15-20 lobes*, or compartments, separated by adipose tissue. The amount of adipose tissue present is the principal determinant of the size of the breasts. However, the size of the breasts has nothing to do with the amount of milk produced. Within each lobe are several smaller compartments, called *lobules*, that are composed of

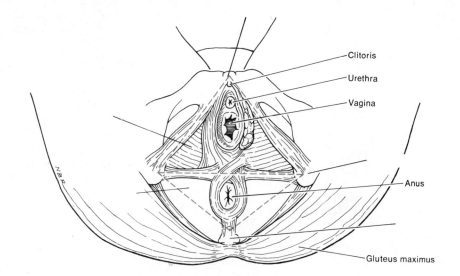

Clitoris

Urethra

Vagina

Anus

Gluteus maximus

**Figure 41-8.** Female perineum.

**415**

connective tissue in which milk-secreting cells referred to as *alveoli* are embedded. Alveoli are arranged in grapelike clusters. They convey the milk into a series of *secondary tubules*. From here, the milk passes into the *mammary ducts*. As the mammary ducts approach the nipple, expanded sinuses called *ampullae*, where milk may be stored, are present. The ampullae continue as *lactiferous ducts* that terminate in the *nipple*. Each lactiferous duct conveys milk from one of the lobes to the exterior. The circular pigmented area of skin surrounding the nipple is called the *areola*. It appears rough because it contains modified sebaceous glands. Label the mammary gland in Figure 41-9.

At birth, both male and female mammary glands are undeveloped and appear as slight elevations on the chest. With the onset of puberty, the female breasts begin to develop, the mammary ducts elongate, extensive fat deposition occurs, and the areola and nipple grow and become pigmented. These changes are correlated with an increased output of estrogen by the ovary. Further mammary development occurs at sexual maturity, with the onset of ovulation and the formation of the corpus luteum. During adolescence, lobules and alveoli are formed and fat deposition continues, increasing the size of the glands. Although these changes are associated with estrogen and progesterone secretion by the ovaries, ovarian secretion is ultimately controlled by FSH.

## CONCLUSIONS

1. What are the functions of the ovaries?


2. How are the uterine tubes adapted to their functions?


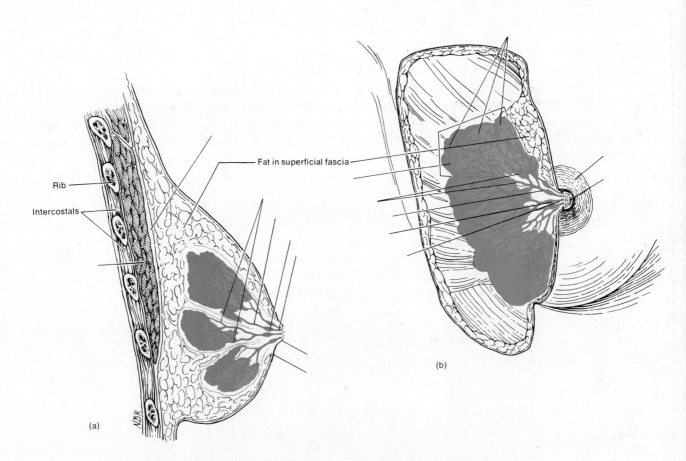

Rib

Intercostals

Fat in superficial fascia

(a)

(b)

**Figure 41-9.** Mammary glands in (a) sagittal section and (b) front view.

3. Define ovulation.

4. What is an ectopic pregnancy?

5. What is meant by retroflexion and anteflexion of the uterus?

6. What are the functions of the myometrium and endometrium of the uterus?

7. Define an imperforate hymen.
   How is the condition corrected?

8. What enables the vagina to stretch?

9. Why is the environment of the vagina slightly acid?

10. Define the function of each of the following:
    a. labia majora
    b. labia minora
    c. clitoris
    d. lesser vestibular glands
    e. greater vestibular glands

## PART III   MENSTRUAL AND OVARIAN CYCLES

The term **menstrual cycle** refers to a series of changes that occur in the endometrium of a non-pregnant female. Each month the endometrium is prepared to receive a fertilized ovum. An implanted ovum eventually develops into a fetus and normally remains in the uterus until delivery. If no fertilization occurs, a portion of the endometrium is shed. The **ovarian cycle** is a monthly series of events associated with the maturation of an ovum.

Gonadotrophic hormones of the anterior pituitary gland initiate the menstrual cycle, ovarian cycle, and other changes associated with puberty in the female (Figure 41-10). FSH stimulates the initial development of the ovarian follicles and the secretion of estrogens by the follicles. Another anterior pituitary hormone, the luteinizing hormone (LH), stimulates the further development of ovarian follicles, brings about ovulation, and stimulates progesterone production by ovarian cells. The female sex hormones, estrogens and progesterone, affect the body in different ways. Estrogens have 4 main functions. First is the development and maintenance of female reproductive organs, especially the endometrium, secondary sex characteristics, and the breasts. Second, they control fluid and electrolyte balance. Third, they increase protein anabolism. Fourth, they cause an increase in the female sex drive. High levels of estrogens in the blood inhibit the secretion of FSH by the anterior pituitary gland. This inhibition provides the basis for the action of one kind of contraceptive pill. Progesterone, the other female sex hormone, works with estrogens to prepare the endometrium for implantation and to prepare the breasts for milk secretion.

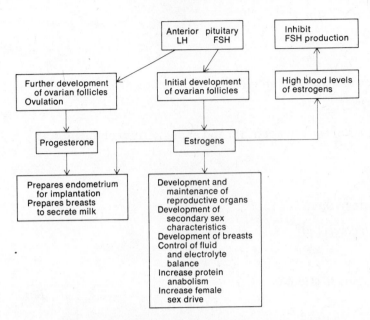

**Figure 41-10.** Hormones of the female sexual cycles.

The duration of the menstrual cycle is variable among different females, normally ranging from 24 to 35 days. For purposes of our discussion, it will be assumed that the average duration of the cycle is 28 days. Events occurring during the menstrual cycle may be divided into 3 phases: (1) the menstrual phase, (2) the preovulatory phase, and (3) the postovulatory phase (Figure 41-11).

The *menstrual phase*, also called *menstruation* or the *menses*, is the periodic discharge of blood (25-65 ml.), tissue fluid, mucus, and epithelial cells. It lasts approximately for the first 5 days of the cycle. The discharge is associated with endometrial charges in which the functionalis layer degenerates and patchy areas of bleeding develop. Small areas of the functionalis detach one at a time (total detachment would result in hemorrhage), the uterine glands discharge their contents and collapse, and tissue fluid is discharged. The menstrual flow passes from the uterine cavity to the cervix, through the vagina, and ultimately to the exterior. Generally the flow terminates by the fifth day of the cycle. At this time the entire functionalis is shed, and the endometrium is very thin because only the basalis remains.

During the menstrual phase, the ovarian cycle is also in operation. Ovarian follicles, called *primary follicles*, begin their development (see Figure 41-6). At the time of birth each ovary contains about 200,000 such follicles, each consisting of an ovum surrounded by a layer of cells. During the early part of the menstrual phase, a primary follicle starts to produce very low levels of estrogens. A clear membrane, the zona pellucida, also develops around the ovum. Later in the menstrual phase (4-5 days) the primary follicle develops into a *secondary follicle* as the cells of the surrounding layer increase in number and secrete a fluid called the follicular fluid. This fluid forces the ovum to the edge of the follicle. The production of estrogens by the secondary follicle elevates the estrogen level of the blood slightly. Ovarian follicle development is the result of FSH production by the anterior pituitary, and during this part of the ovarian cycle, FSH secretion is maximal. Although a number of follicles begin development each cycle, only one will attain maturity.

The *preovulatory phase*, the second phase of the menstrual cycle, is the period of time between the end of menstruation and ovulation. This phase of the menstrual cycle is more variable in length than are the other phases. It lasts from day 6 to 13 in a 28-day cycle. During the preovulatory phase, a secondary follicle in the ovary matures into a *graafian follicle*, a follicle ready for ovulation. During the maturation process, the follicle increases its estrogen production. Early in the preovulatory phase, FSH is the dominant hormone, but, close to the time of ovulation, LH is secreted in increasing

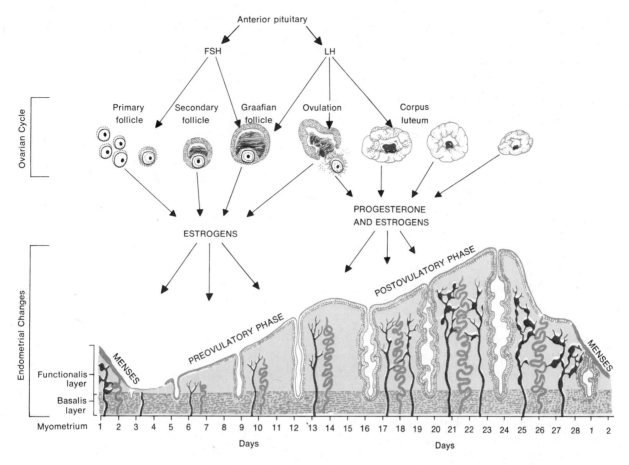

**Figure 41–11.** Menstrual and ovarian cycles.

quantities. Moreover, small amounts of progesterone may be produced by the graafian follicle a day or two before ovulation.

FSH and LH stimulate the ovarian follicles to produce more estrogens, and this increase in estrogens stimulates the repair of the endometrium. During this process of repair, basilar cells undergo mitosis and produce a new functionalis. As the endometrium thickens, the endometrial glands are short and straight, and the arterioles become coiled and increase in length as they penetrate the functionalis. Because the proliferation of endometrial cells occurs during the preovulatory phase, the phase is also referred to as the *proliferative phase*. Still another name for this phase is the *follicular phase* because of increasing estrogen secretion by the developing follicle. Functionally, estrogen is the dominant hormone during this phase of the menstrual cycle.

*Ovulation*, the rupture of the graafian follicle with release of the ovum into the pelvic cavity, occurs on day 14 in a 28-day cycle. Just prior to ovulation, the high estrogen level that developed during the preovulatory phase inhibits FSH secretion by the anterior pituitary. Concurrently, LH secretion by the anterior pituitary is greatly increased. As LH and estrogen secretion increases and FSH secretion is inhibited, ovulation occurs. Following ovulation, the graafian follicle collapses, and blood within it forms a clot called the *corpus hemorrhagicum*. The clot is eventually absorbed by the remaining follicular cells. In time, the follicular cells enlarge, change character, and form the *corpus luteum*.

The *postovulatory phase* of the menstrual cycle is fairly constant in duration and lasts from days 15 to 28 in a 28-day cycle. It represents the period of time between ovulation and the onset of the next menses. Following ovulation, the level of estrogen in the blood drops slightly, and LH secretion stimulates the development of the corpus luteum. The corpus luteum then secretes increasing quantities of estrogens and progesterone, the latter being responsible for the preparation of the endometrium to receive a fertilized ovum. Preparatory activities include the filling of the endometrial glands with secretions that cause the glands to appear tortuously coiled, vascularization of the superficial endometrium, thickening of the endometrium, and an increase in the amount of tissue fluid. These preparatory changes are maximal about one week after ovulation, and they correspond to the anticipated arrival of the fertilized ovum. During the postovulatory phase, FSH secretion gradually increases and LH secretion decreases. The functionally dominant hormone during this phase is progesterone.

If fertilization and implantation do not occur, the rising levels of progesterone and estrogens inhibit LH secretion, and as a result the corpus luteum degenerates and becomes the *corpus albicans*. The decreased secretion of progesterone and estrogens by the degenerating follicle then initiates another menstrual period. In addition, the decreased progesterone and estrogen levels in the blood bring about a new output of the anterior pituitary hormones, especially FSH, and a new ovarian cycle is initiated. A summary of these hormonal interactions is presented in Figure 41-12. Fill in the key words that have been deleted from some of the boxes.

If, however, fertilization and implantation do occur, the corpus luteum is maintained for nearly 4 months, and for most of this time it continues to secrete progesterone. Maintenance of the corpus luteum is accomplished by *chorionic gonadotrophin*, a hormone produced by the developing fetus, until the placenta can secrete estrogen to support pregnancy, and progesterone to support pregnancy and breast development for lactation.

The menstrual cycle normally occurs once each month from *menarche*, the first menstrual cycle, to *menopause*, the complete cessation of menstruation. Menopause typically occurs between 45 and 50 years of age and results from failure of the ovaries to respond to the stimulation of gonadotrophic hormones from the anterior pituitary. The onset of menopause may be characterized by "hot flashes," copious sweating, headache, muscular pains, and emotional instability. Ultimately, menopause results in some degree of atrophy of the ovaries, uterine tubes, uterus, vagina, external genitalia, and breasts.

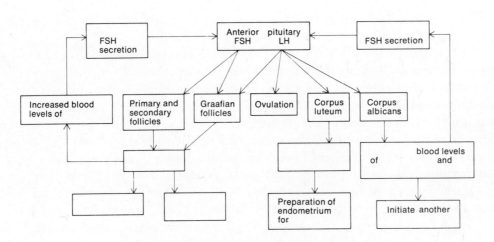

**Figure 41-12.** Summary of hormonal interactions.

## CONCLUSIONS

1. How are the ovarian and menstrual cycles related?

2. What is the importance of anterior pituitary hormones in the female sexual cycles?

3. What causes ovulation to occur?

4. What hormonal changes bring about a new menstrual cycle?

5. What hormonal changes occur if fertilization does take place?

**STUDENT ACTIVITY**

Shown below and on the next page are composite diagrams of the male and female reproductive systems. Label all of the structures shown in Figure 41–13.

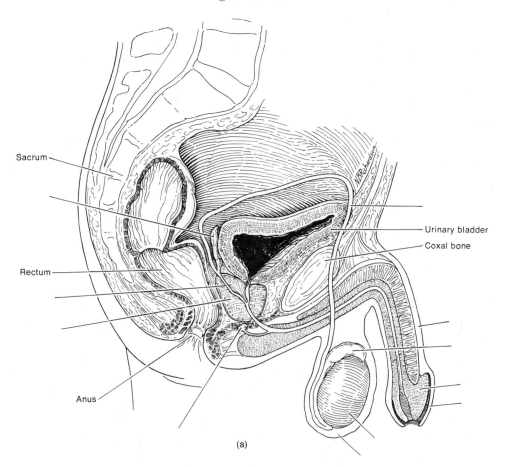

Sacrum

Rectum

Anus

Urinary bladder

Coxal bone

(a)

**Figure 41–13.** (a) Male reproductive system. (b) Female reproductive system.

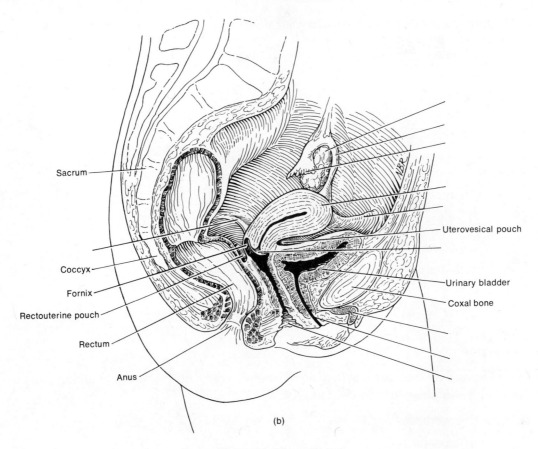

Sacrum

Coccyx

Fornix

Rectouterine pouch

Rectum

Anus

Uterovesical pouch

Urinary bladder

Coxal bone

(b)

**Figure 41–13 continued.**

# MODULE 42
# REPRODUCTIVE DISORDERS

**OBJECTIVES**

1. To define selected prostate disorders, impotence, and male infertility

2. To contrast amenorrhea, dysmenorrhea, and uterine bleeding as menstrual disorders

3. To define ovarian cysts, female infertility, breast tumors, and cervical cancer

4. To describe gonorrhea and syphilis as communicable venereal diseases

The intent of this module is to describe some of the common disorders of the male and female reproductive systems. In addition, a section on gonorrhea and syphilis has been included.

## PART I  MALE DISORDERS

Acute and chronic infections of the prostate gland are common in postpubescent males, many times in association with inflammation of the urethra. In *acute prostatitis* the prostate gland becomes swollen and very tender. Appropriate antibiotic therapy, bed rest, and above-normal fluid intake are effective in treatment.

*Chronic prostatitis* is one of the most common chronic infections in men of the middle and later years. On examination, the prostate gland feels enlarged, soft, and extremely tender. The surface outline is irregular and may be hard. This disease frequently produces no symptoms, but the prostate is believed to harbor infectious microorganisms responsible for some allergic conditions, arthritis, and inflammations of nerves (neuritis), muscles (myositis), and the iris (iritis).

An *enlarged prostate* gland occurs in approximately one-third of all males over 60 years of age. The enlarged gland is from 2 to 4 times larger than normal. The cause is unknown, and the enlarged condition usually can be detected by rectal examination.

*Tumors* of the male reproductive system primarily involve the prostate gland. Both benign and malignant growths are common in elderly men. Both types of tumors put pressure on the urethra, making urination painful and difficult. At times, the excessive back pressure destroys kidney tissue and gives rise to an increased susceptibility to infection. Therefore, even if the tumor is benign,

**423**

surgery is indicated to remove the prostate or parts of it if the tumor is obstructive and perpetuates urinary tract infections.

*Impotence* is the inability of an adult male to attain or hold an erection long enough for normal intercourse. Impotence does not mean infertility.

*Infertility*, or *sterility*, is an inability to fertilize the ovum and does not imply impotence. Male fertility requires viable spermatozoa, adequate production of spermatozoa by the testes, unobstructed transportation of sperm through the seminal tract, and satisfactory deposition within the vagina.

## CONCLUSIONS

1. How is acute prostatitis treated?


2. What is a secondary consequence of an enlarged prostate gland?


   Why is this dangerous?

3. How are prostate tumors correlated with urinary tract infections?


4. Distinguish between impotence and infertility.



5. List some possible causes of impotence.



6. How may sterility arise?



7. Describe how a sperm analysis is performed if sterility is suspected.




## PART II  FEMALE DISORDERS

Disorders of the female reproductive system frequently include menstrual disorders. This is hardly surprising because proper menstruation reflects not only the health of the uterus but the health of the glands that control it, that is, the ovaries and the pituitary gland.

*Amenorrhea* is the absence of menstruation in a woman. If the woman has never menstruated, the condition is called *primary amenorrhea*. Primary amenorrhea can be caused by endocrine disorders, most often in the pituitary gland and hypothalamus, or by genetically caused abnormal development of the ovaries or uterus. *Secondary amenorrhea* is cessation of uterine bleeding in women who have previously menstruated.

*Dysmenorrhea* is painful menstruation caused by contractions of the uterine muscles. A primary cause is believed to be low levels of progesterone.

*Abnormal uterine bleeding* includes menstruation of excessive duration and/or excessive amount, too-frequent menstruation, intermenstrual bleeding, and postmenopausal bleeding.

*Ovarian cysts* are tumors of the ovary that contain fluid. Follicular cysts may occur in the ovaries of elderly people, in ovaries that have inflammatory diseases, and in menstruating females. They have thin walls and contain a serous albuminous material. Cysts may also arise from the corpus luteum or the endometrium. *Endometriosis* is a painful disorder characterized by endometrial tissue or cysts in abnormal locations, such as in the uterine tubes, ovaries, vagina, peritoneum, or any other place in the body outside the uterus.

*Leukorrhea* is a nonbloody vaginal discharge that may occur at any age and affects most women at some time. It is not a disease; it is a symptom of infection or congestion of some portion of the reproductive tract. It may be a normal discharge in some women. If it is evidence of an infection, it may be caused by a protozoan microorganism called *Trichomonas vaginalis*, a yeast, a virus, or a bacterium.

*Female infertility*, or the inability to conceive, occurs in about 10 percent of married females in the United States. Once it is established that ovulation occurs regularly, the reproductive tract is examined for functional and anatomical disorders to determine the possibility of union of the sperm and the ovum in the oviduct.

The breasts are highly susceptible to *cysts* and *tumors*. Men are also susceptible to breast tumor, but certain breast cancers are 100 times more common in women than in men. Usually these growths can be detected early by the woman who inspects and palpates her breasts regularly. To *palpate* means to feel or examine by touch. Unfortunately, so few women practice periodic self-examination that many growths are discovered by accident and often too late for proper treatment.

In the female, the benign *fibroadenoma* is the third most common tumor of the breast. It occurs most frequently in young people. Fibroadenomas have a firm rubbery consistency and are easily moved about within the mammary tissue. The usual treatment is excision of the growth. The breast itself is not removed.

*Breast cancer* has the highest fatality rate of all cancers affecting women, but it is rare in men. In the female, breast cancer is rarely seen before age 30, and its occurrence rises rapidly after menopause.

Breast cancer is generally not painful until it becomes quite advanced, so often it is not discovered early, or if it is noted, it is ignored. Any lump, be it ever so small, should be reported to a doctor at once. If there is no evidence of *metastasis* (the spread of cancer cells from one part of the body to another or from one organ to another), the treatment of choice is a modified or radical mastectomy. A *radical mastectomy* involves removal of the affected breast, along with the underlying pectoral muscles and the axillary lymph nodes. Metastasis of cancerous cells is usually through the lymphatics or blood. Radiation treatments may follow the surgery to ensure the destruction of any remaining stray cancer cells.

Another common disorder of the female reproductive system is cancer of the uterine cervix. It ranks third in frequency after breast and skin cancers. *Cervical cancer* starts with a change in the shape of the cervical cells called *cervical dysplasia*. Cervical dysplasia is not a cancer in itself, but the abnormal cells tend to become malignant.

Early diagnosis of cancer of the uterus is accomplished by the *Papanicolaou test*, or "Pap" smear. In this generally painless procedure, a few cells from the vaginal fornix (that part of the vagina surrounding the cervix) and the cervix are removed with a swab and examined microscopically. Malignant cells have a characteristic appearance and are indicative of an early stage of cancer, even before any symptoms occur. Estimates indicate that the Pap smear is more than 90 percent reliable in detecting cancer of the cervix. Treatment of cervical cancer may involve complete or partial removal of the uterus, called a *hysterectomy*, or radiation treatments.

## CONCLUSIONS

1. Distinguish between primary and secondary amenorrhea.

2. Describe some possible causes of secondary amenorrhea.

3. Why are low progesterone levels implicated in dysmenorrhea?

4. What may cause abnormal uterine bleeding?

5. Define an ovarian cyst.

6. Why is leukorrhea considered to be a symptom rather than a disease?

7. Why is palpation of the breasts important?

8. How are fibroadenomas of the breast treated?

9. What is metastasis?

10. Define a radical mastectomy.

11. How is cervical cancer detected?

## PART III   VENEREAL DISEASES

The term venereal comes from Venus, the goddess of love. *Venereal diseases* represent a group of infectious diseases that are spread primarily through sexual intercourse. With the exception of the common cold, venereal diseases are ranked as the number one communicable diseases in the United States. Gonorrhea and syphilis are the 2 most common venereal diseases.

*Gonorrhea*, more commonly known as "clap," is an infectious disease that primarily affects the mucous membrane of the urogenital tract, the rectum, and occasionally the eye. The disease is caused by the bacterium *Neisseria gonorrhoeae*. Examine a microscope slide containing the bacteria and note their size and shape.

*Syphilis* is an infectious disease caused by the bacterium *Treponema pallidum*. Like gonorrhea, it is acquired through sexual contact. The early stages of the disease primarily affect the organs that are most likely to have made sexual contact—the genital organs, the mouth, and the rectum. Following the initial infection, the bacteria enter the bloodstream and are spread throughout the body. In some individuals, an active secondary stage of the disease occurs, characterized by lesions of the skin and mucous membranes, like those of the first, and often fever. The signs of the secondary stage go away without medical treatment. During the next several years, the disease progresses without symptoms and is said to be in a latent phase. When symptoms again appear, anywhere from 5 to 40 years after the initial infection, the person is said to be in the tertiary stage of the disease. Tertiary syphilis may involve the circulatory system, skin, bones, viscera, and the nervous system. Examine a microscope slide containing the *Treponema* organism and compare its size and shape to that of the *Neisseria* bacterium.

## CONCLUSIONS

1. What are the male symptoms of gonorrhea?


2. How are females affected by gonorrhea?


3. How is the gonorrhea bacterium transmitted to a newborn baby?

4. How does the gonorrhea bacterium affect a newborn baby?


   How is the condition treated?

5. What is the drug of choice for the treatment of gonorrhea in adults?

6. What is a chancre?


7. Where do chancres appear in males infected with syphilis?


   Females?

8. What are some of the effects of tertiary syphilis on the body?




## STUDENT ACTIVITY

What are some of the ways that gonorrhea and syphillis may be prevented?

# MODULE 43
# DEVELOPMENT AND PREGNANCY

**OBJECTIVES**

1. To contrast spermatogenesis and oogenesis

2. To describe fertilization and implantation

3. To explain the formation of the primary germ layers and their fates

4. To describe the growth of an embryo

5. To identify the embryonic membranes

6. To discuss the functions of the placenta and umbilical cord

7. To explain the functions of the hormones of pregnancy

8. To describe parturition, labor, and lactation

Now that we have studied the organs of reproduction, we shall look at developmental processes. The term *developmental processes* refers to a sequence of events starting with the fertilization of an ovum (egg) and ending with the formation of a complete organism. As we look at this sequence, we shall consider how reproductive cells are produced, and events associated with pregnancy, birth, and lactation.

## PART I  MEIOSIS: GAMETE FORMATION

Each human being develops from the union of an ovum and a sperm. Ova and sperm are collectively called *gametes*, and they differ radically from all the other cells in the body in that they have only half the normal number of chromosomes in their nuclei. *Chromosome number* is the number of chromosomes contained in each nucleated cell that is not a gamete. Chromosome numbers vary from species to species. The human chromosome number is 46, which means that each brain cell, stomach cell, heart cell, and every other cell contains 46 chromosomes in its nucleus. In other words, there are 23 pairs of chromosomes in each cell other than a gamete. The ovum or sperm has only one member of each pair. Of these 46 chromosomes, 23 contain the genes that are necessary for programing all the activities of the body. In a sense, the other 23 are a duplicate set. Another word for chromosome number is *diploid number* (the prefix *di-* means two), symbolized as *2n*.

Suppose that a sperm containing 46 chromosomes fertilizes an egg that also contains 46 chromosomes. The offspring then would have 92 chromosomes, the grandchildren 184 chromosomes, and so on. In reality, the chromosome number does not double with each generation because of a special kind of nuclear division called **meiosis**. Meiosis occurs in sex cells before they become mature. It causes a developing sperm or ovum to relinquish its duplicate set of chromosomes so that the mature gamete has only 23. This is called the *haploid number*, meaning "one-half," and symbolized as *n*.

In the testes, the formation of haploid spermatozoa by meiosis is called **spermatogenesis**. The seminiferous tubules are lined with immature cells called *spermatogonia*. Spermatogonia contain the diploid chromosome number and are the precursor cells for all the sperm that the man will produce. During childhood these cells are relatively inactive. When the male reaches puberty, the spermatogonia embark on a lifetime of active division. Some of the spermatogonia undergo developmental changes and become known as *primary spermatocytes*. Each primary spermatocyte then undergoes a special type of cellular divison. As a result, the 2 daughter cells, called *secondary spermatocytes*, contain only half the chromosomes that the parent cell contained. This is the unique nuclear division called *meiosis*. Each secondary spermatocyte then goes through a mitotic division to produce 2 cells called *spermatids*. The spermatids contain the same number of chromosomes as the secondary spermatocytes—thus, they are also haploid. Without further cell division each spermatid develops a head with an acrosome and a flagellum. It is now a mature sperm or *spermatozoan*. Label these various cells in Figure 43-1a and place the appropriate chromosome number next to each cell.

Now examine a prepared slide of spermatogenesis and see how many different cells you can identify.

In the ovary, the formation of a haploid ovum by meiosis is referred to as **oogenesis.** The precursor cell in this sequence is a diploid cell called the *oogonium*. Starting with puberty, one oogonium undergoes developmental changes each month and becomes a *primary oocyte*. The primary oocyte undergoes meiosis and forms 2 cells of unequal size, both of which are haploid. The larger cell is called a *secondary oocyte,* and the smaller is referred to as the *first polar body.* The secondary oocyte

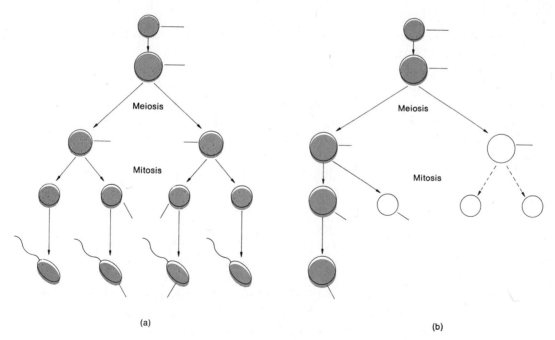

**Figure 43-1.** Meiosis. (a) Spermatogenesis. (b) Oogenesis.

undergoes a cell division and forms 2 more cells of unequal size. The larger haploid cell is an *ootid*, and the smaller is the *second polar body*. In time, the ootid develops into a mature *ovum* and the polar bodies disintegrate. Thus, in the female one oogonium produces a single ovum, whereas in the male each spermatogonium produces 4 sperm. Label these various cells in Figure 43-1b and place the appropriate chromosome number next to each cell.

Examine a prepared slide of oogenesis and see how many different cells you can identify.

## CONCLUSIONS

1. Why does meiosis occur?

2. What is the importance of the haploid chromosome number?

3. What is the principal difference between spermatogenesis and oogenesis?

## PART II   PREGNANCY

Once spermatozoa and ova are developed through meiosis and the spermatozoa are deposited in the vagina, pregnancy can occur. **Pregnancy** is a sequence of events including fertilization, implantation, embryonic growth, and, normally, fetal growth that terminates in birth.

The term *fertilization* is applied to the union of the sperm nucleus and the nucleus of the ovum. It normally occurs in the uterine tube when the ovum is about one-third of the way down the tube, usually within 24 hours after ovulation.

Sperm must remain in the female genital tract for 4-6 hours before they are capable of fertilizing an ovum. During this time, the enzyme hyaluronidase is activated and secreted by the spermatozoa. Hyaluronidase apparently dissolves parts of the membrane covering the ovum. Normally, only one spermatozoan fertilizes an ovum because once penetration is achieved, the ovum develops a fertilization membrane that is impermeable to the entrance of other spermatozoa. When the spermatozoan has entered the ovum, the tail is shed and the nucleus in the head develops into a structure called the *male pronucleus*. The nucleus of the ovum also develops into a *female pronucleus*. After the pronuclei are formed, they fuse to produce a *segmentation nucleus*—a process termed fertilization. The segmentation nucleus contains 23 chromosomes from the male pronucleus and 23 chromosomes from the female pronucleus. Thus, the fusion of the haploid pronuclei restores the diploid number. The fertilized ovum, consisting of a segmentation nucleus, cytoplasm, and enveloping membrane, is referred to as a *zygote*. Label the structures in Figure 43-2a.

Immediately after fertilization, rapid cell division of the zygote takes place. This early division of the zygote is called *cleavage*. The progressively smaller cells produced are called *blastomeres*. Successive cleavages produce a solid mass of cells, the *morula*, which is only slightly larger than the original zygote. Label the blastomeres in Figure 43-2b.

As the morula descends through the uterine tube, it continues to divide and eventually forms a hollow ball of cells. At this stage of development the mass is referred to as a *blastocyst*. The blastocyst is differentiated into an outer covering of cells called the *trophectoderm* and an *inner cell mass*, and the internal cavity is referred to as the *blastocoel*. Whereas the trophectoderm ultimately will form the membranes composing the fetal portion of the placenta, the inner cell mass will develop eventually into the embryo. About the seventh day after fertilization, the blastocyst enters the uterine cavity. Label the structures in Figure 43-2c.

**430**

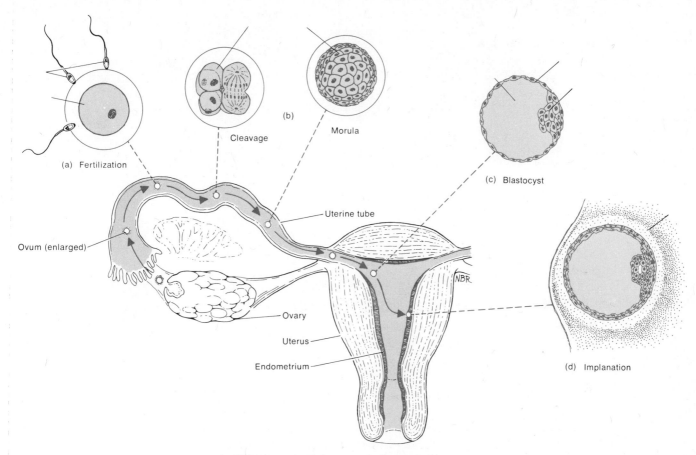

**Figure 43–2.** Fertilization and implantation.

The attachment of the blastocyst to the endometrium occurs 7 to 8 days following fertilization and is called **implantation**. At this time, the endometrium is in its postovulatory phase. The portion of the endometrium to which the blastocyst adheres and in which it becomes implanted is the basalis layer. Implantation enables the blastocyst to absorb nutrients from the glands and blood vessels of the endometrium for its subsequent growth and development. Label Figure 43-2d.

The first 2 months of development are considered the **embryonic period**. During this period, the developing human is called an **embryo**. After the second month it will be called a **fetus**. By the end of the embryonic period the rudiments of all the principal adult organs are present, the embryonic membranes are developed, and the placenta is functioning.

Following implantation, the inner cell mass of the blastocyst begins to differentiate into the 3 primary germ layers: the ectoderm, endoderm, and mesoderm. The **primary germ layers** are the embryonic tissues from which all tissues and organs of the body will develop. The fetal membranes, structures that lie outside the embryo and protect and nourish it, also develop from these 3 germ layers.

In the human being, the formation of the germ layers happens so quickly that it is difficult to determine the exact sequence of events. Before implantation, a layer of *ectoderm* (the trophectoderm) already has formed around the blastocoel. Label these structures in Figure 43-3a. Also label the *inner cell mass*.

The trophectoderm will become part of the chorion—one of the fetal membranes. Within 8 days after implantation, the inner cell mass moves downward so a space called the *amniotic cavity* lies between the inner cell mass and the trophectoderm. The bottom layer of the inner cell mass develops into an *endodermal* germ layer. Label these structures in Figure 43-3b.

**431**

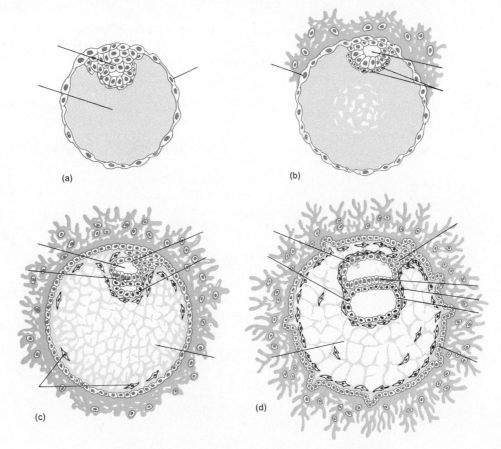

**Figure 43–3.** Formation of the primary germ layers.

About the twelfth day after fertilization the striking changes shown in Figure 43-3c appear. A layer of cells from the inner cell mass has grown around the top of the *amniotic cavity*. These cells will become the *amnion*, another fetal membrane. The cells below the cavity are called the *embryonic disc*; these cells will form the embryo. The embryonic disc contains scattered ectodermal, mesodermal, and endodermal cells in addition to the endodermal layer observed in Figure 43-3b. Notice in Figure 43-3c that the cells of the endodermal layer have been dividing rapidly, so groups of them now extend downward in a circle. This circle is the *yolk sac*, another fetal membrane. The *mesodermal* cells also have been dividing, and many have left the area of the embryonic disc and can be seen around the structures that are becoming fetal membranes. Label Figure 43-3c.

About the fourteenth day, the scattered cells in the embryonic disc separate into 3 distinct layers: the upper *ectoderm*, the middle *mesoderm*, and the lower *endoderm*. At this time, the 2 ends of the embryonic disc draw together, squeezing off the yolk sac. The resulting cavity inside the disc is the endoderm-lined *primitive gut*. The mesoderm within the disc soon splits into 2 layers, and the space between the layers becomes the *coelom*, or *body cavity*. Label these structures in Figure 43-3d.

Refer to Table 43-1. Under each major heading—endoderm, mesoderm, and ectoderm—list some of the structures produced by the primary germ layers.

During the embryonic period, the *embryonic membranes* form. These membranes lie outside the embryo and will protect and nourish the fetus. The membranes are the yolk sac, the amnion, the chorion, and the allantois.

The human *yolk sac* is an endoderm-lined membrane that encloses the yolk. In many species the yolk provides the primary or exclusive nutrient for the embryo, and consequently, the ova of these animals contain a great deal of yolk. However, the human embryo receives its nourishment from the

Table 43–1. STRUCTURES PRODUCED BY THE THREE PRIMARY GERM LAYERS

| ENDODERM | MESODERM | ECTODERM |
|---|---|---|
|  |  |  |

endometrium. The human yolk sac is small, and during an early stage of development it becomes a nonfunctional part of the umbilical cord.

The *amnion* is a thin protective membrane that initially overlies the embryonic disc. As the embryo grows, the amnion entirely surrounds the embryo and becomes filled with a fluid called *amniotic fluid*. The amnion usually ruptures just before birth and it and its fluid constitute the so-called "bag of waters."

The *chorion* derives from the trophectoderm of the blastocyst and its associated mesoderm. It surrounds the embryo and, later, the fetus. Eventually the chorion becomes the principal part of the placenta, the structure through which materials are exchanged between the mother and fetus. The amnion also surrounds the fetus and eventually fuses to the inner layer of the chorion.

The *allantois* is a small vascularized membrane. Later its blood vessels serve as connections in the placenta between the mother and fetus. Label these membranes in Figure 43-4.

Development of a functioning placenta is the third major event of the embryonic period. This is accomplished by the third month of pregnancy. The **placenta** is formed by the chorion of the embryo and the basalis layer of the endometrium of the mother. It provides an exchange of nutrients and wastes between the fetus and mother and secretes the hormones necessary to maintain pregnancy.

During embryonic life, fingerlike projections of the chorion, called *chorionic villi*, grow into the basalis layer of the endometrium. These villi contain fetal blood vessels, and they continue growing until they are bathed in the maternal blood in the sinuses of the basalis. Thus, maternal and fetal

**433**

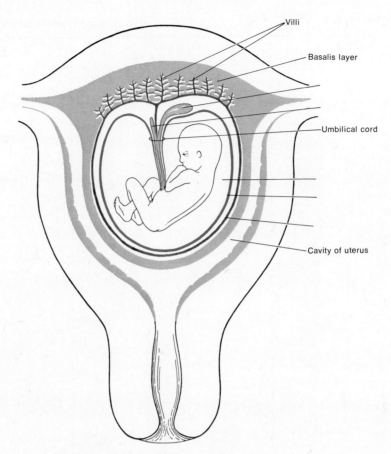

Figure 43–4. Embryonic membranes.

Labels in figure:
Villi
Basalis layer
Umbilical cord
Cavity of uterus

blood vessels are brought into close proximity. It should be noted, however, that maternal and fetal blood do not mix. Oxygen and nutrients from the mother's blood diffuse across the walls and into the capillaries of the villi. From the capillaries the nutrients circulate into the umbilical vein. Wastes leave the fetus through the umbilical arteries, pass into the capillaries of the villi, and diffuse into the maternal blood. The **umbilical cord** consists of an outer layer of amnion containing the umbilical arteries and umbilical vein, supported internally by mucous connective tissue called Wharton's jelly. At delivery, the placenta becomes detached from the uterus and is referred to as the "after-birth." At this time, the umbilicus is severed, leaving the baby on its own. Label the parts of the placenta and umbilical cord in Figure 43–5.

During the **fetal period**, rapid growth of organs established by the primary germ layers occurs. No new structures are formed, but there is accelerated growth and maturation of structures that were laid down during the embryonic period. At the beginning of the fetal period, the organism takes on a human appearance.

## CONCLUSIONS

1. How does an ovum move down the uterine tube?

2. How do spermatozoa reach the uterine tube?

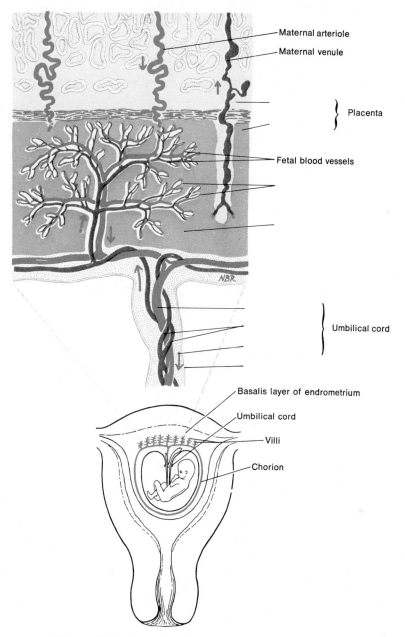

Maternal arteriole
Maternal venule

Placenta

Fetal blood vessels

Umbilical cord

Basalis layer of endrometrium
Umbilical cord
Villi
Chorion

**Figure 43-5.** Placenta and umbilical cord.

3. Define fertilization.

4. What is a zygote?

5. Distinguish between a morula and a blastocyst.

6. What is the ultimate fate of the trophectoderm?
   Inner cell mass?

7. How does implantation occur?

8. Distinguish between an embryo and a fetus.

9. What is a primary germ layer?

10. Define an embryonic membrane.

11. Briefly list the functions of the following:

    a. amnion

    b. amniotic fluid

    c. chorion

    d. allantois

12. Describe the composition of the placenta.

13. What is its function?

14. Why are chorionic villi important?

15. Describe the parts of the umbilical cord.

16. What is the fetal period?

## PART III   HORMONES OF PREGNANCY

Following fertilization, the corpus luteum is maintained until about the fourth month of pregnancy. For most of this time, it continues to secrete estrogens and progesterone. Both these hormones maintain the lining of the uterus during pregnancy and prepare the mammary glands to secrete milk. The amounts of estrogens and progesterone secreted by the corpus luteum, however, are only slightly higher than those produced after ovulation in a normal menstrual cycle. The high levels of estrogens and progesterone needed to maintain pregnancy and initiate lactation are provided by the placenta.

During pregnancy, the chorion of the placenta secretes a hormone called the *chorionic gonadotrophic hormone,* or *CG.* This hormone is excreted in the urine of pregnant women from about the

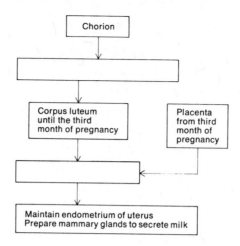

**Figure 43-6.** Hormones of pregnancy.

middle of the first month of pregnancy, reaching its peak of excretion during the third month. The primary role of CG seems to be to maintain the activity of the corpus luteum, especially with regard to continuous progesterone secretion—an activity necessary for the continued attachment of the fetus to the lining of the uterus. Excretion of CG in the urine serves as the basis for some pregnancy tests.

As noted earlier, the placenta provides the high levels of estrogens and progesterone needed for the maintenance of pregnancy. The placenta begins to secrete these hormones no later than the sixtieth day of pregnancy. They are secreted in increasing quantities until they reach their maximum levels at the time of birth. Once the placenta is established, the secretion of CG is cut back drastically at about the fourth month. This decrease results in disintegration of the corpus luteum, which is no longer needed because the placenta supplies the levels of estrogens and progesterone needed to maintain the pregnancy. Following delivery, levels of estrogens and progesterone in the blood decrease to their nonpregnant values. Fill in the key words that have been deleted from some of the boxes in Figure 43-6.

## PART IV   PARTURITION, LABOR, AND LACTATION

The time that the embryo or fetus is carried in the uterus is called *gestation*. It is assumed that the total human gestation period is 280 days from the beginning of the last menstrual period. The term *parturition* refers to birth. Parturition is preceeded by a sequence of events commonly called *labor*. The onset of labor stems from a complex interaction of many factors, especially hormones. Because we already have considered hormonal changes during pregnancy, it will be necessary to review only a few significant changes at this point. Just prior to birth, the muscles of the uterus contract rhythmically and forcefully. Both placental and ovarian hormones play a dominant role in these contractions. You may recall that estrogen stimulates uterine contractions, whereas progesterone inhibits them. Until the effects of progesterone are effectively diminished, labor cannot take place. At the end of gestation, however, there is barely sufficient estrogen in the mother's blood to overcome the inhibiting effects of progesterone, and labor commences. Coupled with this, oxytocin from the posterior pituitary gland stimulates uterine contractions. This mechanism is summarized in Figure 43-7. Label the principal activities in the cycle.

The term *lactation* refers to the secretion of milk by the mammary glands. During pregnancy, the mammary glands are prepared for lactation by estrogens and progesterone. When the levels of estrogens and progesterone in the mother's blood decrease at birth, the hormone prolactin stimulates the secretory cells of the mammary glands to produce milk. Lactation starts 3 to 4 days after delivery. Once initiated by prolactin, it is stimulated and maintained by the suckling action of the infant. Suckling initiates impulses to the posterior pituitary via the hypothalamus. In the posterior pituitary, the

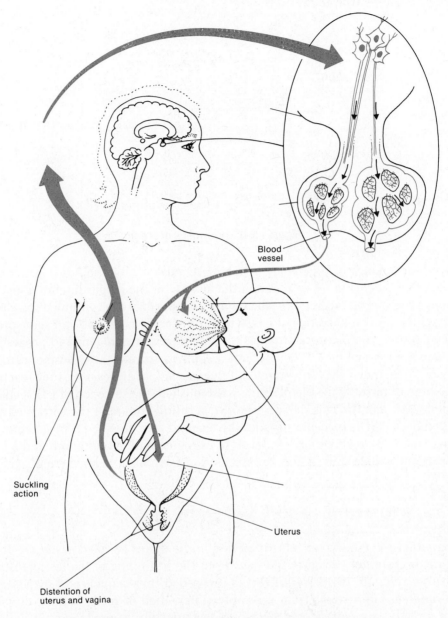

**Figure 43-7.** Hormones involved in labor and lactation.

Blood vessel

Suckling action

Uterus

Distention of uterus and vagina

impulses stimulate the release of oxytocin, causing the ejection of milk. Prolactin formation is inhibited by progesterone. Lactation usually prevents the occurrence of the female sexual cycles for the first few months following delivery. Label these relationships in Figure 43-7.

## CONCLUSIONS

1. Describe the actions of the principal hormones of pregnancy.

2. How does labor start?

3. Distinguish between true and false labor.

4. What is meant by the "show" prior to birth?

5. Briefly describe what happens during each stage of labor:
   a. dilatation
   b. expulsion
   c. placental
6. What is the "afterbirth"?
7. How is lactation initiated and maintained?

## STUDENT ACTIVITY

Below is a Table of representative changes associated with embryonic and fetal growth. Next to each time period indicated, list the approximate size and weight and the principal changes that occur in the respective columns.

| END OF MONTH | SIZE AND WEIGHT | CHANGES THAT OCCUR |
|---|---|---|
| 1 | | |
| 2 | | |
| 3 | | |
| 4 | | |
| 5 | | |
| 6 | | |
| 7 | | |
| 8 | | |
| 9 | | |

# MODULE 44
# PRINCIPLES OF INHERITANCE AND BIRTH CONTROL

**OBJECTIVES**

1. To describe the inheritance of several traits from one generation to another.

2. To distinguish color blindness and hemophilia as sex-linked traits

3. To describe various methods of birth control

**Inheritance,** as you probably know, is the passage of hereditary traits from one generation to another. It is the process by which you acquired your characteristics from your parents and will transmit your characteristics to your children. The branch of biology that deals with inheritance is called *genetics.* In this module we will consider the mechanism of inheritance and birth control.

## PART I   MECHANISM OF INHERITANCE.

The nuclei of all human cells except gametes contain 23 pairs of chromosomes, that is, the diploid number. One chromosome from each pair comes from the mother, and the other comes from the father. The 2 chromosomes that belong to a pair are called *homologous chromosomes.* Homologs contain genes that control the same traits. For instance, if a chromosome contains a gene for height, its homolog, or mate, will also contain a gene for height.

To explain the relationship of genes to heredity, we shall look at the disorder called *PKU* or *phenylketonuria.* People with PKU are unable to manufacture the enzyme phenylalanine hydroxylase. It is believed that PKU is brought about by an abnormal gene, which can be symbolized as *p.* The normal gene will be symbolized as *P.* The chromosome that is concerned with directions for phenylalanine hydroxylase production will have either p or P on it. Its homolog will also have p or P. Thus, every individual will have one of the following genetic makeups, or *genotypes:* PP, Pp, or pp. Although people with genotypes of Pp have the abnormal gene, only those with genotype pp suffer from the disorder. The reason is that the normal gene dominates over and inhibits the abnormal one. A gene that dominates is called the *dominant gene,* and the trait expressed is said to be a *dominant trait.* The gene that is inhibited is called the *recessive gene,* and that trait expressed is called a *recessive trait.*

**440**

By tradition, we symbolize the dominant gene with a capital letter and the recessive one with a lowercase letter. If an individual has the same genes on homologous chromosomes (for example, PP or pp), he is said to be *homozygous* for the trait. If, however, the genes on homologous chromosomes are different (for example, Pp), he is said to be *heterozygous* for the trait. The word *phenotype* refers to how the genetic makeup is expressed in the body. A person with Pp has a different genotype than one with PP, but both have the same phenotype, which in this case is normal production of phenylalanine hydroxylase.

Before moving on, refer to Figure 44-1 and see if you can write the possible genotypes and phenotypes for the offspring.

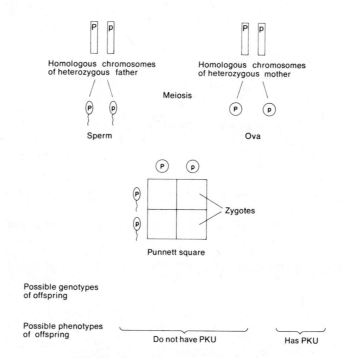

**Figure 44-1.** The inheritance of PKU.

To determine how gametes containing haploid chromosomes unite to form diploid fertilized eggs, special charts called *Punnett squares* are used. Usually, the male gametes (sperm cells) are placed to the side of the chart and the female gametes (ova) are placed to the top of the chart. The 4 spaces on the chart represent the possible combinations of male and female gametes that could form fertilized eggs. Possible combinations are determined simply by "dropping" the female gamete on the left into the 2 boxes below it, and by "dropping" the female gamete on the right into the 2 spaces under it. The upper male gamete is then moved across to the spaces in line with it, and the lower male gamete is moved across to the 2 spaces in line with it. Complete the Punnett square in Figure 44-1.

Microscopic examination of the chromosomes in cells reveals that one pair differs in males and in females. In females, the pair consists of 2 rod-shaped chromosomes designated as X chromosomes. One X chromosome also is present in males, but its mate is a hook-shaped structure called a Y chromosome. The XX pair in the female and XY pair in the male are called the *sex chromosomes*. All other chromosomes are called *autosomes*.

As you have probably guessed, the sex chromosomes are responsible for the sex of the individual. When a spermatocyte undergoes meiosis to reduce its chromosome number, half (or 2) of the daughter cells will contain the X chromosome and the other half (or 2) will contain the Y chromosome.

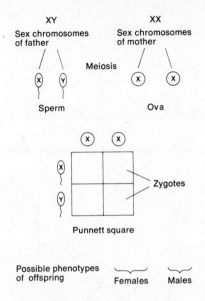

**Figure 44–2.** Sex determination.

Oocytes have no Y chromosomes and produce only X-containing ova. If the ovum is subsequently fertilized by an X-bearing sperm, the offspring will be female (XX). Fertilization by a Y sperm produces a male (XY). Complete the diagram of sex determination in Figure 44-2 by putting X's and Y's in the appropriate places.

The sex chromosomes also are responsible for the transmission of a number of nonsexual traits. Genes for these traits appear on X chromosomes, but many of these genes are absent from Y chromosomes. This produces a pattern of heredity that is different from the pattern we described earlier. Let us consider *color blindness* as an example. The gene for color blindness is a recessive one, and we shall designate it as *c*. Normal vision, designated as *C*, dominates. The C/c genes are located on the X chromosome. The Y chromosome, however, does not contain the segment of DNA that programs this aspect of vision. Thus, the ability to see colors depends entirely on the X chromosomes. The genetic possibilities are:

$X^C X^C$    Normal female
$X^C X^c$    Normal female that carries the recessive gene
$X^c X^c$    Color-blind female
$X^C Y$    Normal male
$X^c Y$    Color-blind male

As you can see, only females who have 2 $X^c$ chromosomes are color-blind. In $X^C X^c$ females the trait is inhibited by the normal, dominant gene. Males, on the other hand, do not have a second chromosome that would inhibit the trait. Therefore, all males with an $X^c$ chromosome will be color-blind. Complete the illustration of the inheritance of color blindness in Figure 44-3.

Traits that are inherited in the manner we have just described are called *sex-linked traits*. Another example of a sex-linked trait is *hemophilia*, a condition in which the blood fails to clot or clots very slowly after a surface or internal injury. It is a much more serious defect than color blindness, because people with severe hemophilia can bleed to death from even a small cut. Like the trait for color blindness, hemophilia is caused by a recessive gene. If *H* represents normal clotting, and *h* represents abnormal clotting, then $X^h X^h$ females will have the disorder. $X^H X^h$ females will be normal but carriers, and $X^H X^H$ females will be normal. Males with $X^H Y$ will be normal, and males with $X^h Y$ will be hemophiliacs, so the condition may be affected by other genes as well.

**442**

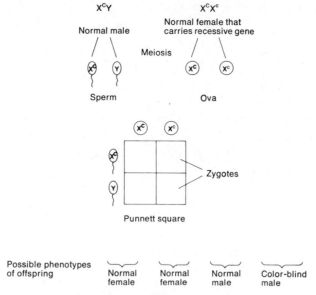

**Figure 44-3.** Inheritance of color blindness.

## CONCLUSIONS

1. What are homologous chromosomes?

2. Define each of the following:

   a. genotype

   b. phenotype

   c. dominant gene

   d. recessive gene

   e. homozygous

   f. heterozygous

3. Why are Punnett squares used?

4. Distinguish between a sex chromosome and an autosome.



5. How is sex determined?

6. Why are color blindness and hemophilia considered to be sex-linked traits?



7. List several other examples of sex-linked traits.



8. What is amniocentesis?

9. List some chromosome disorders that can be diagnosed through amniocentesis.

## PART II  BIRTH CONTROL

Methods of **birth control** include: (1) removal of the gonads and uterus, (2) sterilization, (3) contraception, and (4) abstinence. *Castration*, or removal of the testes in the male, *hysterectomy*, or removal of the uterus, and *oophorectomy*, or removal of the ovaries in the female, are all absolute preventive methods. Once performed, these operations cannot be reversed, and it is impossible for the individuals to produce offspring.

One means of *sterilization* of males is called *vasectomy*, a simple operation in which a portion of each ductus deferens is removed. (Figure 44-4a.) In this procedure an incision is made in the scro-

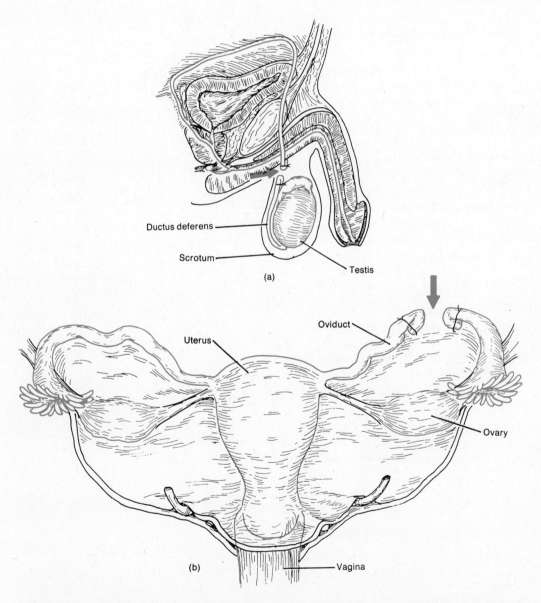

(a)

Ductus deferens

Scrotum

Testis

Oviduct

Uterus

Ovary

(b)

Vagina

**Figure 44-4.** Sterilization. (a) Vasectomy. The ductus deferens is cut and tied after an incision is made into the scrotum. (b) Tubal ligation. The uterine tube is cut and tied after an incision is made into the abdomen.

tum, the tubes are located, and each is tied in 2 places. Then the portion between the ties is cut out. Sperm production can continue in the testes, but the sperm cannot reach the exterior.

Sterilization in females generally is achieved by tubal ligation, a similar operation on the uterine tubes (Figure 44-4b). The tubes are squeezed, and a small loop called a knuckle is made. A suture is tied very tightly at the base of the knuckle and the knuckle is then cut. After 4 or 5 days the suture is digested by body fluids and the 2 severed ends of the tubes separate. The ovum thus is prevented from passing to the uterus, and the sperm cannot reach the ovum. Sterilization normally does not affect sexual performance or enjoyment.

*Contraceptives* include all methods of preventing fertilization without destroying fertility—by "natural," mechanical, and chemical means. The natural methods are complete or periodic abstinence. An example of periodic abstinence is the rhythm method, which takes advantage of the fact that a fertilizable ovum is available only during a period of 3-5 days in each menstrual cycle. During this time the couple refrains from intercourse. One of the difficulties in this method is that few women have absolutely regular cycles. Another problem with the rhythm method is that some women occasionally ovulate during the "safe" times of the month, such as during menstruation.

Mechanical means of contraception include the condom used by the male and the diaphragm by the female. The *condom* is a nonporous, elastic covering placed over the penis that prevents deposition of sperm in the female reproductive tract. The *diaphragm* (Figure 44-5) is a dome-shaped structure that fits over the cervix and is generally used in conjunction with some kind of sperm-killing chemical. The diaphragm stops the sperm from passing into the cervix, and the chemical kills the sperm cells.

Another mechanical method of contraception is an *intrauterine device (IUD)*. These are small objects such as loops, spirals, and rings made of plastic, copper, or stainless steel that are inserted into the cavity of the uterus (Figure 44-6). It is not clear how IUDs operate. Some investigators believe that they cause changes in the lining of the uterus that, in turn, produces a substance which destroys either the sperm or the fertilized ovum.

Chemical means of birth control include the use of various foams, creams, jellies, suppositories, and douches that make the vagina and cervix unfavorable for sperm survival. Of the newer chemical means, oral contraception ("the pill") has found rapid and widespread use. Although several types of pills are available, the most commonly used one is the combination pill. This pill contains a high concentration of progesterone and a low concentration of estrogens. These 2 hormones act on the anterior pituitary to decrease the secretion of FSH and LH. The low levels of FSH and LH are not adequate to initiate follicle maturation or ovulation. Consequently, pregnancy cannot occur in the absence of a mature ovum.

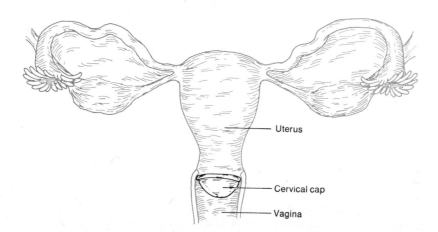

— Uterus

— Cervical cap

— Vagina

**Figure 44-5.** The cervical cap diaphragm. This particular type of diaphragm fits directly over the cervix of the uterus. The thin spring around the margin of the diaphragm opens out, presses against the wall of the vagina, and stretches across the cervix.

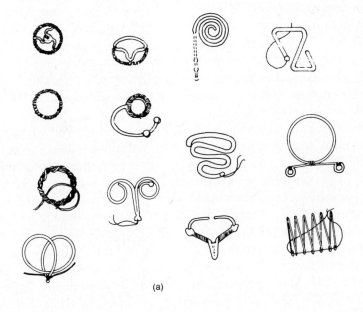

(a)

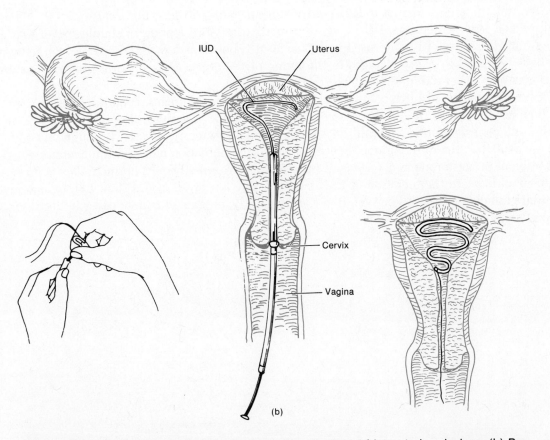

IUD

Uterus

Cervix

Vagina

(b)

**Figure 44–6.** Intrauterine devices. (a) Representative designs of intrauterine devices. (b) Procedure for insertion. IUD's are inserted into a slightly open cervix. The device is threaded into a long, narrow-bore tube that is passed through the cervix. Once in position in the uterus, it spreads out to its former shape. Most IUD's have a thread or chain projecting into the vagina that may be detected by a finger and indicates the device is still in place, and may be used for removal.

Table 44-1 contains a listing of contraceptive methods. Next to each, describe their effectiveness and potential dangers to health, if any.

Table44-1. Contraceptive Methods.

| METHOD | EFFECTIVENESS AND/OR POTENTIAL DANGER |
|---|---|
| Removal of gonads and uterus | |
| Sterilization | |
| Rhythm | |
| Mechanical devices<br>    Condom<br>    Diaphragm<br>    IUD | |
| Foams, creams, jellies, douches, etc. | |
| Oral Contraceptives | |

**STUDENT ACTIVITY**

Prepare a list of about 20 inherited traits in humans. Indicate which are dominant and which are recessive.

# APPENDIX A
# PREPARATION OF SPECIMENS
# FOR EXPERIMENTATION

## I. PITHING OF FROGS AND TURTLES

By definition, *pithing* is the destruction of the central nervous system by the piercing of the brain or spinal cord. This procedure is used in animal experimentation to render the animal unconscious, so that no pain is felt. A singly-pithed frog is one in which only the brain is destroyed, while a doubly-pithed frog has its spinal cord destroyed also. Usually the double-pithing procedure is used.

The frog is held in a paper towel with the dorsal side up and the index finger pressing the nose down so that the head makes a right angle with the trunk. Locate the slight depression formed by the first vertebra and the skull about 3 millimeters behind the line joining the posterior borders of the tympanic membrane. This groove represents the area of the foramen magnum. Carefully insert a long sharp-tipped needle or probe into the foramen magnum and direct it forward and a little downward. Exert a steady pressure and rotate it, moving it from side to side in the cranial cavity to destroy the brain. Destruction of the brain in this manner is a single pith. To double pith the frog insert the needle into the vertebral canal directing it downward until it has reached the end of the canal, moving it from side to side (Figure A-1).

The turtle is pithed in a similar manner. With the head drawn out, a sharp-tipped needle or probe is forced through the brain. This destroys the animal's feeling of pain.

## II. MUSCLE AND NERVE-MUSCLE PREPARATIONS

The gastrocnemius muscle in the frog is commonly used in the laboratory to demonstrate muscle contraction. After pithing the frog, remove the leg by cutting it off two thirds of the way up the thigh. Completely remove the skin at the ankle, gently pulling it loose up over the knee. Cut the Achilles tendon free as far down over the ankle as possible and then gently pull the gastrocnemius free from the lower leg. Cut through the upper leg so as to leave at least half of the femur attached to the gastrocnemius muscle, and carefully cut away the severed muscle of the upper leg. The muscle preparation consists of the gastrocnemius muscle, with its Achilles tendon, attached to the bare lower

**448**

half of the femur (Figure A-2). The other gastrocnemius muscle should be similarly removed and kept in a beaker of cooled saline for future use.

Place the muscle preparation in the recording apparatus with the bone held firmly in a femur clamp, and the tendon connected to a muscle lever by means of a pinhook or S-pin stuck through the tendon (Figure A-3). Remember that isolated skeletal muscle has no blood supply and fatigues

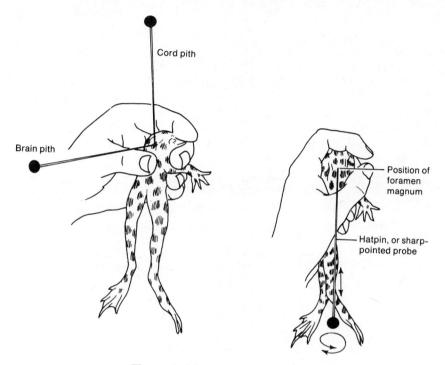

**Figure A-1.** Procedure for pithing a frog.

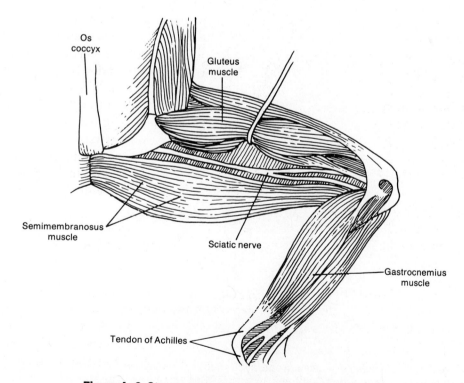

**Figure A-2.** Structures in the lower extremities of the frog.

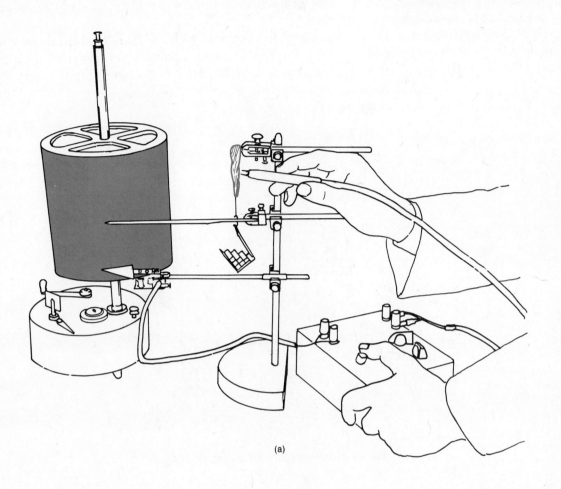

(a)

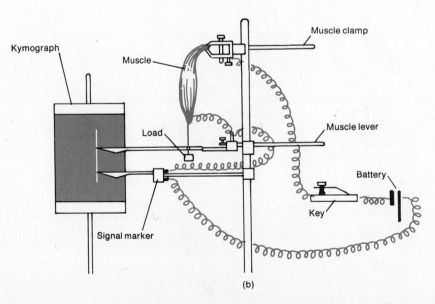

Kymograph

Muscle clamp

Muscle

Muscle lever

Load

Battery

Signal marker

Key

(b)

**Figure A–3.** Kymograph and muscle preparation ready for recording. (a) Redrawn from photograph from Harvard Apparatus Company, Inc. (b) Redrawn and modified from Tuttle, W. W., and Schottelius, B. A., Textbook of physiology, ed. 16, St. Louis, Mo., 1969, The C. V. Mosby Co.)

quickly with very little recovery. A fatigued muscle may give opposite results to those characteristic of a resting muscle, therefore it may be necessary to substitute the preparation with a fresh sample after a period of time.

If a nerve-muscle preparation is necessary, proceed as follows: Make a midline incision in the abdomen of the frog and after removing the viscera expose the sacral plexi. Cut and remove the lower extremities just above where the fibers of the plexus enter the vertebral column. Cut through the pelvis and separate the 2 extremities, continuing the incision all the way through between the plexi. Tie a fine ligature tightly around the anterior part of the plexus on one side. Place the frog in a prone position and expose and dissect out the sciatic nerve from the gastrocnemius muscle through the pelvis to the vertebral column. Do not stretch the nerve with its attached button of bone (Figure A-4). Avoid unnecessary contact between the nerve, metal instruments and other tissue. Glass rods should be used for probes. Then cut the plexus anterior to the ligature and tease the nerve away from adjoining structures.

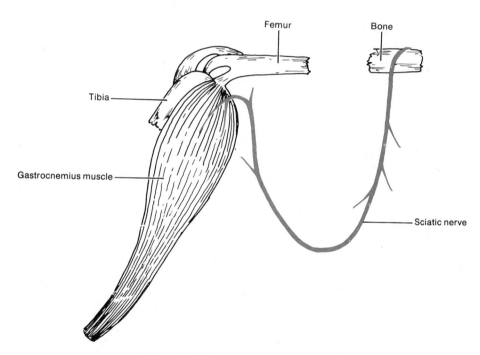

**Figure A-4.** Nerve-muscle preparation.

It is necessary and important to keep both nerve-muscle and muscle preparations constantly moist with 0.7 percent NaCl solution, or Ringer's solution (see Appendix C).

Using the piece of femur as a "handle," clamp it in a flat jaw clamp allowing the gastrocnemius to hang free vertically (Figure A-5). Clamp a glass slide, wrapped with filter paper moistened with Ringer's solution, adjacent to the top of the muscle and gently lay the sciatic nerve on this moist paper. Attach the Achilles tendon to the muscle lever by passing an "S" hook through the tendon or by tying a thread between the 2.

Position the muscle attachment to the lever 1.0 centimeters from the lever fulcrum. Remove any slack in the connection between the muscle and the lever (while it is in the resting position) but do not permit the lever to exert any pull on the muscle. Locate the ring stand so that the writing tip of the muscle lever rests lightly on the drum of the kymograph. Make certain that the tip moves freely across the drum, yet produces a clear continuous line.

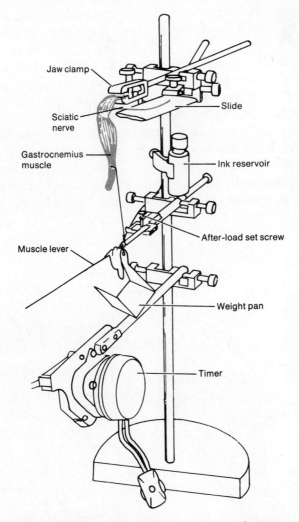

**Figure A–5.** Sciatic nerve-gastrocnemius muscle arrangement.

# APPENDIX B
# USE OF PHYSIOLOGIC EQUIPMENT

## KYMOGRAPH

The kymograph machine consists of a cylindrical drum on a shaft, the shaft being driven by a variable speed motor. The drum will be covered with paper suitable for smoking, for ink writing, or for electric writing.

The drum is rotated in a clockwise direction by either an electric or spring driven motor (Figure B-1). Both types have speed controlled mechanisms and are adequate for muscle contraction experiments. The speed of the kymograph used in an experiment depends upon what phenomenon one wishes to demonstrate. To record heart rate or respiratory movements when amplitude and rate are the only concerns, the speed should be slow; if one wishes to measure latency and the periods of muscle contraction, a fast drum is required.

Kymograph records are made by writing on smoked paper attached to the drum, or by writing on paper attached to the drum with an ink-writing stylus. The latter is mostly used. The stylus is hollow and ink flows through it and out the penlike tip to mark its path on the paper. The kymograph can be used in a variety of experiments whenever recording is required.

In addition to the kymograph, an instrument for providing electrical shocks of variable strength is needed. Either a *stimulator* or an *inductorium* is used, designed so that both the intensity and frequency of the stimulus may be varied. This instrument is arranged to provide a single shock when the signal key is closed. They can also be used to emit multiple shocks at very rapid rates (Figure A-3).

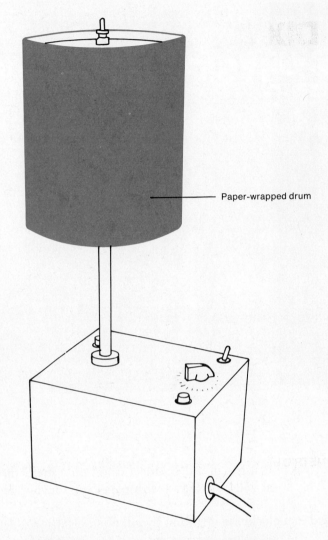

Paper-wrapped drum

**Figure B–1.** An electrically-powered kymograph.

# APPENDIX C
# METRIC UNITS OF LENGTH AND SOME ENGLISH EQUIVALENTS

METRIC UNITS OF LENGTH AND SOME ENGLISH EQUIVALENTS.

| METRIC UNIT (SYMBOL) | MEANING OF PREFIX | METRIC EQUIVALENT |
|---|---|---|
| 1 kilometer (km.) | $kilo = 1000$ | 1000 m. |
| 1 hectometer (hm.) | $hecto = 100$ | 100 m. |
| 1 dekameter (dkm.) | $deka = 10$ | 10 m. |
| 1 meter (m.) | Standard Unit of Length | |
| 1 decimeter (dm.) | $deci = 1/10$ | 0.1 m. |
| 1 centimeter (cm.) | $centi = 1/100$ | 0.01 m. |
| 1 millimeter (mm.) | $milli = 1/1000$ | 0.001 m. |
| 1 micrometer ($\mu$m.) | $micro = 1/1,000,000$ | 0.000,001 m. |
| 1 nanometer (nm.) | | 0.000,000,001 m. |
| 1 angstrom (Å) | | 0.000,000,000,1 m. |

ENGLISH EQUIVALENTS

3280.84 ft. or 0.62 mi.; 1 mi. = 1.61 km.
328. feet
32.8 feet
39.37 inches or 3.28 feet or 1.09 yard
3.94 inches
0.394 inches; 1 inch = 2.54 cm.
0.0394 inches
$3.94 \times 10^{-5}$ inches
$3.94 \times 10^{-8}$ inches
$3.94 \times 10^{-9}$ inches